Low-Carbon Economics

Theory and Application

Low-Carbon Economics

Theory and Application

Editor

XUE Jinjun

Nagoya University, Japan

社会科学文献出版社
SSAP
SOCIAL SCIENCES ACADEMIC PRESS (CHINA)

World Scientific

Published by

World Scientific Publishing Co. Pte. Ltd.
5 Toh Tuck Link, Singapore 596224
USA office: 27 Warren Street, Suite 401-402, Hackensack, NJ 07601
UK office: 57 Shelton Street, Covent Garden, London WC2H 9HE

Library of Congress Cataloging-in-Publication Data
Di tan jing ji xue. English
Low-carbon economics : theory and application / editor, Xue Jinjun.
pages cm
"Originally published in Chinese by Social Sciences Academic Press, copyright 2011, [as Di tan jing ji xue = Low-carbon economics]."
ISBN 978-9814383097
1. Economic development--Environmental aspects--China. 2. Carbon dioxide mitigation--Economic aspects--China. 3. Environmental policy--Economic aspects--China. 4. Climatic changes--Economic aspects--China. 5. Carbon dioxide mitigation--Government policy--China. I. Xue, Jinjun. II. Title.
HC430.E5D52313 2013
363.738'74--dc23

2013000659

British Library Cataloguing-in-Publication Data
A catalogue record for this book is available from the British Library.

低碳经济学
Originally published in Chinese by Social Sciences Academic Press

In-house Editor: Lum Pui Yee

Typeset by Stallion Press
Email: enquiries@stallionpress.com

Printed in Singapore by World Scientific Printers.

ACKNOWLEDGEMENTS

- This book is one of a series of studies conducted by the Institute of Global Low-Carbon Economy (IGLCE).
- We are grateful to the Asian Core Project granted by the Japan Society for the Promotion of Science (JSPS), Japan, Energy Foundation China Office (USA), Kitankai (the alumni of graduate school of economics, Nagoya University) and the Nomura Social Science Foundation for their financial support. We also thank the Economic Research Center of the Graduate School of Economics, Nagoya University, the Development Research Center of the State Council of China, Energy Research Institute of NDRC of China, China Research Center of JST of Japan, Nagoya University, University of International Business and Economics and Hubei University of Economics for their support in conducting the research.
- We thank Professor Soemon Takakuwa and Yuko Arayama (Nagoya University), Professor Yuezhong Zhu (Energy Research Institute of NDRC of China), Professor Zhongxiu Zhao and Bo Wang (University of International Business and Economics) for their encouragement and critical comments.
- Our special thanks are given to Dr. Yoram Bauman (University of Washington, conducted native check), Kaito Yukihara (School of Engineering, University of Tokyo, English translation of Chapter 20), Justin Jonas and Alan Lundell (University of Minnesota, first reading check) for their help and valuable comments.
- We also thank Xiaodong Li (Ph.D. candidate of Graduate School of Economics, Nagoya University), Feng Qi (Ph.D. candidate and research assistant at ERI of NDRC, China), Luo Xiao (Ph.D. candidate of Graduate School of Environment Studies, Nagoya University), Chisato Yukihara (Guest Lecturer of Kinjo Gakuin High School, Japan) for their help in translating the references, proofreading and editing figures and tables.

CONTENTS

PREFACE

TOWARD A LOW-CARBON SOCIETY

As human beings make progresses in economic growth and well-being, growing world population, urbanization and motorization, mass consumption of natural energy, environment pollution, global warming and so on, are becoming serious problems. Among these problems, climate change is especially dramatic due to a rapid increase of carbon emissions and environmental damage. Facing these issues and challenges, many countries are taking actions to deal with the greenhouse effect by reducing carbon emissions. For example, Britain is practicing low-carbon economy aiming at a 70% cut off of CO_2 by year 2050; the United States is conducting a Green New Deal and developing a new energy technological innovation; Germany is developing a low-carbon industry as a new leading sector; in Japan, a low-carbon society is under construction. In 2009, former Japanese Premier Yukio Hatoyama promised at the UN general conference that Japan will cut off 25% CO_2 by 2020 based on the 1990 standard. This is a very hard target but Japan is trying her best to realize it.

Talking about data, wc may mention here that China is facing a serious challenge with regard to carbon emissions because its CO_2 discharge constitutes more than 20% of that of the whole world and becomes the largest carbon emitter in the world, exceeding the U.S.A. in 2009. However, we see that the Chinese government and Chinese people are trying their best to deal with the climate change issue by reducing carbon emission and using new energy-saving technology. We notice that China is working hard in building the largest wind mill generation, diffusing LED lights in cities, promoting low-carbon cities and an eco-economy. In 2010, former Premier Wen Jiabao said in his speech that China will actively take common but different obligations to deal with the climate change issue. In fact, China has accomplished its main targets of the 11th Five-Year Plan: 20% reduction of CO_2 emissions per unite GDP, an increase of the ratio of renewable energy to 10% in total energy and the forest coverage rate to 20% by year 2010. These are very ambitious targets that reflect China's effort to carbon reduction. I believe that China will make great contributions to the world carbon reduction and may create another miracle of low-carbon economic growth in a new world economy in the near future.

Japan had been a poor country in the 1950s–1960s but became the second largest economy in the world after experiencing a high growth period. In the high growth period and the later economic development, manufacturing made a great contribution to the Japanese economy. Even now, Aichi, where the Toyota Motor Corporation is located, is still a center of manufacturing in Japan. Meanwhile, after suffering serious damages from environment pollution in the 1950s–1960s, Japan made great achievements in environment protection and improved the environment significantly. In the process, Japan, especially Aichi prefecture and Nagoya city, accumulated a rich experience of environment regulation, citizen movement, and innovated many kinds of environmental technologies. In 2005, the Aichi World Exhibition and the COP10 were held in Nagoya City in 2010. We hope that Japan's rich experience in environment protection and excellent technologies of carbon reduction can contribute to the world's carbon emission reduction and a low-carbon economic development.

Nagoya University is a research university that has a good academic reputation. We are very proud that in 2008, our alumni and distinguished University Professors, Toshihide Maskawa and Makoto Kobayashi, won the Nobel Prize in Physics, and our former Associate Professor, Osamu Shimomura, won the Nobel Prize in Chemistry. This great honor followed the Nobel Prize in Chemistry presented to Nagoya University Professor Ryoji Noyori in 2001. These four Nobel Laureates truly represent what we call "courageous intellectuals". It is our belief that their sincere attitude and solid commitment to research, developed and fostered in the free and vibrant academic culture of Nagoya University, finally led to this most prestigious commendation.

Besides these fields, we also have advantages in the fields of environmental studies and economic theory. The Wild Flying-gees Model, a well known economic hypothesis which has been used to explain the rapid economic growth in Asia, was originated from the Economic Research Center of the Graduate School of Economics of Nagoya University. All these provide a very good background for us to study the essence of the Japanese style of management of environment and of a low-carbon economy. It is my hope that the publication of the *Low-carbon Economics* makes new contributions to economic research and gains glory for Nagoya University.

Michinari Hamaguchi
President of Nagoya University
October 1, 2012

FOREWORD

This is the first economic theoretical monograph studying the low-carbon economy.

The book is written by members of the Club of Rome, experts from the United Nations Intergovernmental Panel on Climate Change (IPCC), renowned scholars, economists and environmental scientists, experts on the issue of climate change, policy makers and designers of low-carbon society and low-carbon city.

The book covers a variety of disciplines that discusses the definition of low-carbon economy, growth models, analysis methodologies, measures and indicators, the international economic system, international trade, carbon tax, development strategy, policy design, and systematically explores the theoretical framework for the low-carbon economy. As a new discipline, it may seem very naive, and the whole system is still in formation, but the book will impact on the innovation of economic theory, and become reference to the undergoing development of low-carbon economy. So we sincerely hope that it would draw reader's attention.

The concept of "low-carbon economy" originated in England in 2003, since then countries all over the world responded, especially after the British government released a substantial reduction target and leading the trend of international emission reduction. Germany developed a national development plan which regards low-carbon industry as the biggest leading industry. Japan enacted the low-carbon society development plan, and adopted several basic laws to promote low-carbon economy. At the beginning, low-carbon economy is just another important means of responding climate change and stimulating the economy, but after the Copenhagen conference (COP15) together with the world financial crisis, low-carbon economy is going though unprecedented speed of rapid development, and gradually forming a new model of development.

A few years ago, low-carbon economy is a very sensitive topic in China. However, after the heated argument and the media hype during the United Nations climate conference in Copenhagen, and Cancun conference

(Mexico), the concept of low-carbon economy is gradually getting popular and going deeply into its society. Developing low carbon economy and a low-carbon life have become a part of life in China. Consequently the understanding of low-carbon economy has also changed from the negative and being suspected of to positive and actively promote. In particular, on various occasions in recent years, Chinese leaders have declared to actively respond to global climate change, promoting development of low-carbon economy, and making a public commitment of reduction in carbon intensity of GDP unit by 40% to 45% by the year 2020 on the base of 2005, which made China's low-carbon economy as the focus of international concern and a hot topic in the domestic economy. During the National People's Congress and National Committee of the Chinese People's Political Consultative Conference in 2010, low-carbon economy became the topic of most proposals. All levels of government regard low-carbon as a new growth point of the local economy. In 2010, the National Development and Reform Commission specified 8 cities and 5 provinces as pilots of low-carbon city and low-carbon industry development; resulting from that, low-carbon economy attracted further attention from the people, and gradually formed a new model of economic development. Low-carbon projects start to be applied by cities all around, low-carbon economic industries are also booming up, and some aspects even have developed faster than foreign countries. In March 2011, as an outline and national strategy, China issued its 12th FYP in which the new social and economic development plan is described as "Green Development", and the targets relating to GHG emissions are set to 17% reduction of carbon emissions, 16% decline of energy intensity (energy consumption per unit GDP) in the next 5 years (2011–2015). All these show not only China's positive response to climate change, but also the determination and action to realize international promise and national economic development goals through low-carbon development.

Why low-carbon economy is so popular? There are several reasons.

Firstly, low-carbon economy is a new industrial revolution and the new industrial civilization of mankind, if you do not actively deal with that, you will lose the opportunity, lag behind or even be ignored. It can be said that the development of low-carbon economy is to adapt to international climate change requirements, and to greet a new industrial revolution and human civilization. One of the pioneer of China's environmental protection,

Geping Qu said, the future civilization will be built on the basis of new energy, and low-carbon economy is the only way. A key member of the Club of Rome, one of the founders of quantum mechanics, Professor Peter-Hans-Duerr profoundly pointed out that the significance of low-carbon economy is to reposition the meaning of our human life, and rethink the purpose of growth, that is, we are not living for high GDP growth, but for the pursuit of quality of life. This revolution of civilization and philosophy is the root cause of global low-carbon economy boom. Referring to the case of worsening environment, increasing carbon emissions and their huge costs in China, people can't over-emphasis on the "Great Leap Forward of high growth", our own environment, climate, quality of life. In short, "life" should be the target of development, and the new inspection standards for development of productive forces and social progress.

Secondly, since the world financial crisis, the strength of the power of the United States has greatly weakened; the world is facing foremost economic and political restructuring. In this process, every country wants to share a place in the restructuring, and compete for world leadership. This is not sensational prophecy, but some countries have already developed national strategies for this target. In other words, the low-carbon economy is not only a new world economic competition, but also a competition for new international political forces. This competition is reflected in countries around, the pilot cities for low-carbon development has been already competing fiercely. In other words, every country is in the race for capturing the commanding heights.

Thirdly, the world financial crisis has had huge impact on the world economy, especially the United States, Japan and the European countries. How to recover from the crisis and revive the economy has become a national priority. Low-carbon economy, while not a savior, is considered as a new model of economic development, a new growth engine. Some countries have already taken the first step and formed its own characteristics. For example, when explaining about the concept of low-carbon economy, we will surely mention the UK; when talking about new energy technologies and industries, we will associate it with Germany; while the low-carbon technologies and low-carbon society is the new nickname of Japan.

In addition to the above reasons, for China, there are also other issues such as shortage of energy, environmental protection, economic restructuring, energy efficiency, employment security, etc. According to a study by the Development Research Center of the State Council of China, China's economic restructure is predicted to complete in 2030. This means that before 2030, China's industrial structure will still be manufacturing-based, thus environmental pollution will worsen, and carbon emissions will increase.

Low-carbon economy is now quite popular, and someone says it is overheating or even a "speculation". In my opinion, low-carbon economy requires a certain degree of "heating" because firstly, the concept of low-carbon economy need to be spread. Many people do not understand what is truly low-carbon economy. Some regions and local areas are touting their low-carbon economies, but when we have a look at those areas, they are not truly low-carbon economies, so we need publicity. The deputy director of Environmental and Resources Protection Committee of the CPPCC National Committee, Wang Yuqing pointed out that the low-carbon economy need to engage in mass popularity of some of the things that make people understand, because everyone should involve in low-carbon economy, we should all be in action. Secondly, there is the need for theoretical research. Low-carbon economy is a new discipline that needs the joint research of many people. The first report by the Club of Rome, "The Limits to Growth", had put forward some sensational views seemed at the time, but now it appears to be right. At the same time, *Environmental Economics*, a culmination of researches by experts in various disciplines aimed at creating a new discipline, had also caused great controversy at its first photograph. Some people thought it was ironic that environmental study itself is not science, economics is neither. Moreover, some people suspect that these are two completely different subjects, how can they be mixed together? However, at present environmental economics has become an important discipline and is a main course taught at universities. Another example is development economics. Development economics originated in the 1970s. At that time, many neo-classical economists criticized that it was nonsense; there were also people who predicted that development economics should be dead. But now it seems that Development economics is popular in many countries, especially in China, because it studies issues of developing countries, such as environmental, income distribution, education and others that are relating to the "quality" of life but neglected by economic growth. We now have to inherit the history and commit to innovation and introduce new subjects. In 2009, at a seminar in Fudan University, I had argued that today we are talking about low-carbon economy, at seminars at the moment while, in the next 3–5 years, low-carbon economics will appear in the program of academic courses at Universities. It seems my prediction is slower than the actual, as of March 2010, the Ministry of Education of China had introduced "low-carbon economics" as a new subject, and some universities had begun to offer low-carbon economics in their teaching courses, especially courses for doctorate and masters programs. Therefore, the birth of this new discipline — low-carbon economics would be just matters of time.

Low-carbon economy was just a new concept all around the world before 2009. It was COP15 held in Copenhagen that had given a popular lesson to people around the world through the difficult negotiations, serious confrontation between developed and developing countries, and heated exchange opinions at the conference. After the COP15, the low-carbon economy is developing rapidly in many countries. However, the low-carbon economy is still in infancy and lacking in theoretical guidance and experiences, so it is necessary to study the experience of low-carbon economy all over the world, sum up the experience and lessons to provide ideas for future low-carbon economic development. This book is written based on this demand.

This book is the first economic theoretical monograph study on low carbon economy. It brings together experts and government officials from various fields examines the international economic situation after the world financial crisis, compares the experience and policies of development of low-carbon economy around the world, summarizes the new national trend of climate change after COP15, and prepares policy opinion for the United Nations climate negotiations.

The main thrust of this book is as follows.

(1) Mobilize international research efforts, by using international comparative study. This book was authored by the world's leading scholars, experts in environmental economics and low-carbon economy, government officials, policy makers and legislative specialists. The book is written by interdisciplinary researchers, covering economics, physics, sociology, psychology, diplomacy, political science. By taking

the advantages, the book can provide theories and reference for decision making on the low-carbon economic development;

(2) The creation of a new discipline cannot be done by one or two persons, it needs many peoples' efforts; after a long period of exploration, it would gradually form a theory. We hope this book serves as a catalyst for researchers to provide a theoretical basis for a low-carbon economy, exchange the outcomes of various researches and finally establish low-carbon economics with the efforts of all.

(3) The starting point and the focus of the book is global climate change and low-carbon economic development, compare low-carbon economic development models and experience of other countries in order to provide references for low-carbon economic development of countries around the world. For these reasons, the book focuses on the introduction and exploration of low-carbon economic theory, analysis methods, empirical studies in order to provide some operational experience and practical policy tools. We sincerely hope this book will become a reference for government officials to formulate policy options, basic information for researchers, and teaching assistant for students and general readers to learn about the knowledge of low-carbon economy.

This book is an attempt to popularize the concept of low-carbon economy, and low-carbon economic development is still at an initial stage. The study on the theory of low-carbon economy in this book is neither very superficial nor in-depth. In other words, low-carbon economics is a 'developing discipline' which needs more development. If so, we welcome readers' criticism and comments.

In the History, every big event gave birth to a new discipline: the economic competition between the United States and former Soviet Union during the Cold War in the 1950s gave birth to growth theory; the rise of the developing countries in the 1960s induced development economics; the population explosion and environmental issues in the 1970s derived environmental economics. This time, a new low-carbon industrial revolution caused by the energy crisis and global warming has inevitably given birth to low-carbon economics. We believe that low-carbon economics will soon embark on University's pulpit.

Finally, we need to state here that this book is integrated collection of academic researches. Some of the contents of this book that relate to policy research and advocacy represent authors' academic point of view, thus they do not reflect the government's position, are not policy-oriented and only serve as a reference for you.

Jinjun Xue (editor)
Nagoya University, Japan
October 1, 2012

ABOUT THE EDITOR

Dr. Jinjun Xue is a Professor at Economic Research Center, Graduate School of Economics of Nagoya University, Japan. He is also one of the founders and Joint Director of the Institute of Global Low Carbon Economy. He has working experiences as a Professor of Economics School of Wuhan University, China; Fulbright Scholar at Economic Growth Center of Yale University; Associate Professor at Institute of Economic Research, Hitotsubashi University; Visiting Professor at Department of Economics, University of Oxford. He is a Visiting Researcher at Energy Research Institute of National Development and Reform Committee (NDRC) of China, visiting Professor of Wuhan University, Tsinghua University, Beijing Institute of Technology, Xian Jiaotong University, University of International Business and Economics. Dr. Xue has many publications in English, Chinese and Japanese including Growth with Inequality (editor, World Scientific, 2012), Annual Report on China's Low-Carbon Economic Development (editor, Social Science Academy of China, 2012), Inequality in China (editor, Nippon-Hyoronsha, 2008). He is also involved in energy policy discussion, Low-Carbon City Planning, Low-Carbon Campus Design of Universities, Chief Scientist of Emission Trading Scheme of Hubes Province in China, and the EU ETS project. He is the first scholar publishing the book entitled *Low-carbon Economics.*

LIST OF CONTRIBUTORS

(By chapter order)

Jinjun XUE: Please refer to "About the Editor".

Ross GARNAUT: Vice-Chancellor's Fellow and Professorial Fellow in Economics, the University of Melbourne and Distinguished Professor of Economics at the Australian National University. He is an economist whose career has been built around the analysis of and practice of policy connected to development, economic policy and international relations in Australia, Asia and the Pacific. He conducted a review on Climate Change and Policy Response to Climate Change for the Federal, State and Territory governments of Australia (2007–2008), provided an independent update to the 2008 Garnaut Climate Change Review in 2011 for the Australian Government. He has served a Australian Ambassador to China (1985–1988), chairman of Primary Industry Bank of Australia Ltd (PIBA) (1989–1994), chairman of Bank of Western Australia Ltd (1988–1995), etc.

Hans-Peter DUERR: Professor, University of Munich, Germany. He is a German physicist in the field of nuclear and quantum physics, elementary particles and gravitation, epistemology, and philosophy. With others, he published the Potsdam Manifesto and the Potsdam Denkschrift as a follow-up to the Russell-Einstein-Manifesto of 1955. He has won the Right Livelihood Award, the Alternative Nobel Prize. Since 2006 he has been a founding councilor at the World Future Council.

Kunmin ZHANG: Professor, Tsinghua University, China; former deputy director general of the China Environment Protection Agency (CEPA), and current vice chairmen of the Chinese Society of Sustainable Development.

Angang HU: Professor, School of Public Policy and Management at Tsinghua University, China; Director of the Center for China Study at Tsinghua-CAS (Chinese Academy of Sciences), an expert of the Consultant Committee for the 12th FYP of China. He has more than 1,000 publications in Chinese, English and Japanese.

Yoichi KAYA: Professor Emeritus, University of Tokyo, Japan; Deputy Council of Institute for Global Environmental Advanced Industrial Science and Technology. He is former member of the IPCC and the initiator of the Kaya Identity. He had been the youngest Japanese member of the Club of Rome.

Kejun JIANG: Research Professor, Energy Research Institute of National Development and Reform Committee of China. He is a key member in the area of climate change and energy prediction modeling in China.

Shijin LIU: Vice minister of Development Research Center of the State Council, China, executive director of China Industrial Economic Federation and chairman of Academic Committee, deputy director of the National Industry Policy Advisory Committee, member the Experts Commission of the Defense Science, Technology and Industry Committee, member of the Academic Committee of the China Reform Foundation of the China Development Research Foundation governing.

Yongsheng Zhang: Deputy Director General, Department of Development Strategy and Regional Economy, Development Research Center of the State Council, China. Zhang has some journal papers and is involved in many policy-oriented studies, including a joint work done by the center and the World Bank, "China 2030: building a modern, harmonious and creative high-income society", and an official joint report for the BASICs countries, "Equitable Access to a Sustainable Development".

Hidefumi IMURA: Professor, Yokohama city University, Japan. He has headed a project of the Environment Leaders Education in Asia granted by the Ministry of the Environment of Japan.

Mukul SANWAL: Visiting Professor at the University of International Business and Economics, Beijing. He has held senior policy positions in the Government of India and the United Nations. He was closely involved with the Rio Conference, 1992, and the World Summit on Sustainable Development in 2002, and is associated with the South Centre, Geneva, and with think tanks in India.

Libo WU: Professor, Institute of World Economy, Executive Director of Center for Energy Economics and Strategies Studies at Fudan University, China.

Weiqi TANG: Ph.D. Candidate, Institute of World Economy, Fudan University, China.

Soo-cheol LEE: Professor, Faculty of Economics, Meijo University, Japan.

Kazuhiro UETA: Professor and Dean of Graduate School of Economics and Graduate School of Global Environmental Studies, Kyoto University, Japan. He is a key person and one of the promoters of the low-carbon society and environment tax in Japan.

Hideaki ITOH: Professor Emeritus of, EcoTopia Science Institute, Nagoya University, Japan.

Satoshi WATANABE: Assistant Professor, Graduate School of Economics, Nagoya University, Japan.

Makoto NISHIMURA: Professor, Graduate School of Economics, Nagoya University, Japan. He has been heading the Project for the Environment Leader Education in Asia at Nagoya University.

Shin-ichi YAMAMOTO: Professor of bio-agricultural sciences and Vice President of Okayama University, Japan. He was a professor of bio-agricultural sciences and vice president of Nagoya University, Japan.

Kazuhiko TAKEMOTO: Senior Fellow, Institute of Advanced Studies (UNU-IAS) of United Nations University. He has served as Vice-Minister for Global Environmental Affairs at the Ministry of the Environment, Japan, and involved in designing of the Kyoto Protocol, developing policies on international environmental cooperation and global environment and climate change.

Shuzo NISHIOKA: Professor Emeritus, Tokyo Institute of Technology, one of designers of the low-carbon Society in Japan. He has served as a member of the Environment project of Science and Technology Council of Japan, and of Central Council for the Environment of Japan. In the international arena, he has served as coordinating lead author of IPCC Reports since 1988 and Review Editor of Millennium Ecosystem Assessment since 2004.

Keisuke HANAKI: Professor, Graduate School of Engineering, the University of Tokyo, Japan.

Yong-Ho CHOI: Emeritus Professor, KNU, Korea: Director-General of Industry Research Institute, Republic of Korea. He served as chairman of the presidential committee of Low-carbon Green Growth for Korea.

Peter HEINDL: Research Fellow, Centre for European Economic Research (ZEW), Department of Environmental and Resource Economics, Environmental Management, Mannheim, Germany.

Andreas LÖSCHEL: Head of Department, Environmental and Resource Economics, Environmental Management, Centre for European Economic Research (ZEW), Germany; Professor at Alfred–Weber-Institute for Economics, University of Heidelberg, Germany. He is also heading the German government's expert group to monitor the energy transformation in Germany.

Thomas GRIES: Professor and Director, Center for International Economics, University of Paderborn, Germany.

Ziqi LIN: Graduate, Graduate School of Economics, Nagoya University, Japan.

Lin GUO: Research Fellow of Institute of Global Low-Carbon Economy; International University of Business and Economics, China and Nagoya University, Japan.

PART 1

THEORIES AND METHODOLOGIES FOR THE LOW-CARBON ECONOMY

Chapter 1

INTRODUCTION TO LOW-CARBON ECONOMICS

*Jinjun Xue**

1.1. Introduction

The fourth report of the Intergovernmental Panel of the United Nations Climate Change (IPCC) posed a critical concern that climate change and human activities affect each other simultaneously. The nature of the issue is not a scientific argument of whether the temperature rises 1 degree or 2 degrees but a warning to human beings that global warming is one of the most serious challenges in the 21st century, much like the warnings from the environmental crisis in the 1970s. It demonstrates to us that the storage of primary energy is being exhausted and cannot meet the unlimited needs of the human population. Therefore, the development mode of relying on large energy consumption, high carbon emission and large scale environment devastation and the lifestyle of uncontrolled mass consumption, namely the inherent economic and human development mode, should be fundamentally changed. In this sense, the issue of dealing with climate change is the common responsibility of mankind.

The low-carbon economy has been one of the themes of stimulus policy used to cope with climate change issues and to overcome the current global financial crisis. However, it gradually began to form a new kind of development model after the Conference of the Parties of the 15th United Nations Framework Convention on Climate Change (COP15) in 2009. While the low-carbon economy (LCE) theme has expanded rapidly, very few theoretical analyses of low-carbon economy have been conducted and even the definition, the target of analysis and the methodology remain unclear.

*Jinjun Xue, Professor of Economic Research Center, Graduate School of Economics, Nagoya University, Japan.

Therefore, the formation of a theoretical framework for the LCE is urgent and highly expected. This book will meet the challenge and respond to that expectation by creating a basic theoretical scheme for low-carbon economy.

This book is the first monograph of research themes on the economic theory of the low-carbon economy. It brings together experts, scholars, and government officials from all over the world to explore the current situation of the international economy after the financial crisis. Then compares international experiences in low-carbon economic development and summarizes the policies responding to climate change since COP15. Finally, it provides an analytic framework for the future development of the LCE which will aide The United Nations with climate negotiations.

This chapter aims to present a basic explanation for low-carbon economics. Section 6.2 introduces the background of the LCE and describes the concept of low-carbon economics. Section 6.3 analyses low-carbon economics from interdisciplinary aspects and makes clear the relations of low-carbon economics with other related subjects. Section 6.4 gives some expectations on the development of the LCE.

1.2. The Origin of the Low-Carbon Economy

1.2.1. *The origin of the low-carbon economy*

The concept of "low-carbon economy" first appeared in the *Energy White Paper* of the United Kingdom in 2003, and has been spreading around the world one country after another since the COP15. In 2006, British economist Nicholas Stern, former vice president of the World Bank, published *The Economics of Climate Change — The Stern Review*, in which the measures of carbon emissions and the methodology of the costs-benefit analysis of climate change were discussed. In 2008, the British government announced a plan to cut carbon emission by 70–80% and to build a low-carbon economy by 2050. This made Britain one of the world's leading figures in climate change talks and carbon reduction actions. Focusing on the development of renewable energy, in 2008, Germany issued a national strategy in which the low-carbon industry would take over as the leading industry in future economic development. In Japan, the central government created a systemic scheme for the creation of the LCE and made the concept of the "the low-carbon society" (LCS) known worldwide. To ensure the LCS,

Japan instituted many policies and laws on energy conservation and greenhouse gas emission reduction. For example the "Renewable Energy Law" established in November 2011 gave legal support to the development of the LCE. Under the slogan of, "Green New Deal" in the United States, President Barack Obama has been promoting energy technology innovation and the development of a renewable energy industry. As one of the world's largest carbon emitters, China had rejected the idea of the LCE for political reasons, at first, but later accepted it conditionally. Prime Minister Wen Jiabao stressed that China would continue to maintain its policy, "common but differentiated responsibilities," established by the UN Framework Convention on Climate Change, and work hard to fulfill the target of reducing its energy consumption per unit GDP by 20%, raise the proportion of renewable energy to 10% of total, and achieve a forest coverage of 20% by 2010, the last year of the 11th Five-year Plan of National Economic and Social Development (the 11th FYP). At the same time, China has been using carbon emission reductions as a strategy to fight against climate change as well as a target of economic development (Xinhua Net News, August 12, 2009). In 2009, at the COP15, Chinese leaders promised the international community that China would cut 40–45% of its unit green house gas emissions, relative to 2005 levels by 2020. To realize this target, China made national plans for the LCE and began to institute experimental Low-Carbon city planning in six provinces and eight cities since 2010. As an outline of its national strategy, China issued its 12th Five-Year Plan (FYP) in March 2011 in which the new social and economic development plan described as "Green Development", and the targets relating to green house gap (GHG) emissions were set to a 17% reduction of carbon emissions and a 16% reduction of energy intensity (energy consumption per unit GDP) over the next 5 years.

1.2.2. *The low-carbon economy as a new industrial revolution*

The low-carbon economy is crucial because it is the new industrial revolution and will create a new civilization for human society. You may lose the opportunity, lag behind the era or even be eliminated if you do not respond to it positively. It can be said that the development of a low-carbon economy is to adapt to the needs of international climate change in order to usher in a new industrial revolution and human civilization. As Geping Qu, one of the pioneers of China's environment protection activities, said in a speech, "that the new civilization would inevitably be built on the basis of new energies, and the development of the low-carbon economy is the only way to realize it." Pitter-Hans-Durre, a key member of the Club of Rome and one of the founders of quantum mechanics, profoundly pointed out that the significance of the low-carbon economy is to reposition the significance of human life, which is, "life is alive", and the reason we live is not for a high GDP growth rate but for the pursuit of an increased quality of life (refer to Chapter 3). Revolutions in human society and civilization are the fundamental reasons of the rise of the carbon economy and the dawn of carbon economics. In summary, we cannot continue pursuing the "Great Leap Forward" style of high growth. Our own environment, air, climate, quality of life, in short, our life itself, should be the goal of development as well as the new inspection standard for the development of economic and social progress.

After the global financial crisis induced by the Leman Brothers' shock, the power of the United States has been greatly weakened and the world is facing major economic and political restructuring, in which every country wants a place for itself and is competing for world leadership. This is not a myth, but the national strategies which some countries have pursued. The low-carbon economy is not only a new world economic competition, but is also a hunting ground for new international political power. Evidences of the development of low-carbon pilot cities in the EU, USA, Japan, BRICs (Brazil, Russia, India and China) or BASICs (Brazil, South Africa, India and China) and other countries shows that this is a new rivalry for the high ground from which a country can control the direction of the international economy and lead the world into the future.

Meanwhile, the global financial crisis has been seriously damaging the world economies, especially the economies of the United States, Japan and the European countries. Therefore, how to recover from the crisis and revive the economy has become the most important task. And while instituting a low-carbon economy is not a cure-all, it can be seen as a new development model as well as an engine for future economic growth.

For China, in addition to the above-mentioned reasons, the LCE is necessary because of the country's emerging issues of energy shortage as a result of rapid economic growth, the needs for economic restructuring, environmental protection, energy conservation, job creation and so on. A study issued by the Development Research Center of the State Council of China predicted that the transition point of economic structure will be appearing in 2030. If this prediction is true, the Chinese economy, continually relying on manufacturing and its current industrial structure, will continue to increase environmental pollution and carbon emissions.

The Chinese government promised to cut off 40–45% of its carbon per unit GDP by 2020 (base year 1990). However, for a country of 1.3 billion people with an average income per capita of US$4,000 (2010), high growth rate and low efficiency, it is difficult for China to achieve these targets without equivalent experience and external support.

China shows a positive attitude toward the low-carbon economy and responsing to climate change. It has been advancing its low-carbon economic developments in some aspects, for example, having built the largest wind and solar power generation systems in the world. However, overall, China is still in an initial stage of development and thus, needs to learn from other countries' experiences in developing a low-carbon economy. Furthermore, it is also important to embark on a low-carbon development path with Chinese characteristics in order to create a sustainable Chinese model of low carbon development.

1.2.3. *Primary characteristics of the low-carbon economy*

Based on the above discussion, we define low-carbon economy from eight aspects:

(1) Low-carbon production;
(2) Low-carbon energy;
(3) Low-carbon technology;
(4) Low-carbon transportation;
(5) Low-carbon consumption and life style;
(6) Low-carbon housing and building;
(7) Low-carbon farming;
(8) Low-carbon city.

Of course, as the LCE develops, the coverage of the LCE will be expanded.

1.2.4. *Some problems facing the LCE*

The low-carbon economy (LCE) has become a popular term in recent years. Many cities and local governments are making great efforts in building a LCE or low-carbon cities. However, after surveying the related literature and reviewing some model cities, we found that the contents and interpretations of the LCE are quite confusing. The main problems are as follows:

(1) There is no clear definition of a LCE. This may lead to the creation of a poorly designed LCE or model city.
(2) There are no clear studies on the relationship between the LCE and GDP growth, or on the impact of LCE on employment. Therefore, instituting a LCE may hurt the economy and cause serious unemployment problems by misleading economic growth and restructuring labor-intensive industries.
(3) There is no coordination between policies. This may result in a conflict among economic policies and a distortion of resource allocation and cause an unbalance of industrial structure.
(4) There are no clear targets, measures and indexes for the LCE. This may cause problems in identifying the LCE and in creating an LCE standards.
(5) There is no proper national development policy or environmental governance. This may cause new environmental problems.
(6) There is no clear definition of proper low-carbon technology. This may sap the driving force of sustainable development and create a barrier to further economic growth or it may lead a country to rely on the import of technology.

Due to the above problems, we have reasons to worry about a new "LCE bubble" and an unsuccessful transition in the next few years. The Chinese transition to a LCE should be of special concern because of the increasing size of its economy, which includes its emerging LCE, and its influence on the world economy.

Consequently, it is necessary to study the theories of the LCE and compare the experiences of both developed economies and developing economies.

1.3. Principles of the Low-Carbon Economics

1.3.1. *The definition of low-carbon economy*

The concept and idea of the low-carbon economy vary from country to country. In the UK, it is referred to as the low-carbon economy or low-carbon cities while in the United States, it is referred to as "green growth" and "low-carbon energy technology." In South Korea, combined with the principles of environmental protection, it is called "low-carbon green growth". Japan created the concept of "low-carbon society" by infusing the concept of low-carbon into every aspect of society; from its consumption, construction and transportation to its production activities, energy systems and environmental technologies. These systemic actions led Japan to create the nicknames of "low-carbon technology" or "low-carbon society".

In China, there are several unique names for the LCE but no unified name or definition yet exists. For example, in official government documents or scientific paper, it is called a "low-carbon city", or "low-carbon economy" and "low-carbon development", or "low-carbon technologies". In *China Annual Report on Low-carbon Economic Development* (Xue, 2011), it is given the name of "low-carbon development". We prefer to use low-carbon development because China is still a developing country with a large population and an average income per capita, based on the World Bank's ranking. This means that economic development and poverty reduction are still priorities to China. Therefore, the definition should take both economic development and carbon reduction into account, in order to be the most suitable definition for China's situation. In addition, the concept of "development" covers economic growth, environmental protection, education and human resources, income distribution, quality of life and a wide range of other content, which emphasizes the "quality" of an economy, so the connotation of "low-carbon development" is more broad than "low-carbon economy", "low-carbon society", "low-carbon energy" and "low-carbon technology". Therefore, it can avoid the trade-off between economic development and environment problems and thus be more in line with China's national conditions.

However, this book will use the names of "low-carbon economy" and "low-carbon economics" as they are frequently used and well-known terminologies.

1.3.2. *Methodologies of the low-carbon economy*

Every economic branch has its measures, indexes and analysis methodologies; the LCE is no exception.

The methodology of the LCE should be a combination of the analytic methods of natural science, political and environmental economics, econometrics, statistical analysis, international trade theory, public finance, taxation and international relations and politics. In this book, we focus on mathematical analysis and econometrical method for estimation.

In Chapter 7, Professor Yoichi Kaya, the famous scientist who innovated the *Kaya Identity*, provides a mathematical measure for carbon emission estimation using factor analysis methodology. In Chapter 8, some popular econometric models, carbon emissions, like IPAC Model, IPAC/AIM Technological Model etc., are discussed as practical quantitive analysis methodologies for the LCE.

1.3.3. *Classification of low-carbon economics*

Relating to the theory of environment preservation, academic subjects such as the theory of biological diversity, sustainable development, circular economy, ecological economics, eco-city, smart city and energy-saving economics have also been emerging in recent years. In order to clarify the relationship of these subjects to the LCE, we briefly discuss a descriptive classification.

Sustainable development: This is an idea, a kind of philosophy, and an objective for the development of an economy and a society, but not a development model and carrier. The low-carbon economy, however, is not only a development model but also a realistic carrier of development.

Circular economics: It aims to save natural resources and reduce various pollutants and carbon dioxide emissions through the reuse of waste. It is not a development model but an important method in achieving a low-carbon economy.

Ecological economics (including biological diversity): This is the result of high-carbon economic growth. Thus, controlling the ecological problem is only a temporary solution, while the low-carbon economy is a fundamental way of solving the problem in a sustainable way.

Low-carbon technology: This is also not a development model but an important means to reduce energy use and carbon emissions, as well as the key to achieving a low-carbon development. In other words, low-carbon technology is the method while the LCE is the goal.

Environmental economics: Environmental economy is a large category in which low-carbon economy should be included, under certain conditions. However, since carbon dioxide is, or will be, defined in the category of traditional environment pollutants, there still exist some differences between the terms low-carbon economy and environmental economy. Furthermore, the contents and origins of the emissions of carbon are different from that of environment pollutants, as are the solutions of reduction.

1.3.4. *Primary characteristics of low-carbon economics*

As a new discipline, we propose that low-carbon economics covers the contents described in the following:

(1) A background and a general explanation on why the climate change issue is important and why we need a LCE and an economic theory for the LCE. This is the basis for developing a low-carbon economy and creating a new discipline of low-carbon economics. Chapters 1 and 2 will discuss the above topics from the aspects of scarcity of energy, increasing carbon emissions and their damage on climate change and humankind, the importance of development mode transformation.

(2) A new worldview. To achieve the new industrial revolution of LCE, we need to think about the nature and the purpose of economic growth, the meaning of our life and our attitude toward traditional development modes and each of their relations to the LCE. Chapter 3 deals with these issues with a wide view of natural science. It will discuss the changes that have occurred in nature and their sustainability, while also providing us with a philosophical new world view. Professor Hans-Peter Duerr suggests in Chapter 3 that the low-carbon economic development can be applied through "creativity" (innovation of new technology), "change" (a fundamental change of our lifestyle) and

co-ordinations between nature and human beings, economic activity and ecosystems, energy use and the natural order of nature.

(3) Other related subjects and their relation to the LCE. There have been other subjects related to the LCE, e.g., sustainable development, environmental economics, circular economics, eco-economics, the economics of climate change and so on. To develop a new discipline, it is necessary to make clear the relations between these subjects and the LCE and to make a definition of the LCE itself. These issues will be discussed in Chapters 1 and 4. One basic idea is that the LCE is a new evolution of a combination of environment economics and the principle of sustainable development. If so, we can define the LCE as a development mode while the others are simply a conceptual idea and means to conduct the LCE.

(4) The political thinking of the LCE. Climate change itself is a natural phenomenon but the climate change "issue" is a political affair. The latter cannot be decided by a single person or an enterprise but must be decided by the governments and interested parties. However, how should a country deal with this issue? How much should each country reduce its emissions by? What actions a country should take in order to do the reduction? These are affairs relating to a nation's interests and can only be determined through international negotiation, development strategies and government policies. These questions are discussed in Chapter 5 in a scheme about the political economy. Professor Angang Hu, a consultant of the Chinese government, gives his explanation by describing the position of a country in terms of its political consideration, political decision making and international relationship toward climate change talk by using the case of China.

(5) A growth theory for the LCE. It is necessary to establish a theoretical framework and basic principle for the LCE. Chapter 6 provides a new growth model for the LCE and then develops it into a new Kuznets curve — Carbon Kuznets Curve (CKC). Meanwhile, Chapter 6 also defines the LCE, distinguishes the LCE from other related subjects and creates a growth model for the LCE. These analyses build a basis for understanding the LCE and provided a methodology to shape an Environment Kuznets Curve (EKC) or Carbon Kuznets Curve by applying available datasets. This chapter also gives an explanation of the

tunnel effect, by which a developing country can catch up with a developed country through technological progress and carbon financing (foundation).

(6) The analysis methodology for the LCE (refer to Section 1.3.2 of this chapter for the methodology).

(7) The international factors of the LCE. The LCE is not only a domestic issue, referring to energy consumption and the environment, but also an international issue concerning economic and political power restructuring, new international order and future development trends of the world economy. Therefore we need to discuss the issue in an international framework, which calls for the need of international economics and trade theory. Responding to this need, a research group from the State Council of China created an international system for the carbon emission distribution for both developing and developed countries. Chapter 9 makes a detailed explanation of the framework's common but different responsibilities for carbon emission reduction. Based on this, the research provides some global solutions and demonstrative examples of calculations of carbon emissions. These analyses give us a theoretical explanation of why the dispute on emission reduction targets between developing and developed countries exists.

(8) Carbon emission and the environment. Carbon emission has a negative effect on climate change and in turn influences human activities. In this sense, carbon emission should be included as a pollutant and be taken as a kind of new environment pollution. If so, we can classify carbon emission into the environment scheme. In the same principle, environmental issues that occur in international trade, e.g., cross-border pollution and transferred pollution through industrial transformation, are aligned with carbon emission issues such as carbon leak. Consequently, analyses of international trade, investment and global environment problems should be included in the LCE studies.

(9) Low-carbon finance and taxation. There are two aspects that matter to the LCE. One is investment which boosts the LCE and the other is a carbon tax which prevents the increase of carbon emissions. Chapter 13 applies a case study from Japan concerning energy taxes. Professor Kazuhiro Ueda, an environment economist and a major promoter

of the LCE, studied the features and outcomes of energy taxes and suggested changing the current energy tax system to an environment tax system in order to response to climate change issues as well as to reform fiscal policy in Japan.

(10) Indexes for the LCE. Chapter 15 provides some indexes and integrated standards for judging the LCE as well as measuring the effects of it. The components of the indexes are designed by adapting the ideas of sustainable development, eco-life, resource saving and moderate growth. A clear statement should be made that the aim of social and economic development is not about growth rate but about improving human kind's happiness. Therefore the low-carbon economic index should not only contains traditional indexes, like growth rate and income per capita, but also environmental indexes, like quality of life, degree of happiness, social progress, participation rate of citizens and process of policy decision. There are many indexes for the LCE and Chapter 15 introduces the *Ecotopia Indicator*, an index combining the above ideas.

(11) A low-carbon business model. Climate change talks and the LCE have been perpetuated by a top-down business style and are consequences of international negotiation and government policies. However, to apply the LCE, without government intervention, market function is fundamental. In the market place, business model and carbon emission trading market are the most important parts. Chapter 16 demonstrates that designing a business model is the key to materializing a low-carbon economy since an open market is the only reliable method of doing so consistently. Using the case study of flat-screen TV sales and strategies in Japan, the authors implement the idea of the LCE into a low-carbon business model and discuss the possibility of applying it in China. In Chapter 21, the carbon trading system (ETS) will be discussed and we will show the importance of the role it will play in carbon emission reduction and as a support system for developing countries.

(12) Strategies and policies for the LCE. The LCE is a practical science with real actions. In fact, the practice of the LCE is much more advanced than the formation of its theory. Since 2008, the Japanese government has issued several policies for the LCE (Chapter 17). At the same time, Japan has designed the low-carbon society plan and applied it in

several cities for evaluation. Following this plan, Japan has been promoting green technological innovation and extending the low-carbon city planning into Smart Cities (Chapter 18). In South Korea, the government has issued a low-carbon green growth strategy and incorporated it into a National Five-Year Plan for Green Growth. Meanwhile, the Korean government has been encouraging enterprises to stimulate green industry by improving energy efficiency, promoting the development of renewable energy and creating a green industry (Chapter 22). In China, energy conservation and carbon emission reduction have been written into the 12th Five-year Plan of National Social and Economic Development. Another important country of the BRICs, India, has been implementing its own uniquely developed low-carbon economy through designs, strategies and policies fit specifically for itself (Chapter 11).

(13) Empirical studies on the Carbon Kuznets Curve. To test the model and theory presented in Chapter 6, it is necessary to conduct an empirical study. Chapters 23 and 24 raise a trade-off question between economic growth and carbon emissions using the framework of the Environment Kuznets Curve (EKC) and show that as the EKC indicated, economic growth (measured by GDP per capita) and carbon emissions are correlated and can be shaped as an inverted-U curve. Meanwhile, evidences from developed countries and developing countries shows that income level and carbon emissions level are closely associated and that the elasticity of carbon emission is larger in developing countries than in developed countries. In Chapter 23, analyses show that the turning points of carbon emission are quite different between developed countries and developing countries due to different stages of economic growth and income level. These studies give an empirical explanation as to why developing countries conflict with developed countries on the issue of carbon emission reduction.

1.4. Conclusion

Economic growth is the quantitative increase in GDP, or potential output of a country. The main contents of the growth theory explain growth factors, inducing economic growth, the paths of growth and the ways to stabilizing growth. In the 1940s, capital accumulation and labor imput were stressed as the important factors in economic growth, which in turn brought forward the Cobb-Douglas production function and the Harold-Dammar Model. In the 1950s, during the Cold War, American economists compared the productivities and factors which promoted economic growth between the United States and the Soviet Union. They concluded that it was not the inputs of capital and labor but that technological progress that was the core element of modern economic growth. This gave birth to Solow Model economic theory of growth. From the 1960s, T. W. Schultz, G. Becker and J. Mincer *et al.* created the Theory of Human Capital, which states that education, or human capital (which differs from physical capital), played the most important role in improving productivity. In the 1970s, facing a population explosion and environmental pollution worldwide, the Club of Rome issued a report, *Limits to Growth* (1974), and warned the world that environment destruction would result in a great cost to economic growth and would threaten human society. During the 1980s economists began paying closer attention to the growing scarcity of natural resources and the constraints being placed on economic growth by the environment. This research into growth theory and ecological and environmental science and technology led to the creation of a synthesis branch of economics called environmental economics. In the 1990s, institutional factors were introduced into growth theory and the issue of externalities (especially the cost of environment pollution) were evaluated in the solutions of environment problems from the Pigovian tax scheme to the Coase Theorem (a basis of the climate change talks). At the beginning of the 21st century, IPCC reported on the emergency of fossil energy depletion and the threat of global warming to human survival. The report linked global warming with a rapid increase of greenhouse gases which is thought to be caused mainly by carbon emissions originating from human activities. Therefore, it sparked great concern on issues of climate change, and the reduction of carbon emissions was thought of as the most efficient way of stopping climate change.

With this background, the development of a low-carbon economy has become an important scheme for the reduction of carbon emissions as well as the promotion of economic growth. This is how, low-carbon economics came into being.

Reference

Xue, J. (ed.) (2011). *China Annual Report on Low-Carbon Economic Development*. China: Social Science Academy Press.

Chapter 2

LOW-CARBON NATIONAL ECONOMICS

*Ross Garnaut**

2.1. Introduction

2.1.1. *The Chinese interest in climate change mitigation*

China has an immense interest in effective global mitigation of climate change. China's strong science community has shown that, like Australia, and many of our neighbors in Asia and the Pacific, China faces risks of damage that are larger than the considerable risks faced by most established developed countries.

One example of the special risks which has been the focus of Chinese science is associated with the deglaciation of the Tibetan Plateau. This could destabilize what for thousands of years have been steady flows in the great rivers, first of all the Yangtze and the Yellow, that have nurtured Chinese and human civilization from the cradle. Another is the effects on run-off for irrigation and therefore the threat to agriculture in the warming of the North China plain. A third is the threat that rising sea levels poses to the huge concentrations of economic activity that have developed in the era of reform in low-lying areas of coastal China, including through the Pearl River delta and around Shanghai, Ningbo, Tianjin and elsewhere. Like Australia, China would share the trauma in Southeast and South Asia that could be associated with rising sea levels, disruption of the flows of the other great rivers flowing from the Tibetan Plateau and the Himalayas and any dislocation of the poorly understood South Asian monsoon.

*Ross Garnaut, Vice-Chancellor's Fellow and Professorial Fellow in Economics, University of Melbourne.

We are too late to avoid substantial damage from climate change, so part of the response of each country and the international community must be to adapt efficiently to the change. But we are not too late substantially to reduce the likely costs of climate change. As countries facing large potential costs, Australia and China share interests in ambitious and effective mitigation.

2.2. Assessing the Costs and Benefits to China of Mitigation

The National Development and Reform Commission (NDRC), China, is to be commended for the research that it is doing to identify the costs and benefits to China of varying levels of mitigation.

The costs are associated with a reasonable Chinese contribution to a global effort to reduce greenhouse gas emissions. These are relatively straightforward, and are amenable to general equilibrium economic modeling. They can be reduced by international cooperation in the development and dissemination of new technologies.

The benefits come from reduced risks of various climate change damage to China. Some of the damage to China will be direct. Some will be indirect, through its effects on economic development and political stability in countries which are important to China's own prosperity and stability.

Only some of the benefits from avoided climate change are amenable to standard economic modeling, based on scientific assessments of the "average" or "median" of the possible impacts on China's industries and infrastructure. Three categories of costs of climate change and therefore benefits of mitigation are not amenable to standard economic modeling, and therefore must be assessed qualitatively. One drives from the increasing difficulty of modeling economic effects as we go further into the future. It is necessary to define some cut-off date after which formal modeling is considered to be too unreliable to be useful, and to recognize, and to assess qualitatively, benefits from reduced damage from climate change beyond that date.

A second comes from the uncertainty of impacts, and the possibly catastrophic damage from the outer ends of the probability distribution of impacts. These also need to be assessed qualitatively.

The third comes from the non-market dimensions of many of the benefits of, avoided climate change, such as greater environmental amenity and longevity and better health. When I did all of these assessments for Australia in the Garnaut Climate Change Review, it turned out that the benefits of strong global mitigation that could be calculated by economic modeling exceeded the costs of Australia playing its proportionate part in a global mitigation effort, even though Australia's proportionate part, as a developed country, was relatively large. This meant that the benefits after this century, the insurance value of reducing the chances of extreme outcomes and the non-economic benefits of mitigation were all "bonuses". Moreover, the net surplus of benefits was greater for stronger (450 ppm) than weaker (550 ppm) mitigation objectives.

2.3. Why Early Participation by China and Other Developing Countries is Essential?

The world is moving rapidly towards the concentrations of greenhouse gases in the atmosphere that the science tells us are associated with high risks of dangerous climate change. Until late in the 20th century, the increase in concentrations came mainly from changes in economic activity in what are now the developed countries. This is the source of a special responsibility on the high-income countries to take the lead in mitigation, developing low-emissions technologies, and assisting poor developing countries to adapt to climate change.

But most of the increase in global emissions in recent times and as far into the future as we can see is expected to come from the advanced developing countries, led by the three most populous of them: China, India and Indonesia. Even if the developed countries now dropped their emissions to zero — so long as this occurred without diminution of growth in the developing countries — the growth in emissions from developing countries alone would take the world beyond the thresholds of dangerous climate change in a relatively short period. Effective mitigation will require participation of all developed and all substantial developing countries. Within the framework of common and differentiated responsibility, it will be necessary for developing countries to reduce emissions substantially below business as usual from an early date, and to put them on a declining path before too long.

This is unfamiliar territory for developing countries following successive discussions within the United Nations framework at Rio de Janiero, Kyoto and Bali. These discussions did not contemplate early binding commitments to mitigation from developing countries. While it is unfamiliar territory, it is territory that we will have to cross if any of us is to avoid high risks of dangerous climate change.

There is a need to define a set of mitigation responsibilities across all countries including developing countries that is widely recognized as being fair. This, in turn, requires concrete proposals for the distributing of mitigation burden across countries, that "adds up" to concentrations of greenhouse gas that avoid high risks of dangerous climate change.

2.4. The Elements of a Fair International Agreement

The Garnaut Climate Change Review put forward for discussion one approach to distribution of the burden of mitigation across countries. It could be described as "Modified Contraction and Convergence Plus Compensation" (MCCC). The starting point is the classic Contraction and Convergence, favored and much discussed in India and many developing countries over a long period. I note that this is the core of an approach favored by scholars in the NDRC. Under Contraction and Convergence, each country's per capita emissions entitlements move in a linear way from current levels to some low level at a specified future date. This approach reconciles the fairness of equal per capita entitlements, with the practicality of moving entitlements towards that level at a rate that was not seriously disruptive of global economic growth.

The Garnaut Climate Change Review suggested that the specified convergence date could be 2050. Per capita emissions entitlements would be equal and low in 2050.

The "modified" part of MCCC allows the entitlements of the rapidly growing economies for a period to grow faster than the rate that would be allowed under Contraction and Convergence. This recognizes the principle that countries that are successfully lifting their people from poverty at a rapid rate should not be required to compromise their development ambitions if these can be accommodated within a global agreement. It also recognizes the reality that successful developing countries will not be willing to compromise their development objectives. The Garnaut Climate Change Review allowed for this principle and reality by proposing that the emissions entitlements of rapidly growing developing economies should be allowed to grow at half the rate of their economic output until their per capita emissions had reached the rapidly declining levels of the developed countries.

Note that we are talking about entitlements to emit and not actual emissions. Any country can emit more than its entitlement if it purchases unutilized entitlements from another country. The "compensation" part of MCCC recognizes that contraction and convergence allows the currently developed countries to emit more greenhouse gases than developed countries for some time to come, and also that the developed countries in the past have used much more than their shares of the capacity of the earth to absorb greenhouse gases. The Garnaut Climate Change Review suggested two elements to the compensation: high-income countries accepting responsibility for high levels of investment in research, development and commercialization of low-emissions technologies; and developed countries accepting responsibility for providing assistance to developing countries for adaptation to climate change.

The form, quantum and distribution across countries of the support for low-emissions technologies and adaptation are at an early stage of international discussion. The Garnaut Climate Change Review suggested that economies with per capita incomes exceeding $11,000 should contribute a total of $100 billion per annum of public funding to research, development and commercialization of new, low-emissions technologies. A proportion — perhaps half — of the low-emissions technology commitment would be deployed in developing countries. The $11,000 corresponds to the World Bank's definition of a high-income country. This threshold would require contributions from the Special Administrative Region of Hong Kong, but not from the mainland of China at this stage. The contributions of high-income countries would be in proportion to the excess of their incomes over $11,000.

The Review stressed the importance of developed country commitments to adaptation to climate change in poor developing countries, in addition to current development assistance, but did not place numbers on those commitments. In the lead-up to this month's G20 meeting in Pittsburgh, and to the Copenhagen conference in December, there has been increased focus on developed country support for low-emissions technologies and for adaptation in developing countries. These discussions are crucial to the agreement at Copenhagen and subsequently.

2.5. The Suitability of MCCC to China

There has been considerable discussion of the Garnaut Climate Change Review's approach in developing countries, including China. Scholars based here at the NDRC have supported the broad principles underlying the approach, but have suggested that equity would require convergence on equal per capita entitlements before 2050. Similar sentiments have been expressed in India.

This is a discussion that we have to have. It raises important questions for research. In particular, it raises important questions for research in relation to the Chinese interest in effective global action on climate change being undertaken at China's NDRC. Or rather, these are important questions for research, as there are several inter-related matters that require investigation. What workable rules for distributing the global mitigation effort across countries have a chance of being accepted within the international community, for each of several levels of ambition in an international mitigation effort? What is the cost and benefit of each of these for China, presuming that China will need to play its proportionate part within some agreed set of rules for allocating the international mitigation effort?

Unfortunately, these are urgent matters. The answers have to come in time for China to play a productive part in international negotiations on the post-Kyoto international arrangements. It is not going to be easy to find a basis for international agreement. The difficulty means that none of us will be able to hold out for what is most in our own narrow national interest: there will be no agreement without early compromise and accommodation of the perspectives of others.

The Garnaut Climate Change Review's suggestions took into account China's domestic mitigation policies. In particular, the proposal to limit emissions growth to half the GDP growth rate is comfortably within China's ambitions to reduce the energy intensity of GDP by four percentage points per annum, and to rapidly increase the roles of renewable and nuclear energy. As a failsafe, China's strong fiscal and external payments positions would allow it to purchase permits from abroad if there were any overshooting of emissions above entitlements.

2.5.1. *The importance of international commitments*

It is important for China not only to achieve strong domestic mitigation outcomes consistently with current policy, but also to enter firm international agreements to do so. The difference between China implementing its own announced policies for slowing the growth in emissions, and entering a commitment to that same outcome within an international agreement, may well be the difference between success and failure in the effort to reduce the risks of dangerous climate change. An international commitment by China to implement its own domestic policies would also have important beneficial effects on the stability of the global trading system and on global recovery from the Great Crash of 2008.

Why is it so important for China to bind by international agreement the emissions outcomes to which it is already committed in domestic policy?

Chinese domestic policies are relatively unknown outside China and outside China are not treated with the seriousness that they warrant. In truth, Chinese domestic policies have already moved the trajectory of emissions growth substantially below business as usual, but the rest of the world considers that China has done nothing. This false impression can be removed, and China's major mitigation efforts brought to account internationally, through China entering an international agreement to do what it has already committed itself to do. If China entered a binding international agreement, it would be in a position to accept a cap on emissions (albeit, determined differently to developed country targets). This, in turn, would allow Chinese participation in a productive way in international trade in entitlements. This would lower the costs to some extent of China and the rest of the world meeting their respective emissions reduction targets.

It would also establish comparable emissions pricing in China and other countries, mainly the developed countries, which have accepted a cap on emissions. This would weaken the claims that firms producing emissions-intensive tradable products in other countries make for subsidies, as free permits within emissions trading schemes in other ways. It would remove any valid basis for developed countries to impose emissions-related tariffs on imports from China. It would weaken a major negative influence on implementation of effective mitigation policies in the developed countries.

The removal of the case for free permits or other forms of subsidy in developed countries would free large amounts of revenue-raising capacity — so large that it would materially improve the prospects for restoring sound fiscal arrangements in the developed countries after the Great Crash of 2008. That, in turn, would materially improve the prospects for restoring global economic stability and sustainable growth.

2.6. China and Global Leadership

China is now an essential participant in any effective global action in any area of policy. This follows simply from China's importance in the international political and economic systems. That importance has been growing steadily since the reforms began, and more rapidly since the turn of the century.

The rise in the relative importance of China in the international system has been accelerated by the global financial crisis. The Great Crash of 2008 and its recessionary aftermath have placed long-term growth in the United States and Europe on a lower trajectory. They have had relatively little effect on the long-term growth trajectory of China and such other large developing countries as Indonesia and India. A long term tendency towards increased Chinese influence in the international system has accelerated.

We are now in a world in which China's international commitments are centrally important to good international outcomes. China's domestic policies affecting climate change mitigation broadly meet the current requirements of strong global outcomes. Turning those same policies into international commitments could make all the difference to effective global action on climate change.

Chapter 3

THE PHILOSOPHY OF LOW-CARBON ECONOMY

*Hans-Peter Duerr**

3.1. Introduction

Our main concern today is the problem of sustainability and this focused on the question of survival of humankind. This problem has caught presently our attention on the approaching danger of a dramatic climate change apparently caused by the continuous emission of carbon dioxide into the atmosphere in the process of burning fossil fuels. With carbon dioxide being only a trace gas of about 0.03% in the atmosphere, it was rather unexpected that it should have such a sensitive green house effect on the climate with large consequences for the biosphere, including the global weather conditions critical for human settlements. This may be less surprising if we realize that with sun energy it took the plants on our earth million of centuries to photosynthesize and accumulated the carbon containing fossil fuels which we are burning now in two centuries, i.e. million times faster, returning carbon back into the atmosphere. The climate change problem, however, represents only the "tip of an iceberg". Nevertheless it is a very good starting point for caring about sustainability. Since everybody is capable to experience the weather and its changes and its better and worse consequences for their way of life, the climate change problem offers an excellent example

*Hans-Peter Duerr, Professor at University of Munich, Germany. He is a German physicist in the field of nuclear and quantum physics, elementary particles and gravitation, epistemology, and philosophy. With others, he published the Potsdam Manifesto and the Potsdam Denkschrift as a follow-up to the Russel-Einstein-Manifesto of 1955. He has won the Right Livelihood Award, the Alternative Nobel Prize. Since 2006 he has been a founding councillor at the World Future Council.

for establishing the necessary cooperation for a successful solution. The low-carbon economy hence is an excellent start in this direction.

3.2. Re-orientation and Sustainability

My present contribution is meant to remind us, that this initiative of a low-carbon economy trying to defuse or resolve the climate change problem, will turn out to be only a first step with others following being much more complex. I claim: sustainability can only be established if we shake off the fetters of our presently overly narrow world view. For this, we need a new orientation and the ability to act and shape in a broader and more creative sense than in the framework of our old conception of the world, in which we imagine ourselves as or are forced into the state of little cogwheels in a machine. To indicate this I have added the subtitle: New Thinking as Challenge to face Global Dangers.

Stating this I realize that I have mainly a Western point of view in mind with its tendency to perceive the world primarily as a thing-like material Reality (*res* lat. = thing) in complicated but well-determined motion dictated by strict natural laws. It is my impression that in China's long philosophical tradition there is less temptation to quickly press the world into a simplified model, which can be more easily manipulated, with the consequence that complicated changes or something like "liveliness" play only a secondary role. Modern physics has revealed that there is basically an indetermination which softens the Reality into a "Potentiality" leading to continuous change and action, emphasizing connectedness rather than substance. What I call "new thinking" in our perception may, from a Chinese point of view, be much closer to your common thinking. Perhaps I could have formulated my subtitle as: Chinese Thinking as Challenge to face Global Dangers.

If we possess such new abilities, and want to change our society in this way, the question arises: where are the actors who can achieve this? This question is by no means easy to answer. Many people have quite good ideas of what should be done, but who will actually do it? How can these be achieved? What kind of people do we need for this? But the answer is no great secret: it is all of us who have to do it! We cannot rely on some outside forces, say from government or the business world. These are more likely to be obstacles. But we may be able to convince a few of them to participate in activities where human beings are the focus, and not technology. If we chase after technology, we will forget to live.

The social challenges can be divided into external and internal ones. The external ones are shown by economic activity that leads to destabilization. This is not done intentionally, but unintentionally we get into a vicious circle from which we can hardly escape. A true vicious circle is characterized by the fact that within it, the tools to get out of it no longer exist. So we have to come up with some brilliant or even seemingly paradoxical new ideas in order to escape from the vicious circle.

One open problem is that we are in the process of destroying the vital natural basis of life. Many people still believe that humankind hovers, as it were, over nature, and that we can do whatever we like. They do not grasp that we human beings are embedded in a larger whole. If we destroy that larger whole, we will be the first to crash. We have to understand that.

Another problem is that justice and peace are more remote today than ever before, and we don't really see how we can get out of this situation. Most people regard the establishment of justice and peace as a completely unrealistic vision. They think that reality is the way it is, and that things will always be unjust and unpeaceful in the end, i.e. a nature in which only the most powerful survive in the end and the rest simply perish. But these are fundamentally false ideas, which I think reflect our false image of humankind and the world.

3.3. New Technology But Old Ways of Thinking

It appears to me that in the West dominantly rational and analytical approach to problems we are in a schizophrenic situation. There exists an inconsistency between the way of our thinking, the perception of our technology, and the intentions of our future requirements. We have a way of thinking that is still the old Classical Descartes-Newton world view — namely the way celebrated in the late 19th century — in which the world is regarded as a material-energetic, mechanistic reality.

But our present-day technology is based on an entirely different conception, namely on the achievements of this new revolutionary way of looking at things which physicists developed about one century ago, called Quantum

World, a holistic worldview, where matter is replaced by a Connectedness, an immaterial Texture, Potentiality instead of Reality, a Process rather than a thing an isolated object, in German "Wirklichkeit" (reality) meaning continuously acting and changing. I am sure that there will be a better word in Chinese: "Daoism".

But we, the West, are not aware of this. And now we are trying to shape the 21st century — with modern technology based on the new Quantum Texture, combined with the old classical way of thinking. This can only go wrong. But we hear the statement: "There is no alternative" to the world in which we live now, we simply have to accept it as it is. Of course, Ms Thatcher meant the neoliberal theory of economics, to which there is supposedly no alternative. That is pure fatalism. We have to struggle against it. Fatalism means a lack of imagination, and not only a lack of imagination, but a lack of understanding of what actually lies behind this world.

Why do we cling so much to this old materialistic-energetic, mechanistic worldview? Because we can grasp it, comprehend it. Matter is something that I can grasp. What I can hold in my hands, so to speak, I can own and hence manipulate.

This is all related to the old worldview. As the basic principle of our development we have competitiveness. This has become the leitmotif in the West today. We have to start choosing valid goals again, as the actual reason for doing things. Only people who can orient themselves, who have insight into their environment, in which they are existentially embedded, can determine such goals. This is the difficulty facing us at the moment in the West and running the risk to infect the others. Due to obsolete worldviews, we human beings feel to be apart from nature, something different from nature. Religions also have to do with this disregard for or forgetting of nature. The ways in which some humankind have been raised above nature and said to be in God's image have degraded nature. People in this case regard nature as nothing more than a tool or building blocks — although we know that all of us are embedded in nature. It leads on the one hand to an overestimating what we can do, and on the other hand underestimating the possibilities that are actually available to us. We must overcome this idea of separation. The new natural sciences have demonstrated that the strict laws of nature, as we used to conceive of them, do not actually exist. Behind things there is a linkage which is not strict, but possesses certain openness.

This enables us to develop ways of life in which creativity also plays a role. This applies not only to us human beings. Creativity is a property which we share with everything in the world.

3.4. Openness

Thus we have a situation that is much more favorable to us than we have imagined. Instead of being tied to natural laws, there is a certain openness which we can utilize in order to exert influence ourselves. But this openness is not at all easy to experience. It is related to the fact that, when we act, we are always acting from a background of some vague knowledge, with "Ahnungen" (presentiments), of dark hunches or "dawnings" that give rise to new ideas. This "Ahnung" (idea) or dark hunch is in another world than the old world in which we grasp things in the literal sense, and can say, "I have understood that, I am holding it here in my hand".

In this new other world, we experience this openness. The new insights show us that we do have creative abilities. But is that only characteristic of human beings? No, it is a general feature, and we can rely on it being necessary and important to be able to partially shape the future. The future is not entirely predetermined by laws of nature.

In this context, we can also ask: what future do we want? That is not a pointless question; we can actually shape the future — not generally, of course, but as possible in principle. There is not much point in waiting to see what the future will bring. No, we should not wait but consider what kind of future we want. What visions of the future do we have? If we have these visions, then we also know that we can move towards this future as a possibility. Not alone, but together with others.

We all have the predisposition to be a *homo sapiens sapiens* — the wise human being — this is not restricted to just a few people. But we make no effort to get people to that state. Who talks about human beings these days? We talk about jobs, but never about human beings who want to act, who want to shape their lives.

Talking only about jobs is a sterile language in which one loses sight of the people, the actual people. Instead of technology being there to help people to fully develop, we chase after technology so as to meet the demands which this new technology puts on us. This misses real life entirely. We all are that *homo sapiens*. Every one of us is unique, and can make a contribution in that direction.

3.5. Sustainability

We also need to face up to the question of who is really concerned about the future. What must we do in order to arrive where we all really want to get to? It is a matter of sustainability — and the word contains "ability". But this concept does not mean that we want to sustain in the sense to maintain the world the way it exists now. This world contains dynamism, vitality, and productivity. We want to keep its robustness and resilience, but not its current state. Things should continue in this direction, with this vitality. In my mind sustainability should be interpreted in the dynamical sense: "Let the Living become more alive!" Reminding me of the wonderful exclamation of Albert Schweitzer: "I am Life who wants to live amidst of Life who wants to live!"

But this means that we can never include only human beings in this concept of sustainability. Sustainability refers to the vitality of the entire bio-system in which we are embedded. We can distinguish three levels of this: we do not want to destroy the natural foundations of life; we want people to live together in peace and justice; and we also want to lead a good and worthwhile life.

This sounds a bit like the "three pillars" theory which is commonly used in the Western world: Threefold Sustainability — "Ecology, Society, and Economy". Except that the "Economy" in this common context does not characterize the full character and development of human beings as *homo sapiens sapiens*, but only their material basis, as the reduced *homo oeconomicus (economist)*. But in my mind this is an important issue: it is not just a matter of the physical survival of humankind, but also of human beings being able to develop their emotional and intellectual potentialities, which shape their personalities and make up their particular characters, optimally. Everybody has this gift, and can demand it for themselves. This denotes an essential precondition for human beings that cannot be covered by the concept: economy.

In addition, introducing these three pillars it is inadequate to suggest by this image equal priority of each pillar and even then completely wrong as usually presented put the Economy first, because one thinks one has to provide optimum conditions for competition first. Then, second, attention is directed to social problems, and finally, third, at the very end, if there are still intentions and enough resources left, one turns to nature, putting on the final decorative touches as it were.

No, it will not work that way. The priorities must be just the opposite. Nature, the natural foundations of life, is the basis in which humanity as a species among all others is embedded. We have to ensure that the natural foundations of life are not destroyed, and then, that humanity can develop in this world with healthy foundations of life, and that human beings are not mere cogwheels in the social order, but can develop their own nature as individuals, and as responsible supports of the community find their appropriate places in it.

3.6. Differentiation and Co-ordination

How can we live together and reach a new whole? This requires a positive-sum game, in which one does not abandon one's specific characteristics, but everyone can essentially retain their different qualities, similarly as an orchestra manages to give a concert, in which different instruments play together constructively, such that the whole becomes richer in its expression and qualitatively more sophisticated, than the sum of its parts.

This is the paradigm of life: two creative steps — Differentiation and Cooperative Integration on a higher level. The first process is the individualization of human beings, their emancipation; and we must support it. We want people to be varied. For if I only have legs that are tied to each other, I will fall over just as quickly as with one leg. We must become increasingly different, so that we can provide more and more supporting functions, which permit more and more dynamic stabilization of statically instable subsystems, a situation with a lot of cooperative balances. Only then will we be able to walk through the world in a crazy alive way without falling. This however requires a second creative step to successfully perform a cooperative interplay to succeed finding an optimum win-win solution. We may believe this might be impossible, but it is not, because, despite our variability we are all basically connected and will always remain so. We never are and were actually separated and hence can already look back on a lengthy common evolution, on which we can build.

Therefore, we can imagine our bio-system as a big house of cards, on the peak of which we are dancing and even jumping around such as if we did not know that we were balancing on a house of cards. The house of cards is to symbolize that all life is built basically on instability. However, the house of cards is only a defective image, because its final stability depends on the friction of the cards against one another, in contrast to the final stabilization of the biosphere which is achieved dynamically. To imitate this living case a better image would be the pyramid of people we sometimes can watch in a circus growing to higher levels by person after person climbing up where a clap of the pyramid is prevented by a continuous balancing of all the people involved.

The Sun employs about 450 billion "energy slaves", as I call them, to re-adjust the cards constantly and prevent a collapse of the house of cards, the biosphere. An "energy slave" is the equivalent of a quarter of a horsepower working uninterruptedly for twelve hours a day (in electric units 100 Watt). I derived this conversion from experience after the end of the Second World War, when the farmers lacked horses, and four strong adult men were needed to pull a plow. But this was only for brief periods e.g. one hour, after which they had to take a breather, and by no means for twelve hours at a stretch, as my energy slaves are assumed to represent.

So the Sun has 450 billion energy slaves for stabilizing the entire biosphere. It is important that we ask ourselves to what extent and in what ways we can jump around on this house of cards without it collapsing. We not only have six- and -a-half billion people on this Earth; these people also substitute a great number of powerful machines for their weak bodily strength. And for this we need energy, not just the energy that we take in ourselves by eating for our living existence, but also that with which we power our machines. If we convert the primary energy we need for this, then the present population of six- and -a-half billion people employ about 130 billion energy slaves. This is well above a quarter of the energy slaves that the Sun needs to stabilize the entire biosphere. How much is allowed?

This is a matter of life and death for us and other creatures. Observations on the resilience of our biosphere, which is reflected in changes in its biodiversity, suggest that the maximum sustainable burden on the biosphere from our interventions is less than a quarter of the stabilization input of the Sun, about 100 billion energy slaves. So the load we are presently putting on the biosphere is already above this, which appears to be indicated by a frightening reduction in biodiversity.

I would like to emphasize another point, as well. The evolutionary progress of life by the combination of differentiation and cooperative interaction of different beings results in a new whole, a new "Holon", as it was named by Arthur Koestler to distinguish it from a really isolated whole which has no "outside". A human being is also such a Holon having still a certain degree of interaction with the outside. But when we look at how many opposing forces act within us, at how many levels there are on which we have to achieve this equilibrium again and again, a highly-integrated, global structure of the system becomes apparent. So globalization is not per se something bad; on the contrary, it is a necessity that the differences stage a positive-sum game, which is not only mutually compatible, but where an advantage to one is on average to the advantage of the others, as well.

A group of people cannot say, this or that property is more important or valuable, and should therefore be globalized, and everything else is suppressed and treated as a troublemaker. We must ensure that the substance of all the cultures of this world is preserved, and incorporated into the desired overarching culture. For only the sum of all these cultures can create a world culture in this interaction, forming a new higher sophisticated and qualified Holon that exhibits a higher level of evolution. If one culture believes it is the one culture to which all the others must be subordinated, it would mean that we are tying all these legs together by force, and then, as a result of violating the paradigm of life, will all trip and fall together.

3.7. The New Worldview and the World of Everyday Life

"Wirklichkeit" or Potentiality contrary to Reality is a non-divisible, a-material system of relationships, a sort of "expectation field" for future possible energy-material manifestations. The future is essentially open, which is to say that it is not quite randomly open, but rather restricted by certain general conditions, which are related to what are called the conservation laws, and result from symmetry properties of dynamics. These include in particular the conservation of energy, which includes "mass" as a concentrated form of energy, according to Einstein, but also the conservation of electrical charge and others. They ensure that, in large-scale averaging, properties, the characteristic parameters of classical physics, are left and become graspable.

Non-living and living nature are no longer fundamentally different, but instead appear, respectively, as statically stable or dynamically stabilized statically unstable (metastable) agglomerates, or better "articulations", formed participants in the holistic one. An important consequence is that humankind and nature are not separated in principle in this way of seeing things, like everything else. This does not mean that qualitative differences cannot emerge on the large scale, such as, in particular, the possibility of creative shaping in the framework of the conditionally open future.

The openness in principle of the future has substantial consequences for our understanding of the world, its development, and our relationship to it. It is very important that we do not experience ourselves as parts of, but as participants in this cosmos, which emphasizes the relationship and not the substance, and that it is therefore up to all of us. We are co-creators. Future development depends on us. Of course, we cannot change the world arbitrarily, but we should realize that our decisions are always also contributing to the whole. On the other hand, however, our decisions themselves are always embedded and incorporated in something that we have in common with everything, which modifies the strictly private nature of each personal decision. Based on these ideas, we find that if we want to change the world, it is not necessary to enter into a dialogue with six- and -a-half billion people. We hardly need this dialogue to convince others, but only to remind them of what they already actually know.

For we have three- and -a-half billion years of the same evolution behind us. This development is not related to our energy and material realization, our bodies; because what is our "software", so to speak, is not enclosed in our bodies, but is everywhere, as it were, in other spaces, and we are all connected to it. This is a kind of Internet version that I can poll. I can find out where the others are, and also orient my own decisions accordingly. We are not alone in the world like a clump of matter, only connected to our surroundings by interactions; rather, we are embedded in the holistic one, so that we already know something which we can pass on, and in addition can reinforce certain processes that are genuinely sustainable.

3.8. The Finiteness of the Terrestrial Ecosystem — Energy and Order

In our simplified way of speaking, we can speak of parts of the cosmos, such as our solar system with its planets and our Earth as one of them. The separation of our Earth is indeed valid to a large extent, but of course it is not true in the strict sense. Gravitational attraction forces our globe into an elliptical orbit around the Sun. The biotope in which we live, the terrestrial ecosystem, is only a tiny part of this Earth, a thin skin about 20 km thick, about 10 km of atmosphere above the Earth's surface and 10 km below it; and in the narrower sense really on where there is habitable land. Most is oceans and mountains. We regard this biotope as relatively closed. This applies to a large degree to its material basis. Now and then, volcanoes still eject something from the Earth's interior onto the surface, providing it with a vital resource, heavy metals, which has also shaped human culture, if we think of the development of weaponry and industrialization. What is of predominant importance, however, is that our ecosystem is open with respect to the effective energy that irradiates it from the Sun. This forms the necessary precondition for all life, including us human beings. This source of effective energy (exergy) is available to us every day, a real gift without debt.

So in the background, there is the possibility of growth, but only due to this effective energy. We have to deal with matter in its various forms, for example the material we obtain from mines, carefully; otherwise we will lose it by dissipation. Copper, for example, does not really disappear, but is merely scattered all over the place by its multitude of uses. In principle, it could be collected together again, with a large input of energy; this is made enormously easier by careful recycling. The irradiated ordered solar energy is also reradiated again as less ordered thermal energy, so that in the end, what matters is its property of order, its syntropy or usually called negentropy (negative disorder).

Due to these circumstances, the demand for sustainability becomes vitally relevant to our ecosystem. We see a violation of sustainability most clearly in the economy of raw materials. We have fewer and fewer materials, some materials are running out, and in the end this means we have to collect them again. It becomes even more obvious in our energy economy.

We are constantly using up available power. Where can we get this energy from, if not from the Sun? How do we use these forms of energy? If we use non-renewable fuels, what happens to the non-recycled waste? We have our carbon-dioxide problem with climate change, but also the question of "safe" ultimate disposal of the spent radioactive fuel rods of our nuclear reactors.

What are our sources of energy today? In the past, we lived at first only from the Sun, by using solar energy for food and building materials indirectly, via photosynthesis by plants, just like the animals. Later, it was doubly indirectly, via animals as food and labor power. Then we discovered the fossil fuels coal, petroleum, and natural gas. These are solar energy collected over millions of centuries, which we are now mainly consuming, exhausting it in only two centuries. Since the middle of the last century, nuclear fission has been added as a new source of energy, specifically of the heaviest element occurring naturally, radioactive uranium. This was created many billion years ago in supernova explosions, and has survived because of its half-life of billions of years. So these are the sources of energy currently available to us.

By using fossil fuels, we are obviously doing something that is irreversible, and thus not sustainable. So we are living in a sort of bank-robbery society: we manufacture welding torches on our own, and use them to crack one of Nature's safes after another, and take out their riches and energy in order to manufacture more welding torches. Of course, this allows us to lead the good life. It is more lucrative to be a chauffeur for the Mafia than for an NGO, because they can pay a lot better. Clearly, such shortsighted exploitation is much more profitable and competitive than genuine adding of value.

So it is not just the resources that are the problem. At the moment, the fossil fuels are running out. Now, in particular in Germany, people are saying what should we do with nuclear power? We must have it, because the oil is running out. The same discussion is going on that we had thirty years ago. I don't want to go into that now, just one sentence: that is something which we absolutely must not do!

So first of all: how is solar energy disposed of? Sunlight shines in, and the solar energy consumed is taken back into outer space as thermal radiation. Fortunately, the night sky is black, and therefore makes an ideal garbage dump, as it were, to get rid again of the energy as low effective thermal radiation compensating the incoming high effective radiation of the Sun, otherwise we here on Earth would boil. The night sky is black because we live in an expanding universe.

Then we have the carbon dioxide, as an end product of fuels containing carbon, which is giving us problems. We take the coal from the ground, and blow it off burned as a gas into the atmosphere. And now we have discovered that, although these are very small quantities, they influence the world's climate in a very critical manner. Humanity will not perish if the climate changes, but we will have real difficulties. I would like to point out that it is not just these 24 billion tons of carbon dioxide which we are putting into the atmosphere every year. We are also ruining the micro-flora and micro-fauna in humus with chemical fertilizers, causing a further quantity of carbon dioxide, comparable to the total emissions of all cars, to be emitted into the atmosphere. This is only meant as an indication of all the other things we need to watch out for, such as protecting the earthworms which, in the temperate zones, pull the leaves underground, so that they do not come in contact with oxygen creating CO_2, and form carbon rich humus instead. Everything is connected.

3.9. Resilience of the Biosphere

Today, there is a lot of talk about the limits to fossil fuels, especially about shortages of petroleum, whose production peak is expected around now, and we ask ourselves about a suitable replacement, with nuclear power being considered more and more again. Almost nobody talks about how much we can actually carry on in our biosphere without endangering its metastability irreversibly. To put the specific question: how many people can live on Earth, i.e. in our smaller biosphere? Better yet: what is the maximum amount of energy we human beings may use for our purposes, or the maximum number of energy slaves we can employ? For the six- and -a-half billion people are not the critical factor, if compared with their 130 billion energy slaves. We need birth control for energy slaves! I already mentioned that about 100 billion energy slaves represent the limit to the burden on the biosphere. That is more than a quarter of the stabilization energy of the biosphere, after all.

This demonstrates the enormous robustness of the biosphere, a firm that has been in business for three- and -a-half billion years without going bankrupt. It will not break down at the first kick in the shins, and can put up with one in four being a troublemaker, but not more. I don't know how well a human society or a family would work if one in four were a troublemaker, whether it could stand that.

3.10. New Life Styles

It is simply a matter of stability, and therefore, in my opinion, the greatest challenge for us: can we develop life styles that are compatible with not overburdening the carrying capacity of the biosphere, i.e. maintaining its dynamic stabilization, and not exceeding a total of 100 billion energy slaves employed. For six- and -a-half billion people on Earth, this means 15 energy slaves per person — what they call 1 *ecoson* in Japan — and not more. Now you might ask where all these energy slaves are? They are very unevenly distributed, in any case. On average, one inhabitant of the U.S.A. has 105 energy slaves, we in Central Europe have 53, the Chinese have 10, and in Bangladesh, people have only one. In total, we have about 130 billion energy slaves. Our task would be not to have more than fifteen per person. If we have more, then somebody else in the world must have fewer.

This is a limitation that has nothing to do with the availability of the resource, but with the stabilization of the system in which we are embedded. Of course, this is a bitter pill to swallow. But it is also a beneficial pill, because this means it is not necessary to prepare for an energy future in which in the year 2040 perhaps 7.6 billion people will be living on Earth, all of them striving to live like Americans today. That is simply not possible. For then we would have 900 billion energy slaves, and would need to fetch eight more Earths up out of the cellar. Anyhow this would be impossible for many other reasons, such e.g. as the supply of drinking water.

Therefore, we must develop life styles that are feasible within this limit of fifteen energy slaves. This is not a life on bread and water in a mud hut, but that of the average Swiss in 1969 with the present efficiency level. Things were not that bad then. A reduction by a factor of three or four is not such impossibility here in Europe. We can manage with far fewer energy slaves, because in reality, we do not make them all work for us. Actually, they are only banging about the place, characterizing how much energy we waste pointlessly. This is a question of efficiency.

We must develop life styles within the necessary limits that are not only viable, but also fully worth living. This calculated limit of fifteen energy slaves can be observed without any loss in the quality of life, in my opinion. To give but a few examples: if firstly, we use the services of our energy slaves much more efficiently; secondly, throw less energy out the window by means of better thermal insulation of our housing; thirdly, shorten the constantly lengthening, energy-consuming haulage routes for items of daily use substantially, and brake or abandon our increasing tendency to exaggerated speeds. If you tear through the countryside in your car, you actually do not need any energy for your movement per se: any ride where you return to your starting point, your garage, does not consume energy, since your potential energy does not change. The gasoline you have burned in the process simply serves to compensate for the heat which you have emitted on your route, via the internal-combustion engine and a variety of friction forces, to the air, the tires, and via these to the pavement. The faster we drive, the more heat we generate, and the more energy we consume accordingly. A bit slower is quite possible and can be reasonably expected; indeed, this is a source of a better quality of life. We simply have to devote more attention to all these things, and alter some of our habits accordingly.

Some restrictions we may find difficult, especially with regard to mobility. Mobility is a difficult topic, because it has great psychological importance for us. There is a strong desire for mobility, which goes far beyond simply getting from here to there. It should be added that what we understand by mobility is in reality immobile. Getting out and walking is much more mobility in the language of physics than sitting in a motorcar strapped in. It is only the car that is mobile. But I am not concerned with the mobility of the automobile as much with my own. We simply need to use other methods to satisfy this demand. As I said, there is plenty of room for other options.

The central question is not: where will I find sufficient resources in the future for our energy demands? Will it be mainly solar energy, or continue to be fossil energy, or even nuclear power or something else entirely? Rather, the main question is: how do we handle the energy available to us, what do I do with it? The answer to the question of the resources is simple: the exergy, the usable or effective energy, which we can utilize daily under the limits demanded can be obtained entirely from the Sun easily. The total amount of energy which the Sun delivers to the Earth's surface is greater by a factor of 2,000 than what is pumped through the biosphere, and 8,000 times more than the maximum amount we humans can utilize without overburdening the resilience of the biosphere. So it is not true that the exergy we require cannot be satisfied sufficiently from the Sun.

We are not failing because it is impossible. Why we have failed to date? The answer is the system of centralization and power politics. Whenever a source of energy is proposed that is decentralized, we meet solid resistance from the powerful. This is one reason for nuclear power, a concentrated form of energy — quite apart from the fact that nuclear power is unaccept-able for very many reasons. I will mention only one of them, which for me as a nuclear physicist is a killer argument against nuclear power. We human beings should never ever develop technologies which in their worst possible accident, lead to damages that is no longer acceptable, for which we cannot take responsibility. And this demand must apply regardless of what probabilities have been calculated for the occurrence of such an incident. Why is an accident in a nuclear power plant unacceptable? It is not because of the danger that all humankind might perish, which could occur with nuclear weapons in many circumstances, but because the vast majority of the people who are affected could have the nothing to do with it.

Many people counter this argument by saying that humankind must simply learn to take such risks. It is usually men over sixty who say this. They can easily take the risk. The probability that it will affect them personally is not very large. But we must remember that if something happens, the great majority of the people affected will be the ones who have absolutely nothing to do with the nuclear power plant or its benefits, because they were not born until much later. To put it another way: it makes a big difference whether I play Russian Roulette with my own head, or with my child's head. The latter is simply forbidden, unacceptable from the point of view of human rights. I may not assume a risk for which I will not have to take the consequences if it goes wrong. And therefore I must not use a technology which does this in the case of an accident. For I have no control over it in time or space. And anyway — which cannot be emphasized often enough — we do not have our backs to the wall in the energy problem by any means.

It is so easy for such an accident to occur, if we keep all the possibilities in mind. I know the ways in which probabilities are calculated or estimated. Statistical calculations are no help at all in this case. Probabilities are relative frequencies of something happening and once are never frequent. You can see this by the fact that we cannot find any insurance company that is willing to provide full coverage for a nuclear reactor. What significance has the statement that the calculated probability of a certain incident is one in a million? It can happen right now or this year, or not for a million years. What must not ever happen simply cannot be insured. So why should I impose the risk on others, if even an insurance company is not willing to assume it?

3.11. Fault Tolerance, Creativity and Change

And there is another aspect that makes things worse: if something goes wrong, it is usually not what we calculated, but something that nobody thought of. Just imagine that you are required to think of everything that might happen.

Physicists are accustomed to designing apparatus to the best of their knowledge, which they will then use to experiment with for a few years. But when you switch on such an apparatus after it has been completed, as a rule it will not work properly in the beginning, because something was overlooked or misjudged. People do not get excited about this, because nobody can anticipate or calculate everything accurately in advance. But the less imagination somebody has, the more reliable he will think his apparatus is.

However, the shortcomings and hazards of a machine are not always due to its design, but also arise from the people who handle it, switch it on and operate it, and interpret its behavior. Human beings are simply not fault-free. This is why many people recommend handing over the operation entirely to robots. But who is capable of writing an error-free program for such a robot? What does the demand for fault-free human beings really mean? Wouldn't it mean driving all the creativity out of them? For after all, it is people's creative gifts that permit deviations from the norm, which we often designate as errors.

Technology is made for people, and not people for technology. We want to help creative human beings, *homo sapiens sapiens*, achieve their optimum development. We want to live in a world in which everybody can be creative and can and may make mistakes in doing so, without running the risk of endangering the robustness of our terrestrial world, the fruit of three- and -a-half billion years of evolution, in an unacceptable fashion. As participants in the biosphere, it is our responsibility to recognize the inherent instability of live and promote its dynamic equilibrium. Einstein was right when he said, with regard to the atom (and I would add: to nuclear power plants), that a mouse would never have thought of building a mouse-trap. In other words: we need a technology that is compatible with creative human beings, who are allowed to make mistakes. For the mistake they make today may also open the future of tomorrow. This is not a life of renunciation and sacrifice, but rather a life of openness and flexibility, of vitality in the deeper sense. This is the new life style we should strive for, with which our problems will be and remain solvable.

Chapter 4

LOW-CARBON AND SUSTAINABLE DEVELOPMENT

*Kunmin Zhang**

4.1. Introduction

The history of humanity's survival and development is one of interaction, common development and continuous evolution between humans and nature. For what development pattern one should choose is always the vital topic troubling human beings. From the worship of nature to the self-confidence of technologies, to the persistence of the idiom "man can conquer nature", then to the recognition and implementation of sustainable development, this is a hard process of cognition, practice, recognition and re-practices (Zhang, 1997).

All environmental problems, such as local air and water pollution, soil erosion, regional acid rain, global ozone layer destruction, endangered species and climatic change, are almost associated with the process of human development. However, these bring great threats to human survival and development.

4.2. Sustainable Development Strategy as a Researching Target of Human Beings

4.2.1. *A long history searching for sustainable development*

The concept of sustainable development is long standing. The "Tin Lv" (the law of nature) of Chinese Qin Dynasty (2nd century B.C.) clearly reflected

*Kunmin Zhang, Professor at Tsinghua University, former deputy director general of the China Environment Protection Agency (CEPA) and current vice chairmen of the Chinese Society of Sustainable Development.

the ideas of sustainability. This is possibly the earliest sustainability law in China, and even in the world. Some economists from the western countries, such as D. Ricardo, T. R. Malthus and J. Mill, also recognized substance-based restrictions on human consumption, namely that the range of human economic activities had ecological boundaries.

Since the 1960s and 1970s, with the appearance and aggravation of "the public hazard" and the shock of the "energy crisis", a worldwide debate has begun on "the limits to growth". With human cognition deepening that equity (intergenerational equity and intra-generational equity) is the goal of social development and with the confirmation of global problems, the idea of sustainable development was shaped gradually in the 1980s. In November 1983, the United Nations established the World Commission on Environment and Development (WCED). The commission formally proposed sustainable development in the 1987 report on "Our Common Future".

In 1992, the United Nations Conference on Environment and Development (UNCED) adopted the "Rio Declaration", "the 21st Century Agenda" and other documents, which crystallized contemporary people's deepening desire for sustainable development. They were formally accepted by the governments of all countries, which indicated humans' denial of the previous development pattern and also reflected humans' yearning for new development goals and roads. Around that period, many scholars and international agencies proposed many creative views that highlighted the global research and practice for sustainable development strategies. Sustainable development was the goal that developing countries and developed countries were both trying to achieve, and most of the developing countries were willing to join the practice of sustainable development. This was the vital reason that sustainable development theory became fashionable. Table 4.1 shows the chronology of the strategic thought of sustainable development.

4.2.2. *A core framework of sustainable development and sustainable economics*

One of the founders of sustainable development theory, Munasinghe (1996), considered that the core framework of sustainable development

Table 4.1. The chronicle of sustainable development and strategy.

The 18th century	The Western Industrial Revolution began.
The 19th century	Research on sustainable yield of trees. In 1872, the British scientist Smith found acid rain in an industrial city.
The early 20th century	Research on sustainable yield of fisheries.
The middle 20th century	Public hazard incidents appeared constantly in Europe, America and Japan.
1962	U.S. scientist Rachel Carson's "Silent Spring" published.
1970	On April 22, more than two thousand people in the U.S. took to the streets, calling for environmental protection.
1972	Club of Rome's "limits to growth" published. On June 5, the United Nations conference on human environment, held in Stockholm, decided to establish the United Nations Environment Programme (UNEP).
1974	The United Nations conference on human settlements held in Vancouver.
1980	IUCN/WWF issued "World Conservation Program" and first proposed the words "sustainable development".
1981	American Lester Brown's "Building a Sustainable Society" published.
1984	The United Nations set up "World Commission on Environment and Development" (WCED).
1985	UNEP organization concluded the "Vienna convention" on ozone layer protection.
1987	WCED published "Our Common Future" and proposed the definition of sustainable development and a series of recommendations. UNEP organization negotiated and passed the "Montreal Protocol" on ozone layer protection.
1988	UNEP and the World Meteorological Organization (WMO) set up the "Intergovernmental Panel on Climate Change" (IPCC).
1989	Environment Ministers from 69 countries gathered in the Netherlands and presented "Nuodeweike Declaration" for air pollution and climate change issues. The 44th term of the General Assembly adopted resolution No. 228, and decided to prepare the United Nations Conference on Environment and Development (UNCED).

(*Continued*)

Table 4.1. (*Continued*)

	UNEP negotiated and passed the "control of transboundary movement of hazardous wastes and their disposals, the Basel Convention" (effective in 1992).
1990	UNCED 1st substantial preparatory meeting held in Nairobi.
1991	World Bank, UNEP and United Nations Development Program (UNDP) set up "Global Environment Facility" (GEF). "Framework Convention on Climate Change" and "Convention on Biological Diversity" started to be negotiated.
1992	UNCED held in Rio de Janeiro, and passed the "Rio Declaration", "Agenda 21" and the "Declaration Forest Principles". The "Framework Convention on Climate Change" and the "Convention on Biological Diversity" opened for signing. "China's Ten Countermeasures on Environment and Development" published, and China Council for International Cooperation on Environment and Development (CCICED) established.
1993	"Basel Convention", the 1st Conference of the Parties. The United Nations Commission on Sustainable Development (UNCSD) 1st annual meeting, to appraise the process of "Agenda 21" year after year.
1994	"China's Agenda 21" published. "Convention on Biological Diversity", the 1st Conference of the Parties.
1995	"Framework Convention on Climate Change", the 1st Conference of the Parties (COP1). "Desertification Convention" finished negotiations, opened for signing.
1997	United Nations General Assembly 19th special session, made comprehensive review with the progress of "Agenda 21" in 5 years. "Framework Convention on Climate Change", the 3rd Conference of the Parties (COP3) passed the "Kyoto Protocol".
In 1999	"Rotterdam Convention on the Prior Informed Consent Procedure for certain hazardous Chemicals and Pesticides in international trade" (PIC) opened for signing.
2000	"Cartagena Protocol on Biosafety" and "Convention on Biological Diversity" opened for signing. The United Nations "Millennium Development Goals" (MDGs) for the next 15 years launched.
2001	"Stockholm Convention on Persistent Organic Pollutants" (POPs) opened for signing.

(*Continued*)

Table 4.1. (*Continued*)

2002	World Summit on Sustainable Development (WSSD) held in Johannesburg, South Africa, passed the "implementation plan" and emphasized five main points: water, energy, health, biodiversity, and agriculture.
In 2005	"The United Nations Millennium Ecosystem Assessment Report" released.
2007	IPCC published its 4th assessment report, and more data and facts indicated that global climate change "very likely". COP15 passed the "Bali road map".
2009	Copenhagen Meeting (COP17) reached the "Copenhagen Agreement" which was not legally binding.
2010	Reserved Cancun meeting (COP18).
2012	The United Nations decided that the summit conference on sustainable development will be held again in Rio de Janeiro in the year of 2012.

Source: Annual reports of "*China Environment Yearbook*", Japan "*Earth Environment Manual*", and the events that the author personally experienced and sorted out.

depends on the following basic principles and methods:

(1) to make development more sustainable;
(2) sustainable development triangle and balance (see Fig. 4.1);
(3) to surpass the traditional boundaries for better integration;
(4) from data collection to policy implementation and feedback, the whole cycle fully applies practical analysis tools and methods.

Based on sustainable development science and a substantive interdisciplinary framework, the three dimensions of economics, society and environment need to be balanced to achieve sustainable development. In the above, traditional sustainability development, the key point is harmony. There are different conditions in different countries, and the emphasis is placed on different points. In the South, priorities include sustainable development, consumption and growth, poverty reduction and equity. However, the mainstream literature about sustainable development in the North tends to focus on pollution, growth in non-sustainability and population.

Munasinghe *et al.* (1996) determined that sustainable development covered all human activities, including the complex interactions of society,

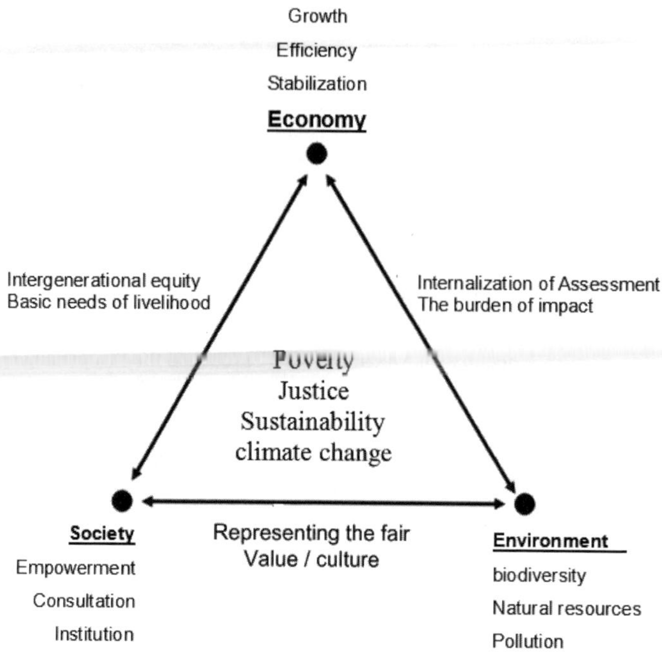

Fig. 4.1. Sustainable development triangle — key elements and consumption relationship.
Source: Munasinghe (1996).

economics, ecology and physical systems. Accordingly, sustainable economics encourages practitioners to surpass the traditional boundaries that are limited by the criteria, space, time and stakeholder viewpoints and operation, and to make great efforts to integrate new solutions. "Sustainable economics" is a new term that emphasizes the focal point on sustainable development and stresses that this is a neutral method with no bias or hegemony.

Sustainable economics is a practical, interdisciplinary framework. It attempts to establish a rich, overall design framework for guiding policies, while various elements (from many other disciplines of the principles, methods and tools) provide a strict "minimalist" approach to the building blocks and the foundation. This integration requires the use of core subjects, such as ecology, economics, sociology, anthropology, botany, chemistry, demography, ethics, geography, law, philosophy, physics, psychology and zoology. Other subjects such as engineering, biology and other technological subjects, are equally important. It is the complement, but not the replacement of other methods for sustainable development.

4.2.3. *Challenges facing human society*

The Former UNDP Director James Speth believed that sustainable development was the most important challenge that human society would face in the 21st century. It affects everyone on the planet, thus we are all stakeholders. Traditional development patterns focus on economic growth with a material basis. Problems such as poverty, hunger, disease and inequality still become increasingly severe in the poor communities of numerous low-income countries, as well as in industrialized countries. At the same time, new challenges are appearing, such as environmental degradation, violent conflict, climate change and runaway globalization, making the already existing problems worse.

At the global level, thousands of scientists in the United Nations Intergovernmental Panel on Climate Change (IPCC) further confirmed that human activities result in the increase of greenhouse gas emissions, which is likely to lead to the disastrous consequences of climate warming. The "Millennium Ecosystem Assessment report", initiated by Kofi Annan, the former U.N. Secretary General, and completed by the world's most important ecologists, showed the ecosystem service functions, which all life on earth depend on, were declining in a process of historical change. These studies invariably call on people to take action as soon as possible to reverse this dangerous situation. Indeed, those who live in extreme poverty and spend less than one dollar each day need to be helped with sustained economic growth which will reduce poverty. Keeping a balanced development of the economy, society and environment is the essence of sustainable development.

4.3. Sustainable Development in the Context of Climate Change

4.3.1. *About the data of global climate change*

Responding to the "conspiracy theory" of climate change, China Meteorological Administration chief Zheng Guoguang recently cited the following data (*People's Daily, June 23, 2010*), IPCC reported that from 1906–2005 global average surface temperature increased by 0.74°C according to a large and long-term observational data set. Furthermore, other observational evidence also verified global warming in different ways. For example, during the 20th century global average sea level rose by 0.17 m and the area of seasonally frozen ground in the northern hemisphere decreased by about 7%. Since 1978, the Arctic sea ice area has been shrinking at a rate of 2.7% per decade. Nearly 30 years of upper air exploration and satellite observations show that temperatures in the atmosphere 8–12 km above the earth's surface (earth's troposphere) also appear consistent with warming of the earth's surface. The latest observation from the World Meteorological Organization indicates that 2000–2009 was the warmest decade since worldwide systemic meteorological observation started in 1880. From December 2009 to February 2010, Canada experienced their warmest winter on record while in China the average temperature was 0.69°C higher than normal. In 2010, the global earth surface temperature was 0.69°C higher than normal from January to April, which is also the highest recorded temperature over that period since 1880. In April 2010, the global earth surface temperature was 0.76°C higher than normal which is also the highest over that period since 1880. During 1906–2005, the earth surface temperature in China rose by about 1.1°C.

Both natural factors and human activities influence climate change. Over geological history, the main factor affecting climate is nature. Since the industrial revolution, however, the concentration of carbon dioxide and other greenhouse gases in the atmosphere have been increasing due to large-scale use of fossil energy such as coal, oil and so on. The global average concentration of carbon dioxide in the atmosphere has increased from 280 ppm before the industrial revolution to 385 ppm in 2008. This is significantly outside the range of natural variability over the past 65 million years. So there is no denying that human activities are the main reason for global warming.

4.3.2. *China has suffered the pain of climate disasters in the past and it will be more vulnerable to climate change*

Pan Jiahua pointed out that the key reason is that the Chinese natural environment is very fragile (Pan, 2009). For example, in 1931 Yangtze floods claimed 14.5 million lives and thousands of people were left homeless. Among a population of 1.3 billion, 800 million are living in the eastern coastal area which is the most economically vital region. The more affluent areas are also concentrated there, especially the Yangtze River delta, Pearl River delta and Bohai rim. Over the past 30 years, sea level has increased 2.6 mm annually and this trend continues. The population density is 890 persons/km^2 in the Yangtze River delta region where the land scale of the 15 largest cities only accounts for 1% of the total land area in China. However, in 2008, the GDP of this area accounted for 17% of the whole of China. In the northwest, people's lives are highly dependent on the Himalayas and the Tian Shan snow. Warming means the agricultural oasis there will gradually disappear.

Since 2008, China has experienced extreme weather events over and over again. The snowstorm during the 2008 Chinese New Year covered 18 provinces and autonomous regions. From July 2009 to May 2010, 6 provinces in the southwest suffered continuous drought, the greatest disaster in 50 years to happen in Guizhou (Renmin Net, June 23, 2010). Since the flood in 2010, 10 southern provinces and regions have suffered 10 large-scale heavy rainfalls in less than two months. In Guangdong, some areas have experienced heavy rainfall, more than any year in the past 500 years (Xinhua New Agency, June 26, 2010).

The experiences of China and other countries all over the world shows that through development people can gradually adapt to the change of climate. In 1998, the Yangtze River had a flood of similar scale to that of one in 1931. However, the losses in 1998 were small compared with 1931. The reason is simply that the dam is stronger and more resources can be used for flood control. Before 2000, extreme weather events caused economic losses accounting for 3% to 6% of China's GDP. However, during the past 10 years, the losses account for only 1% of GDP or less, though the absolute amount of damage may be greater. Before 1978, typhoons claimed many lives and destroyed many houses in coastal areas each year. Now buildings can withstand much stronger typhoons. Furthermore, early warning systems can effectively allow people to make full preparation in advance, and water-saving technology can reduce people's demand for water.

4.3.3. *Chinese leaders' promise over the climate change issue*

On September 8, 2007, at the 15th APEC Leaders Meeting, President Hu Jintao made four proposals about climate change. The second one stressed the need to, "adhere to sustainable development". He said, "basically climate change is a development issue and so it can be resolved only in the context of sustainable development. Sustainable development requires economic growth, social development and environmental protection, all at the same time. Stagnation in response to climate change or the one-sided pursuit of development ignoring climate change are both undesirable. To meet the demand of sustainable development and the challenge of climate change, a new production and consumption system should be established, the energy structure should be optimized, industrial upgrading should be promoted, the low carbon economy should be advocated and a resource-saving, environment-friendly society should be built".

4.4. Low-Carbon Economy is the Key Strategic Point for Sustainable Development

4.4.1. *The origins of low-carbon economy*

Considering climate change, energy security, and international competitiveness, Britain first put forward the concept of "creating a low carbon economy" in the "White Book of Energy". Then *Economics of Climate Change — The Stern Report* (Stern, 2007) was issued in 2006, domestic legislation on climate change was passed in 2008, and the "Plan of Low-Carbon Transition" was issued in 2009.

In order to construct a "low-carbon society", Japan decided to change the way that the emissions reduction of a factory depended only on the factory itself.

In China, the "low-carbon economy" development path was put forward in the "strategy of restraining climate change in China", the 25th chapter in "National Evaluation Report of Climate Change" issued in December 2006 (Science Publisher, 2006). In this report, it was written that there is a need for "technological innovation, developing and extending advanced energy conservation techniques, improving the utilization ratio of energy, developing techniques of renewable energy sources and advanced nuclear energy, as well as optimizing energy structures and reducing CO_2 emission with hydrogen energy which is efficient, clean and low-carbon, improving land management and utilization, reinforcing forest reserves, increasing forest cover area, conserving ecology areas and increasing carbon absorption".

4.4.2. *A positive attitude towards Low-Carbon Economy by Chinese government, academics and the public*

It is not only the results of the political study of "the Scientific Outlook on Development", but also our past experiences with environment and development that has led more and more people to the understanding that many environmental problems result from conventional economic growth patterns: especially the unclean utilization of coal. After entering into the new century, the energy elasticity coefficient is almost 1 because of the rapid increase in energy consumption; this is doomed to not last. Meanwhile, the energy endowment in China (China's per capita carbon emissions) is only half of the world's based on energy reserves source, which means that this kind of rapid energy consumption could not be permanently supported. From the Industrial Revolution to 2005, the emissions load of CO_2 has only has been about 60 tons per capita, which is much lower than the thousands of tons per capita in developed countries. But because of the huge population, the gross emissions load of CO_2 in China has been first in the world, and the emissions load per capita has also approached or even slightly exceeded the world average (although it is still much lower than in OECD countries). China has always paid attention to energy conservation, but it is not enough. It will be very difficult to control, or decrease CO_2 emissions if it becomes too late to make a detailed plan of low-carbon development which emphasizes "climate change, energy security and international competitiveness". If China fails to mobilize the society and international cooperation, more positive actions will be necessary because of the seriousness of the situation.

4.4.3. *Economic growth should gradually decouple from carbon emission*

China stated that "carbon emission intensity" (emissions of CO_2/unit GDP) should be a mandatory control target in the 11th Five Year Plan. Compared with "energy intensity" in the 11th Five Year Plan, "carbon emission intensity" puts forward higher requirements not only about energy efficiency goals but also about the reduction of carbon emissions per unit GDP. The actual aim of this is the realization of decoupling between economic growth and carbon emissions. The reciprocal of carbon emission intensity is carbon productivity (unit GDP/emission of CO_2), which is one of the most

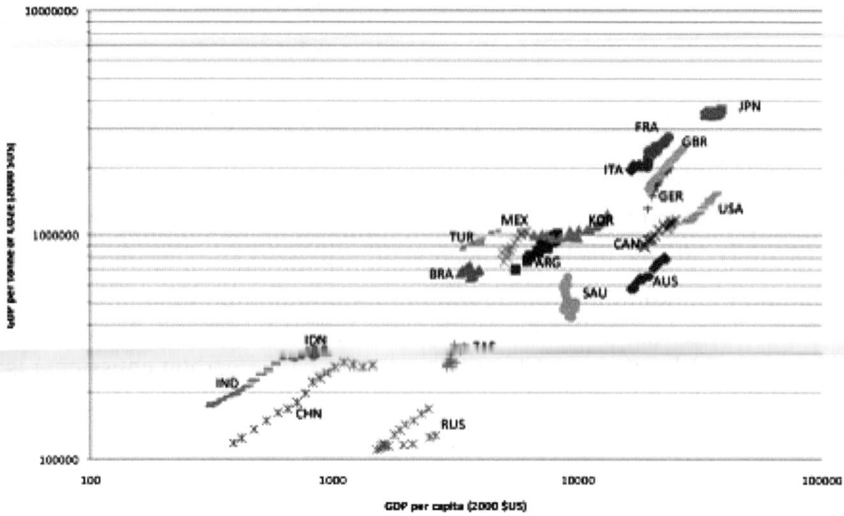

Fig. 4.2. Logarithmic relationship between carbon productivity and GDP per capita of Group of 20 countries.

Source: The Climate Institute and E3G (2009). *G20 Low Carbon Competitiveness*. Vivid Economics, p. 43. www.climateinstitute.org.au or www.e3g.org.

important indicators to measure international competitiveness. The comparison of members of the Group of 20 countries is shown in Fig. 4.2. As shown in the graph, China's GDP is only slightly smaller than Japan's, but the GDP per capita of China, which is larger than only Indonesia and India, is in 18th place. Carbon productivity in China, even less than Indonesia's and India's, puts it in 19th place. The gap between China and developed countries is about an order of magnitude. Larger carbon productivity means the ability to create higher GDP with less carbon emissions. In consideration of climate change, energy security and international competitiveness, larger carbon productivity means more contribution to the mitigation of climate change, less dependency on fossil energy and a higher level and larger scope of new energy application technologies. In short, larger carbon productivity means strong international competitiveness. The starting points, status and responsibilities of developed countries and developing countries in these three areas are different; the low-carbon economy, represented by the carbon productivity index, attracted attention all around the world.

Carbon emission intensity and carbon productivity are two indicators that relate to each other closely. Those two indicators reflect the degree of sustainable development of a nation or an area. Human population on earth keeps increasing, fossil energy is limited and carbon emissions continue to increase. Measuring low-carbon development, which aims at gradually decoupling economic growth from carbon emissions, could show us whether this economic growth model has a qualitative change or not. It can also show us whether the economic growth model is being transformed to sustainable development or not and put an end to the anti-science growth mode that only pursues GDP. At the same time the "recycling economy" that puts an emphasis on the recycling of resources and waste is easily quantified and operated. A low-carbon economy that highlights the key word "carbon" points out the main point of sustainable development. If a necessary and reasonable system is to be established, it must become a development model that is much easier for monitoring, reporting and verification than "Cleaner Production", "green economy", the "ecological economy" and so on.

4.5. Combining Low-Carbon Economy with Sustainable Development

4.5.1. *Interaction of climate change and sustainable development*

In its 3rd assessment report of 2001, the IPCC provided a framework for a comprehensive evaluation model which included the causal relationship between climate change and sustainable development (see Fig. 4.3). In the lower right quadrant, each socio-economic development path (driven by population, economy, technology and governance) gives rise to different levels of greenhouse gas emissions. The solution should include two types of adaptation and mitigation, which are detailed in the economic assessment of Stern.

4.5.2. *Policy instruments for a sustainable development and low-carbon economy*

The consensus at the RIO conference in 1992 was that environmental protection and economic development are inseparable. Macroeconomic policy directly affects the environment. Some effects are positive, such as

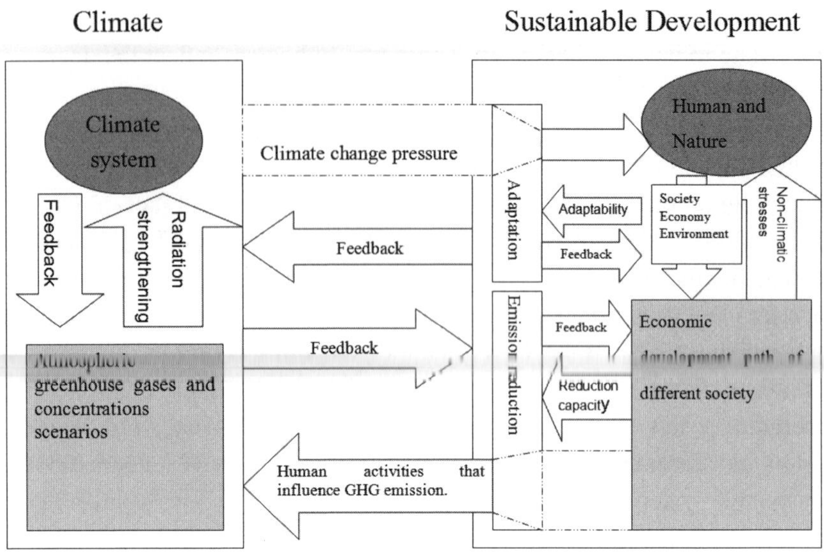

Fig. 4.3. The interaction between climate change and sustainable development.

correcting price distortions to improve resource use efficiency; some are negative, such as economic liberalization that may lead to increased production, increased pollution and the emergence of new environmental problems. In the past, the most common method of environmental regulations (command and control) involved the direct management of market behavior, but this has proven to be both expensive and complex. Use of the market, or creating a market where no market existed is a policy innovation: through reform of subsidies in order to eliminate price distortions; through the use of taxes and fees to raise prices to reflect social costs; to create a new market for sewage permits or resource development rights that can be traded. Informing the public and promoting public participation is an efficient method, which can be done through environmental education, eco-labeling, establishment of a pollutant emissions database or directly involving the public in the resource management system. The World Bank summed up these four types of policy instruments into a policy matrix in Table 4.2. Low-carbon economics and sustainable economics are closely related in theory and application of these methodologies. From a resource perspective, they include fossil fuels, renewable energy and new energy; from a pollution

Table 4.2. Policy matrix — The means of sustainable development.

	Policy instruments			
Theme	Using market	Creating market	Implementing environmental laws and regulations	Encouraging public participation
Resources	Reducing subsidies	Clarification of property rights	Standard	Public participation
Management	Environmental tax charge	Decentralized authority	Ban	Information disclosure
And	for use	Tradable permits	Quota of the permit	
Pollution	Deposit-refund system	Certificate/ development rights		
Control	Special subsidies	International compensation system		

control point of view, they include greenhouse gases, conventional pollutants and special pollutants.

In June 2009, the 17th European Environmental and Resource Economics Annual Conference was held in the Netherlands. More than 700 scholars from Europe, the United States, Japan and China took part in this conference. The themes of the General Assembly were: "Global Crisis and Natural Resources: Developing Outlook," "Wealth, Social Welfare and Sustainable Development," "Environmental Crisis: Past, Present and Future" and "Face the Challenge of Climate Change". Breakout sessions offered 576 academic reports, mainly on economic growth and the environment, trade and the environment, sustainable development and green accounting, environmental value assessment methods, environmental economic policy and environmental management, climate change policy, energy, water and natural resource management, ecosystem services and biodiversity and food and the environment. In particular, climate change has become an important area of research in environmental and resource economics. Vulnerability, adaptation and mitigation, climate policy and the European Union during the post-Kyoto climate policies were discussed in separate arrangements for a sub-venue. This fully reflects the degree of attention paid to climate change and related environmental policy, but it also partly reflects low-carbon economics' relationship with the environment and resources, or sustainable development, in a broader study of the close link between them. Such close contact is not accidental.

4.5.3. *Through combination of low-carbon economy and sustainable development to actively promote the 3rd environmental revolution*

Pearce and Wofford (1993) wrote in their report that the first environmental revolution happened in the late 1960s and 1970s, when people debated the relationship between environmental quality and economic growth, turning to pay more attention to the environment. The second environmental revolution happened in the late 1980s and early 1990s, which redefined and expanded many existing concepts, and proposed sustainable development problems, shifting the discussion to implementing economic growth that would benefit the environment. Obviously, the correct environmental policy will benefit economic growth and, if it is well managed, economic growth will be in favor of environmental protection. Combining development with environment yields the philosophical idea of sustainable development, which is bound to accept applied environmental economics. Environmental economics as a discipline was formed in the 1950s and 1960s and covers the analysis of the economic cost-effectiveness of environmental projects and policy evaluation to the practical implications of absolute resource limits of the traditional economic growth model; the search for reasons and solutions for resource degradation; and efforts to value ecosystem services and address a series of global environmental problems.

In the last 50 years, environmental economics has made considerable progress. Now making the scientific analysis and classification of the related parts of low-carbon economics, environmental economics and sustainable development science is ready for certain conditions.

In practice, the author thinks that the world's population will continue to increase; poverty must be eliminated, the environment must be protected; the level of welfare should be improved; depletion of fossil energy will be just around the corner; a new energy revolution is about to begin. Developing the low-carbon economy in China is not the result of external pressure, but of inherent demand (Zhang, 2009). China's development of the low-carbon economy is not only necessary but also possible. The domestic and international situations give China's low-carbon economy a sense of urgency. On the basis of the last two global environmental revolutions and a large number of pioneers' contributions, coupled with the proposed low-carbon economy in the new century and the promotion of a new energy revolution, it is possible that orderly low-carbon development can gradually solve China's environmental issues (such as coal burning pollution), energy problems (renewable energy and new energy) and the leap-forward development issues (change the advantage of cheap labor force into all-round clean technologies).

Deep research is needed on the low-carbon economy, including learning from other scholars, such as Magner, who recommended that we "surpass the traditional boundaries for better integration" and "from data collection to policy implementation and feedback, the whole cycle fully applies practical analysis tools and methods". In practice, consciously combining low-carbon economics with sustainable development will produce infinite power. If operated properly, it could drive the 3rd global revolution and sustainable development.

References

Munasinghe, M. (1992). "Environmental Economics and Sustainable Development." Paper presented at the UNCED, Rio de Janeiro, Environment Paper No. 3, World Bank. Washington D. C.: USA.

Munasinghe, M. *et al.* (1996). *Defining and Measuring Sustainability*. New York: The Biogeophysical Foundation.

Munasinghe, M. (2007). *Making Development More Sustainable*. Sri Lanka: Vijitha Yapa Press.

Pan, J. (2009). "Solve the Climate Change Issue through Sustainable Development". (in Chinese) Bureau of International Information of the U.S. Department of State, net news, September 2009. http://www.america.gov/mgck/media/pdf/ejs/climate.pdf#popup

Pearce, D. W. and Warford, J. J. (1993). *World Without End: Economics, Environment, and Sustainable Development*. Washington, D. C.: The World Bank.

People's Daily, Renmin Net, June 23, 2010, http://news.163.com/10/0623/10/69RTSAS3000146BD_2.html

Xinhua News Agency, June 26, 2010. http://news.xinhuanet.com/society/2010-06/26/c_13370727.htm

Zhang, K. (ed.) (1997). *Sustainable Development*. Beijing: China Environment Science Press.

Zhang, K. (2009). "Develop Low-Carbon Economy Needs a Sense of Urgency." (in Chinese), *Qiushi*, 23, 29–35.

Chapter 5

LOW-CARBON POLITICAL ECONOMY

*Angang Hu**

5.1. Introduction

5.1.1. *Climate change involves political economy*

The well known saying, "It does not matter whether the cat is black or white, as long as it catches mice," was said by former Chinese leader Deng Xiaoping for the first time in the early 1960s to indicate that he was against the policies of Mao Zedong. At the time, this statement had not been implemented, but it began to take effect after Mao Zedong died. The doctrine of the "Two Whatevers" (*liangge fanshi*) was successfully removed and economic reforms started. If the "cat theory" is used to summarize the characteristics of China's reform and development, then different times require different color cats. Mao Zedong's era was the red cat theory, Deng Xiaoping's era was the black cat theory, and now we need a green cat theory because we need a new reform which will develop a new green economy and reduce carbon emissions. If Mao Zedong and Deng Xiaoping were still alive, they would agree that we must create not only peace but also a green international environment.

Since 1978, the international environment has been critical to the success of China. China has also continued to shorten the income gap with developed

*Angang Hu, Professor at the School of Public Policy & Management, Tsinghua University and Director of the Center for China Study at Tsinghua-CAS (Chinese Academy of Sciences).

countries. Now, China must strive to innovate. However, if China wants to improve its green performance there must first be a better understanding of the relationship between altruism and self-serving behavior. On the issue of carbon emissions reduction, developing an active role fits the interests of China. Although China's per capita income and emissions are lower than those of Western countries', the overall effect is equally damaging. China's rapid growth produces two important effects. On the one hand, it quite successfully makes a positive contribution to global growth, trade and poverty reduction; on the other hand, it brings about large negative externality. Therefore, as China is the biggest beneficiary of globalization, China must assume global responsibility and try to reduce negative impacts.

After the Copenhagen conference, China faced sharp international criticism. China is currently at a crossroads, it can begin active cooperation with both developed and developing countries, or be criticized by both developed and developing countries.

The 12th Five-Year Plan is leading China to a new green path. In it, green development serves as a basic principle in the next Five-Years. If China can realize its target following this path, China will make a global contribution.

This chapter aims to discuss China's response to global climate change. In general, the green economy, adjusting industrial structure, developing green industries, investing in green energy and promoting green consumption, and the green modernization process are obligatory choices for China. Green Modernization will not only affect the long-term economic growth of China, but will also greatly improve the quality of economic growth and social welfare, and get win-win outcomes in economic development, environmental protection, ecological security and adapting to climate change. This chapter is divided into three parts. The first section discusses why the Chinese government needs to publicly promise a target of emissions reduction by 2050. The second part discusses the evolution of China's climate policy from 1998–2009, in order to develop a more long-term, strategic and forward-looking climate policy. Section 3 explains China's green modernization and a roadmap for emissions reduction by analyzing China's green modernization "three-step" strategy.

5.2. Why the Chinese Government Needs to Publicly Promise a Target of Emissions Reduction by 2050.

First, China has already been the world's pollution "superpower". China is the world's largest emitter of sulfur dioxide. Since 1990, China has been the country whose carbon dioxide emissions have grown the fastest in the world.[1] In 2006, China became the world's largest emitter of greenhouse gases.[2] By 2030, China's greenhouse gas emissions will reach 12.8 billion tons of carbon, which will account for 30.5% of the world total (Table 5.1). This is the largest negative externality in the world and is created by China's rise. It is the greatest danger to the future global climate change of the world. It is an important source and the basis of the "China threat" to the world. It makes China a "target of public criticism" and is the biggest liability of "soft power" internationally. It is not that China does not need to develop, but how to develop? Whether its China's national conditions or the world's conditions, China's development in the 21st century must be "self-disciplined" development "self-restrainted" development and green devel-

Table 5.1. The proportion of China's CO_2 emissions to the world total (2005–2020).

	2005	2010	2015	2020	2030
CO_2 (hundred million tons)	51.0	67.0	83.5	100	128
The proportion to the world total (%)	19.2	22.2	24.5	22.4	30.5

Data sources: IEA (2007).

[1]In the period 1990–2005, carbon dioxide emissions in China increased from 2.244 billion tons to 5.101 billion tons, an annual average growth rate of 5.6%. The proportion of the world total rose from 10% to 19%. According to estimates from the International Energy Agency, by 2015 the carbon dioxide emissions of China will reach 8.632 billion tons, which will be 35% greater than the emissions of the U.S., and have an annual average growth rate of 5.4%. Even if the carbon dioxide emissions growth rate in China declined from 2015–2020, reaching only 8.9 billion tons in 2020, it would still account for 52% of the new increment of the world.

[2]Based on data provided by the Union of Concerned Scientists, in 2006 greenhouse gas emissions in China reached 6 billion tons, more than the emissions of the U.S. (5.9 billion tons); however per capita emissions in China were 4.6 tons, while the per capita emissions of the United States were 19.8 tons, 4.3 times that of China.

Table 5.2. Damage and disaster conditions due to floods, drought (1950–2006).

Year	Floods: Damage area (ten thousand hectare/year)	Floods: Disaster area (ten thousand hectare/year)	Floods: Disaster rate (%)	Drought: Damage area (ten thousand hectare/year)	Drought: Disaster area (ten thousand hectare/year)	Drought: Disaster rate (%)
1950–1959	789.13	496.25	57.53	1322.38	416.63	34.11
1960–1966	942.00	585.43	57.74	2164.71	1002.57	45.80
1970–1979	535.70	224.30	39.64	2164.10	750.00	28.02
1980–1989	1042.50	552.90	52.71	2162.80	1176.10	47.56
1990–2000	1459.36	923.00	63.2	2632.27	1331.82	50.60
2001–2006	1064.60	629.00	59.1	2325.80	1363.1	58.60

Data sources: Author's calculation based on the data from *China Statistical Yearbook*.

opment, not "freewheeling" development, "laissez-faire" development or the continuation of black development.

Second, China has been the first major nation impacted by climate change in the world. China, as the nation with the largest population and agricultural production, has been affected by natural disasters most seriously. The scale of both the damage area and the disaster area are the largest in the world, and are growing rapidly (see Table 5.2). At the same time, China is the country where disaster has caused the largest direct food losses: from the 1950s, the average annual direct losses accounted for 2.1% of total grain output, and increased to 7.38% during the period 2001–2006 (see Table 5.3).

According to the information provided by the State Flood Control and Drought Relief Headquarters, in recent years, China's climate has become abnormal due to global climate change, torrential rain, typhoons, drought, freezing and other extreme weather events, which have increased significantly, as well as the problem of uneven spatial and temporal distribution of water resources which have become more prominent. The effect on economic development, social development and ecological systems grows larger each year (Chen, 2009).

From the perspective of the future, China is the biggest beneficiary of the global emissions reduction due to the Copenhagen Summit. However, if the meeting fails, China will be one of the biggest losers.

Table 5.3. The national average annual reduction of food production due to natural disasters (1952–2006).

Period	Annual reduction of food production due to natural disasters (hundred million tons)	Annual total grain output (hundred million tons)	Proportion of loss to total output (%)
1952–1959	379.49 (151.79)	18025.125 (1437.224)	2.1 (0.80)
1960–1966	612.26 (170.73)	17386.143 (2597.466)	3.5 (1.0)
1970–1979	662.72 (356.00)	27612.400 (2896.999)	2.4 (1.3)
1980–1989	1595.12 (325.02)	27699.000 (3254.569)	4.2 (0.9)
1990–2000	3290.89 (641.60)	47036.03 (2903.38)	7.00 (1.37)
2001–2006	3404.31 (757.18)	46522.66 (2378.71)	7.38 (1.91)

Notes: 1. The values in parenthesis are standard deviations.
2. The loss of grain is defined according to disaster area, namely, it is calculated based on 30% of the production in a normal year.
Data sources: Author's calculation based on the annual *China Statistical Yearbook*.

Third, the Earth is an expanding global village, and is the home of all the peoples in the world. The world needs a green China to fulfill their public commitment to the world which can be laid out in the document of 17th National Party Congress: help each other and work together to promote environmental protection and jointly maintaining the earth which is home to all human beings. China has to not only say it, but also to do it. China's policies and attitudes about climate change are vital in reaching world consensus. If China took the initiative to make a public commitment on emissions reduction it would hasten a "global consensus" and help develop a "global action plan" from the Copenhagen Conference, as well as play a leading role in saving the Earth and caring for our homes. Making a public commitment was a successful major policy decision in the 17th session of the Central Committee, like the successful case that the Central Political

Bureau made when China entered into the WTO against all odds in the 5th session of the Central Committee.

Fourth, while China has become the biggest beneficiary and promoter of economic globalization, it has also become the biggest stakeholder of the world's common interests. Any actions of China's, including rapidly developing the economy and pursuing its own national interests, should not be to the detriment of the common interests of mankind. Rather, it should make the economy grow slowly (even if China can grow) or sacrifice some of its own immediate interests (even if they conform to the long-term interests of its own). In terms of global interests as well as the interests of China, emissions reduction is not only altruistic, but also since China has one of the largest acid rain areas in the world[3] and more than 1.3 billion Chinese are the victims of air pollution.[4]

Protecting ecology and the environment conforms to China's own core interests. In the 21st century, because of global climate change, ecological security has become the biggest national security issue. It is even greater than other national security issues and has become the core issue of national security.

Fifth, for the whole world and China, global climate change can be described as unprecedented in two ways. The first is as an unprecedented development challenge, in essence the biggest challenge to the traditional black mode of development. The second, is an unprecedented development opportunity, in essence the largest innovation opportunity for the green development model.

Sixth, the world needs China to take the lead to reduce emissions. Taking the lead to provide global public goods is the best opportunity and long-term investment for China in order to become a brilliant world power with high

---

[3]According to research from the China Academy of Environmental Sciences at Tsinghua University, the area of acid rain pollution caused by sulfur dioxide emissions is one third of total area. In 2005, 695 cities and counties were monitored throughout the whole country and more than half had experienced acid rain.

[4]Coal has become the biggest culprit of air pollution and a major force which harms people's health. According to World Bank database of statistics, China has been the world's largest emitter of sulfur dioxide. Twenty of the world's 30 cities with the most serious air pollution are in China. According to OECD estimates, by 2020 health loss due to air pollution will be equivalent to 13% of China's GDP. According to WHO estimates, about 380 thousand people every year in China die due to air pollution resulting from solid fuel burning.

Table 5.4. Emission amount of carbon dioxide and per capita emissions of selected countries (2006).

	Total emission amount (hundred million tons carbon dioxide equivalent)	Per capita emission amount (tons carbon dioxide equivalent)
China	60	4.6
America	59	19.8
Russia	17	12.0
India	13	1.2
Japan	13	9.8

Data sources: Union of Concerned Scientists.

affinity.[5] While the United States has tried to take the lead to reduce emissions,[6] China is the most critical to global emissions reductions and the most important country. To this end, China, together with the five major greenhouse gas emitting countries, can take the lead to reduce greenhouse gases, even though China's per capita emissions are much lower than those in developed countries (see Table 5.4). In fact, China is one of the most stringent countries on taking actions to reduce emissions. Not one developed country in the world has promised and achieved its targets of saving energy by 20% and reducing the major pollutants by 10% every five years. This gave China a "reputation" and "good fame" in terms of opening a commitment and fulfilling a promise. On the contrary, If China had done something but had not promised it in words, China would become a "target of public criticism" in the eyes of the whole world and end up in an embarrassing position if China does not meet its promise to the international society in reduction target.

[5]Professor Joseph Nye from Harvard University thinks that the U.S. can invest in global public goods to achieve their interests. These include: achieving global economic development, protecting public health, dealing with climate change and maintaining an open and stable international economy. He made no secret of, and claimed all these goals require American leadership. *Diplomacy*, No. 3, 2009, July–August.

[6]On June 26, 2009, the U.S. House of Representatives passed the "Clean Energy and Security Act of the United States"; by 2020, it will reduce greenhouse gas emissions equivalent to 17% of 2005 levels (equivalent to 4% of 1990 levels); by 2050, emission levels will be equivalent to 83% of 2005 levels (equivalent to 80% of 1990 levels).

5.3. China's Climate Change Policy Evolution (1998–2009)

We will now discuss the evolution of climate policy in China: from meteorological resources policy to climate change policy; from energy saving policy to a comprehensive policy reducing greenhouse gas emissions by conserving energy to deal with climate change; from the policy for energy, environmental protection, meteorology sector to national development strategies.

In 1998, the National Climate Change Coordination Office was founded and the "Energy Conservation Law" was formally promulgated.

In 2004, the National Development and Reform Commission promulgated the "Medium and Long Term Energy Conservation Plan".

In February 2005, "Renewable Energy Law" was formally promulgated.

In March 2006, "National 11th Five-Year Plan Sketch" was formally promulgated. It instituted direct indicators of energy saving (3) and indirect indicators (2); proposed to establish a resource-saving, environment-friendly society and develop a circular economy. A "Development and utilization of climate resources" policy was also first proposed,[7] but is not a climate change policy.

In December 2006, the National Science and Technology Ministry published the "National Assessment Report on Climate Change".

In 2007, the "National 11th Five-Year Energy Plan" proposed goals to increase energy efficiency.

In June 2007, the Leading Group of Climate Change and Emissions Reduction by Conserving Energy, headed by Premier Wen Jiabao, was established; the National Development and Reform Commission formulated the "China National Climate Change Program", which marked the formation of a relatively complete climate change policy system and created an example for developing countries. The program first defined targets for dealing with climate change by 2010. To be more exact, the program was only a short-term (2007–2010) action plan, but it also gave a roadmap for long-term targets on limits of total CO_2 emissions and emissions reductions.

In September 2007, the "Renewable Energy and Long-Term Development Plan" was published; it proposed that renewable energy consumption reach 10% of total energy consumption by 2010 and 15% by 2020.

---

[7]Refers to Chapter 3, Section 2 of "The 11th Five-Year Plan of National Economic and Social Development".

In October 2007, at the 17th party congress, General Secretary Hu Jintao proposed the green strategy and objectives for 2020: "building an ecological civilization". He also proposed that, "the proportion of renewable energy should be increased significantly; major pollutants (including carbon dioxide) should be effectively controlled; environmental quality (including air quality) should be obviously improved," the report also proposed, for the first time, to "strengthen building capacity of dealing with climate change, and make new contributions to protecting the global climate" (Hu J., 2007).

In December 2007, "China White Paper of Energy Development and Policy" placed renewable energy development as an important part of the energy development strategy and a priority development area.

In October 2008, "China White Paper of Policy and Action to Deal with Climate Change" was published, and it clearly defined six principles: to deal with climate change under the framework of sustainable development; "common but differentiated responsibilities" principle; to lay equal stress on both mitigation and adaptation; convention and protocol is the main channel to deal with climate change; to rely on technological innovation and technology transfer; public participation and extensive international cooperation.

In 2009, "Energy Revitalization Plan (2009–2011)" proposed that in 2010 the proportion of renewable energy to total energy consumption should reach 10%. The capacity of wind energy was adjusted to 20 million kilowatts by 2010, and adjusted to about 0.1 billion kilowatts by 2020.

In June 2009, the meeting of the National Climate Change Leading Group and the Energy Conservation State Council Leading Group formally proposed that China take carbon dioxide emissions reduction as an important development indicator,[8] which laid a foundation for bringing the indicator into the 12th Five-Year Plan of Economic and Social Development as a core indicator.

Global climate change is not only a crisis that happens daily, but also an increasingly serious long-term crisis. Its influence is far beyond the short-term impact of the global financial crisis, and it is a great constraint to

[8]The National Climate Change Leading Group and the Energy Conservation State Council Leading Group proposed that "We need to bring dealing with climate change and reducing carbon dioxide emissions into the national economic and social development plan" (June 5, 2009).

long-term development in China. We need to institute longer-term, strategic and forward-looking climate change policies and objectives, and give a clearer roadmap for reducing greenhouse gas emissions.

5.4. China's Green Modernization and Emissions Reduction Roadmap (2006–2050)

Our general idea is: One World, One Dream, One Action (emissions reduction). Here, we get rid of narrow "emission reduction" thinking and consider dealing with climate change in terms of the whole historical process of modernization in China.

The theme of China's modernization in the 21st century is green development and scientific development. Namely, from black industrialization, urbanization and modernization to green industrialization, urbanization and modernization; from black manufacturing to green manufacturing; from black energy to green energy; from black trade to green trade; from black urban cities to green urban cities; from black consumption to green consumption.

The three steps strategy of China's Green Modernization[9]:

The first step (2006–2020) is to reduce CO_2 emissions and adapt to climate change. In 2020, carbon dioxide emissions are predicted to reach their peak.[10] According to IEA data, energy-related CO_2 emissions in China were 2.244 billion tons in 1990, and reached 5.684 billion tons in 2006. In 2020 we will strive to keep them within 8 billion tons (even with this, China will account for about 20% of the world total). This requires China to

---

[9]Here, it refers to the thinking about modernization by the "three-step" strategy proposed by Deng Xiaoping in 1987, namely: by the end of this century, China's per capita GDP will reach 800 to 1000 U.S. dollars. If it seems that it is hopeful to reach a thousand dollars, then population is 1.2 to 1.25 billion and GDP is 1000–1200 billion U.S. dollars. Our socialist system is based on public ownership and common prosperity, and then we call it a well-off society which widely improves people's living. More importantly, after reaching this point we shall take that figure as a new starting point and try to quadruple it again, so that per capita GNP will reach 4000 U.S. dollars in another 50 years. Although China will still rank below dozens of countries in the world, China will become a medium-developed country. At that time, the population will be 1.5 billion, GDP will be six trillion U.S. dollars, which is calculated based on parity of the dollar and yuan in 1980, and this figure will certainly be the highest in the world (Deng, 1987, 1993).

[10]According to IEA estimates, greenhouse gas emissions in OECD countries will peak in 2015 and then begin to decline (IEA, 2007).

greatly reduce the emissions rate during the "12th Five-Year Plan" period (referring to 2011–2015); in the "13th Five-Year Plan" period (referring to 2016–2020), emissions are to be stabilized and reach their peak. By then, the proportion of agriculture will only be 8% of GDP, the proportion of industry will have dropped to 38%, and the proportion of the service industry will have increased to 47%; urban population will be about 57%; the proportion of renewable energy will have approached or reached 20%, the proportion of coal consumption drop to 60%, and the efficiency of cleaning coal technology (especially CCS) will be higher; forest coverage rate will be 23%; Human Development Index (HDI) will reach 0.88; patent applications by residents will rank third in the world; China's GDP will rank second in the world and GDP (PPP) will rank first in the world.[11]

The second step (2020–2030) goes into the stage of CO_2 emissions, by 2030 they will have dropped substantially and strive to reach the levels of 2005.[12] By then, the proportion of agriculture will be only 5% of GDP, the proportion of industry will be only about 38%, and the proportion of the service industry will approach 47%; urban population will be about 65%; the proportion of renewable energy will exceed 25%, the proportion of coal consumption will have dropped to 45–50%, the efficiency of clean coal technology will be very high; the forest coverage rate will be 24%; Human Development Index (HDI) will reach 0.93; patent applications by residents will rank second in the world; China's GDP will rank first in the world and GDP (PPP) will rank first in the world.

The third step (2030–2050): carbon dioxide emissions will continue to decline significantly by 2050, and reach half of the emissions of 1990.[13] China will keep pace with the world. By then, the proportion of agriculture

[11]This also shows that the implementation of emission reduction targets does not affect China's overall goal of building a moderately prosperous society in 2020; not only will its economic aggregate rank first in the world, but its human development index will also reach a high level group in the world.

[12]IPCC (2007) reported that by 2050 the concentration of greenhouse gas in the atmosphere was long-term stabilized in the 445–490 ppm range and global greenhouse gas emissions were reduced by half the emissions of 1990.

[13]In the morning of April 26, 1987 during a meeting with Czech Prime Minister Bomir-Strougal, Deng Xiaoping pointed out that by the middle of the next century we could achieve the level of a moderately developed country. If we reach this point, first, a very difficult task was completed; second, a real contribution to mankind was made; third, the superiority of the socialist system will have been reflected (Chinese Communist Party Literature Research Center, 2004).

in GDP will only be 2–3%, the proportion of industry will drop to less than 20%, the proportion of the service industry will approach 80%; urban population will be more than 78%; the proportion of renewable energy will exceed 55%, the proportion of coal consumption will have dropped to 25–30%, completely clean coal will be used; forest coverage rate will be 26%; Human Development Index (HDI) will reach 0.98; patent applications by residents will rank first in the world. China will basically realize green modernization, reach the level of developed countries and make a real green contribution to humanity.[14]

The modernization of China will be on a path of innovation. It will be different from the traditional development model of common growth of economic and greenhouse gas emissions, a model that has been familiar since the British Industrial Revolution in 1750. Instead it will be a green development model with the economy growing and greenhouse gas emissions falling, even decoupling, which will be innovated in the first half of the 21st century.

Green modernization is a necessary path for China. Dealing with global climate change, developing the green economy, adjusting the industrial structure, developing green industries, investing in green energy and promoting green consumption will not affect the long-term economic growth of China, but in fact will greatly improve the quality of economic growth and social welfare, as well as achieve "win-win" outcomes in economic development, environmental protection, ecological security and adapting to climate change. At the same time, because China is the world's most populous country, it will have the largest economy and the most patents for inventions. These will make significant contributions to dealing with climate change, achieving the "450 ppm stabilization" program and reducing half of the greenhouse gas emissions through peaceful, green development and international green collaboration.[15]

Finally, will China be able to achieve these emissions reduction targets by 2050? And will China realize the goal of green modernization? As Deng

---

[14]IEA (2009) reported that during the period (1750–2007), global GDP, dollar value and carbon emissions all showed a growth trend.

[15]In order to ensure that global temperatures rise no more than 2 degrees, before 2020 the corresponding greenhouse gas emissions must be reduced by 25–40%, relative to 1990 levels, and in 2050 achieve the target reduction of 50% of overall emissions (Yang, 2009).

Xiaoping said, "our next generation will not only be more intelligent than us (the older generation), but also own more innovative capacity than our generation" Deng (1987). Now, new leaders need to make an international commitment, just as "a good beginning is half done". If this commitment is made, China will enter the successful path of scientific and green development. It will become the banner of the world's green revolution and make great green contributions for all of humanity.

References

Chen, L. (2009). Speech at the 2009 National Flood Control and Drought Relief Work Conference (in Chinese). January 9.

China Statistical Yearbook (various years). Beijing: China Statistics Press.

Chinese Communist Party Literature Research Center (2004). *Chronicles of Deng Xiaoping (1975–1997)* (in Chinese). Beijing: Central Literature Publishing House.

Deng, X. (1987). Speech to meet with a member of the Drafting Committee for the Basic Law of the Hong Kong Special Administrative Region. (in Chinese). April 16, 1987.

Deng, X. (1993). *The Works of Deng Xiaoping*, Volume III. Beijing: People's Publishing House.

Hu, J. (2007). A Speech at the 17th national meeting of CCP (in Chinese). October 15, 2007.

IEA (2007). *World Energy Outlook 2007*. Paris: Stedi Media.

IEA (2009). "The Impact of the Financial and Economic Crisis on Global Energy Investment." IEA Background paper for the G8 Energy Minister's Meeting, May 2009.

IPCC (2007). *Climate Change 2007: Synthesis Report*. Geneva, Switzerland.

Yang, M. (2009). World Climate Diplomacy and China's Response. Beijing: Shishi Press.

Chapter 6

LOW-CARBON GROWTH THEORY

*Jinjun Xue**

6.1. Introduction

In dealing with the issue of global warming, *a low-carbon economy* (LCE) is thought of as an effective way of stopping climate change. However, at present, all actions are driven mainly by global climate change talk, policies and schemes relating to the targets of emission reduction and energy conservation through government promises without enough economic verification and theoretical explanation. This encouraged us to think about a clear definition and develop a new growth theory for LCE.

This chapter aims to provide a theoretical explanation for the emerging LCE by means of presenting a low-carbon growth model (LGM) and illustrating a Carbon Kuznets Curve (CKC).

To do so, we will take advantage of the Environmental Kuznets Curve (EKC) as the basis for analysis because carbon dioxide emissions can be considered as environmental issues.[1] On the other hand, we will use the Solow Model as an analysis framework because modern economic growth is mainly driven by technological progress and a low-carbon economy is no exception. Based on these traditional theories and models, we develop a low-carbon growth model, which takes into account environmental technology progress and further provide a Carbon Kuznets Curve by importing the concept of the "tunnel effect" in order to give a theoretical explanation of the relationship between economic growth and carbon emissions. Our hope is that this study can supply an introduction to the economic theory of a

*Jinjun Xue, Professor at the Graduate School of Economics, Nagoya University, Japan.
[1] Carbon emissions are or will be classified as a new type of pollutant in more and more countries.

low-carbon economy and induce the birth of a new branch of economics, low-carbon economics.

This chapter is organized as follows: Section 6.2 provides a theoretical explanation for a LCE with a framework of growth models; Section 6.3 develops a sustainable growth model for a low-carbon economy; Section 6.4 forms a Carbon Kuznets Curve. Section 6.5 presents an empirical study to test the theory; Section 6.6 contains a brief conclusion.

6.2. A New Growth Theory for a Low-Carbon Economy

The Kuznets Curve originated as a theory illustrating the relationship between economic growth and income inequality (i.e., the Income Kuznets Curve, shortened to the IKC; Kuznets, 1955). The IKC has been developed into a model showing a process of high growth accompanied with high environmental pollution. In this section, we will discuss a growth model incorporating environment constraints and highlight how pollution influences the path of economic growth.

6.2.1. *The Environmental Kuznets Curve*

Paralleling the IKC, the EKC describes the dynamic relationship between economic growth and pollution emissions as an inverted-U shaped curve. Namely, pollution per capita will increase rapidly as the economy accelerates in the take-off stage, reach a peak (turning point) in the high growth stage, and then decrease in the stable growth stage. Through this process, the EKC assumes that the pollution problem will gradually be solved and the economy will eventually arrive at a stable growth stage (see Fig. 6.1).

The EKC is a market, fundamentalist based, model and it suggests that pollution will be reduced automatically by market mechanisms as an economy matures. Consequently there is no need for government intervention in the economy (e.g., via a strong environmental policy regime).

6.2.2. *A theoretical explanation of the EKC*

We assume a production function as $Y = F(K, L, E)$, then its inverse function is just the Kuznets curve. By the Kuznets hypothesis, in the first stage, the environment damage begins to intensify, the slope of the EKC is

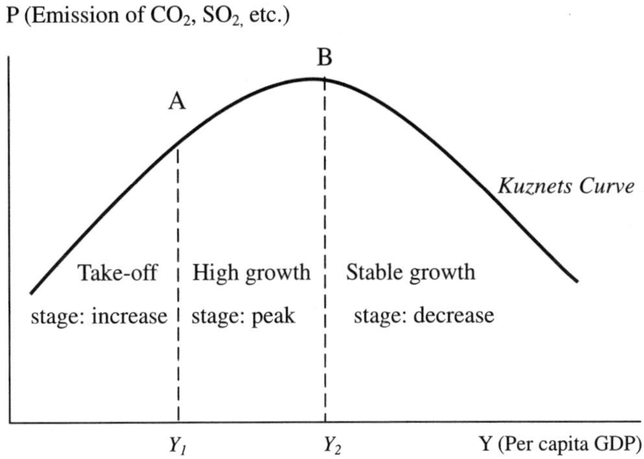

Fig. 6.1. An illustration of the environmental Kuznets Curve.
Source: Drawn by Jinjun Xue based on the Kuznets hypothesis (Kuznets, 1955).

greater than zero, so its reciprocal is also greater than zero. In the second stage, as environmental control technology begins to mature, the position of the environment curve that load is less than zero, so the slope of EKC, as well as the reciprocal are less than zero; namely minus. In the third stage, the economy as a whole reaches a stage in which the environment is not so polluted and economic development is stable (sustainable stage), the slope of the EKC trends toward zero, so its reciprocal tends to be infinite. The EKC can be obtained by the following formula.

$$\begin{cases} \text{1st stage } \dfrac{dE}{dY} > 0 \Rightarrow \dfrac{dY}{dE} > 0 \\ \text{2nd stage } \dfrac{dE}{dY} < 0 \Rightarrow \dfrac{dY}{dE} < 0 \\ \text{3rd stage } \dfrac{dE}{dY} \to 0 \Rightarrow \dfrac{dY}{dE} \to \infty \end{cases}$$

However, most analysis and country-based case studies show that the progression from a high pollution regime to a low pollution regime could be more protracted, for example if there were no strong government policy interventions, if distortions exist that favor inefficient practices, or if technological progress and technology transfer linked to improvement in the environment are not pursued aggressively. Furthermore, damage to the wider environment while the EKC is on its upswing may be so pervasive that it impinges on current and future economic growth.

6.2.3. *A growth model without technological progress*

In the literature of growth theory, the Cobb–Douglas production function illustrates that capital investment and the labor force are the only factors determining the quantity of production. Following the implication of this function, we assume that the economy is driven mainly by capital and that technological progress is absent. In light of this, the Cobb–Douglas production function can be expressed as,

$$Y^N = F(K, L) \quad \text{or} \quad Y^N = K^\alpha L^\beta \tag{6.1}$$

where, K = capital investment, L = labor force, $\alpha = (1 - \beta)$.

We take total differential of the Cobb–Douglas production function to get,

$$dY = \frac{\partial Y}{\partial K} dK + \frac{\partial Y}{\partial L} dL \tag{6.2}$$

Divide both sides of the function by Y,

$$\frac{dY}{Y} = \frac{\partial Y}{\partial K}\frac{dK}{Y} + \frac{\partial Y}{\partial L}\frac{dL}{Y}. \tag{6.3}$$

Then,

$$\frac{dY}{Y} = \frac{\partial Y}{\partial K}\frac{K}{Y}\frac{dK}{K} + \frac{\partial Y}{\partial L}\frac{L}{Y}\frac{dL}{L} \tag{6.4}$$

By the marginal revenue theorem we have the ratio of marginal return to capital,

$$\frac{\partial Y}{\partial K} = \frac{r}{p} \tag{6.5a}$$

By the marginal revenue theorem we have the ratio of marginal return to laborer,

$$\frac{\partial Y}{\partial L} = \frac{w}{p} \tag{6.5b}$$

Substituting (6.5) into (6.4), we get,

$$\frac{dY}{Y} = \frac{r}{p}\frac{K}{Y}\frac{dK}{K} + \frac{w}{p}\frac{L}{Y}\frac{dL}{L} \tag{6.6}$$

Under the conditions of perfect competition, by maximizing corporate profits, we will have,

$$pY = rK + wL \tag{6.7}$$

Divide both sides of (6.7) by pY,

$$1 = \frac{rK}{pY} + \frac{wL}{pY} \tag{6.8}$$

Assume that the contribution of laborer equals θ, then the contribution of capital equals $(1 - \theta)$

$$\theta = \frac{rK}{pY}, \quad 1 - \theta = \frac{wL}{pY} \tag{6.9}$$

Substituting (6.9) into (6.6), we get,

$$\frac{dY}{Y} = \theta\frac{dK}{K} + (1 - \theta)\frac{dL}{L} \tag{6.10}$$

here, Y = output, K = capital, L = labor.

Because savings equals investment in the Keynesian equilibrium of national income, s can be interpreted as investment i (i includes both physical investment and human capital; here we define human capital as the labor force); therefore we can carry the equation into the Harrod Growth Model (Harrod, 1948) as follows,

$$\frac{dY}{Y} = \frac{dK}{K}\alpha \tag{6.11}$$

$$G = s/k \tag{6.12}$$

where G is the growth rate, s is the savings rate, and k is the coefficient of capital, which can be defined as the ratio of incremental input-output.

6.2.4. A growth model with technological progress

The Harrod growth model stresses the absolute role of savings or investment in the process of economic growth but ignores the role of technological progress. To modify the model, we can change the economy into a modern economy by introducing the factor of technological progress, A, into the production function. Therefore the above production function can be re-written as,

$$Y^N = F(K, L, A) \tag{6.13}$$

or,

$$Y^N = AF(K, L) \tag{6.14}$$

where Y^N is natural growth rate.

Harold growth model can be transformed to the Solow model with technological progress (dA/A) based on this new production function. Taking the total differential of (6.14), we get,

$$dY = \frac{\partial Y}{\partial K}dK + \frac{\partial Y}{\partial L}dL + \frac{\partial Y}{\partial A}dA \tag{6.15}$$

Then, both sides are divided by Y,

$$\frac{dY}{Y} = \frac{\partial Y}{\partial K}\frac{dK}{Y} + \frac{\partial Y}{\partial L}\frac{dL}{Y} + \frac{\partial Y}{\partial A}\frac{dA}{Y} \tag{6.16}$$

To sort out the above formula (6.16) divided by K, L, we get,

$$\frac{dY}{Y} = \frac{\partial Y}{\partial K}\frac{K}{Y}\frac{dK}{K} + \frac{\partial Y}{\partial L}\frac{L}{Y}\frac{dL}{L} + \frac{\partial Y}{\partial A}\frac{dA}{Y} \tag{6.17}$$

By the marginal returns theorem, we substitute formula (6.5) into (6.17) and obtain,

$$\frac{dY}{Y} = \frac{r}{p}\frac{K}{Y}\frac{dK}{K} + \frac{w}{p}\frac{L}{Y}\frac{dL}{L} + \frac{\partial Y}{\partial A}\frac{dA}{Y} \tag{6.18}$$

Let the contribution rate of labor be θ and capital contribution rate be $(1-\theta)$,

$$\theta = \frac{rK}{pY}, \quad 1-\theta = \frac{wL}{pY} \tag{6.19}$$

Substituting (6.19) into (6.18), we get,

$$\frac{dY}{Y} = \theta\frac{dK}{K} + (1-\theta)\frac{dL}{L} + \frac{\partial Y}{\partial A}\frac{dA}{Y} \tag{6.20}$$

Let technological progress be $\lambda = \frac{\partial Y}{\partial A}\frac{dA}{Y}$,

$$\frac{dY}{Y} = \theta\frac{dK}{K} + (1-\theta)\frac{dL}{L} + \lambda \tag{6.21}$$

Based on the new production function, we can transfer the Harrod-style model into the Solow Model in which technological progress (designated as dA/A) is included.

$$\frac{dY}{Y} = \frac{dK}{K}\omega_K + \frac{dL}{L}\omega_L + \frac{dA}{A} \tag{6.22}$$

The Solow model explicitly demonstrates that technological progress is the decisive factor in modern economic growth (Solow, 1957). However, because technological progress is an external factor, the Solow model is treated as an exogenous growth model. Implicitly, the economy will grow with constant returns to factors of production.

6.2.5. *A growth model with environmental constraint but without technological progress*

Now we develop an EKC model in which an environmental factor, E, is considered. As a starting point, we assume a new production function as follows.

$$Y^A = F(K, L, E) \tag{6.23}$$

where Y^A is available economic growth and E is the environmental factor, which can be defined as the cost of environmental deterioration and is negatively related to economic growth.

By taking the total derivative of available economic growth formula (6.23), with respect to environmental constraints without technological progress, we can obtain,

$$dY = \frac{\partial Y}{\partial K}dK + \frac{\partial Y}{\partial L}dL + \frac{\partial Y}{\partial E}dE \tag{6.24}$$

here, E is the cost of environmental pollution, measured by the ratio of investment of environmental control in GDP; dK, dL and dE are the elasticity of capital, labor and environment, respectively.

By dividing both sides of formula (6.24) by Y, we get,

$$\frac{dY}{Y} = \frac{\partial Y}{\partial K}\frac{dK}{Y} + \frac{\partial Y}{\partial L}\frac{dL}{Y} + \frac{\partial Y}{\partial E}\frac{dE}{Y} \tag{6.25}$$

Rearranging the equation (6.25), we get,

$$\frac{dY}{Y} = \frac{\partial Y}{\partial K}\frac{K}{Y}\frac{dK}{K} + \frac{\partial Y}{\partial L}\frac{L}{Y}\frac{dL}{L} + \frac{\partial Y}{\partial E}\frac{dE}{Y} \tag{6.26}$$

Using the solution from the last session, we get,

$$\frac{dY}{Y} = \theta\frac{dK}{K} + (1-\theta)\frac{dL}{L} + \frac{\partial Y}{\partial E}\frac{dE}{Y} \tag{6.27}$$

If we set $\omega = \theta\frac{dK}{K} + (1-\theta)\frac{dL}{L}$, we can get the following by rearranging equation (6.27),

$$\frac{\partial Y}{\partial E} = \frac{\frac{dY}{Y} - \omega}{dE/Y} \tag{6.28}$$

This is the mathematical expression of the EKC (Fig. 6.1). The EKC shows that the contribution of the environmental load to the production function is negative, namely $\frac{dY}{Y} - \omega < 0$.

We can divide the EKC into three stages based on the Kuznets Hypothesis.

In the first stage, environmental damage is increasing and the cost to environmental load is negative, namely $dE < 0$. Then,

$$\frac{\partial Y}{\partial E} > 0 \tag{6.29}$$

In the second stage, environmental pollution arrives at a peak, then begins to decline, and environmental damage tends to be decreasing. Therefore we have $dE > 0$, then,

$$\frac{\partial Y}{\partial E} < 0 \tag{6.30}$$

In the third stage, environmental load is stabilized (i.e. enters in a stable stage of low environmental pollution) and environmental load tends to be zero. Therefore, $dE \to 0$, then,

$$\frac{\partial Y}{\partial E} \to \infty \tag{6.31}$$

This equation shows that environmental pollution and economic growth stabilize in the third phase and the partial derivative of economic growth on the environment is infinite. This means the economy can have a sustainable development with lower environmental load.

To understand the above theory more easily, we can change formula (6.31) into the following Solow-style Growth Model.

$$\frac{dY}{Y} = \frac{dK}{K}\omega_K + \frac{dL}{L}\omega_L - \frac{dE}{E}\omega_E \tag{6.32}$$

here, dE/E is the cost of environmental pollution measured by the ratio of investment of environmental control in GDP; ω_E is elasticity of resource consumption.

6.2.6. *The environment-constrained EKC*

To explain the relationship between economic growth and environmental change, we introduced an environmental factor into the EKC model and constructed an *Environment-Constrained EKC Model.* This aims to show that when an economy grows at an accelerating speed, environmental pollution as a byproduct of mass consumption of energy (e.g. the emission of SO_2, CO_2, etc. from coal and crude oil) will become more and more problematic. Furthermore, the pollutants will cause damage not only to nature, people, and society, but also to economic growth because it will induce a huge cost and weigh down the economy with a heavy burden. As a result, the economy will have to invest in environmental measures in order to improve economic performance and the quality of people's lives. In other words, the cost of environmental pollution and investment in pollution control measures will have a significant impact upon GDP growth and induce a constraint, vis-à-vis future economic growth. This is what we defined as the cost of high growth.

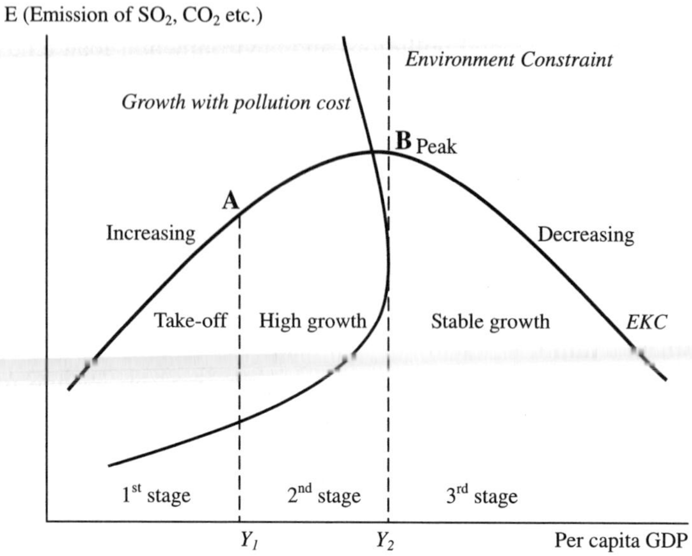

Fig. 6.2. The environment-constrained EKC.

To illustrate the cost of high growth, we consider an EKC shaped as a rebound growth path with environmental constraint in Fig. 6.2. It demonstrates that in the high growth stage, in the absence of technological change, increasing pollution will induce a strong environmental constraint on the growth path of per capita GDP. As a result of environment degradation, the growth path of per capita GDP may not go up but bend in the direction of a low speed, unstable growth position. Accordingly the distorted growth path will delay the time of getting to the turning point (the peak) and not enable the economy to reach the third stage (i.e. stable growth with low pollution as shown by the EKC) but remain in a stage of low growth and high pollution.

6.3. A Sustainable Growth Model for the LCE

In this section, we will develop a new EKC model that incorporates technological progress, which drives positive environmental outcome, into the basic neoclassical growth theories discussed above.

6.3.1. *Economic growth with pollution-suppressing technology*

Here we introduce pollution-suppressing technology into the EKC to derive sustainable economic and environmental development.

To evaluate the applicability of our hypothesis, we presume that the EKC includes a pollution-suppressing technology which can guarantee sustainable development. Therefore, production functions of gross output can be written as,

$$Y^N = AK^\alpha L^\beta, \quad \alpha + \beta = 1$$

and,

$$R = BK^\gamma L^\delta, \quad (\gamma + \delta) > 1$$

here we assume returns to factors $\gamma + \delta$ are exceeding unity. This means that the pollution-suppressing technology possesses the nature of increasing returns to scale.

If we introduce the factors of environmental technology in the EKC, we can acquire a sustainable economic growth model. To this end, we assume that the environmental technology can bring about sustainable development. Based on this assumption, we can rewrite the production function as,

$$Y = F(K, L, E, T_E) \tag{6.33}$$

Take the total differential of (6.33), we get,

$$dY = \frac{\partial Y}{\partial K}dK + \frac{\partial Y}{\partial L}dL + \frac{\partial Y}{\partial E}dE + \frac{\partial Y}{\partial T_E}dT_E \tag{6.34}$$

Divide both sides of (6.34) by Y,

$$\frac{dY}{Y} = \frac{\partial Y}{\partial K}\frac{K}{Y}\frac{dK}{K} + \frac{\partial Y}{\partial L}\frac{L}{Y}\frac{dL}{L} + \frac{\partial Y}{\partial E}\frac{dE}{Y} + \frac{\partial Y}{\partial T_E}\frac{dT_E}{Y} \tag{6.35}$$

Using the solution of formula (6.5a and 6.5b), we can get,

$$\frac{dY}{Y} = \theta\frac{dK}{K} + (1-\theta)\frac{dL}{L} + \frac{\partial Y}{\partial E}\frac{dE}{Y} + \frac{\partial Y}{\partial T_E}\frac{dT_E}{Y} \tag{6.36}$$

Let $\omega = \theta\frac{dK}{K} + (1-\theta)\frac{dL}{L}$, (6.36) can be rewritten as,

$$\frac{\partial Y}{\partial E} = \frac{\frac{dY}{Y} - \omega - \frac{\partial Y}{\partial T_E}\frac{dT_E}{Y}}{\frac{dE}{Y}} \tag{6.37}$$

Because environmental technology's contribution to GDP is positive, that is $\frac{\partial Y}{\partial T_E}\frac{dT_E}{Y} > 0$ and $\frac{dY}{Y} - \omega$ is less than zero, namely, the environmental technology can promote economic growth. In summary,

$$\frac{dY}{Y} - \omega - \frac{\partial Y}{\partial T_E}\frac{dT_E}{Y} < 0 \tag{6.38}$$

6.3.2. *A sustainable growth model (SGM)*

Due to reduction of environmental load induced by environment technology, $dE > 0$ or $dE \to 0$. At the same time, $\frac{\partial Y}{\partial E} < 0$ or, $\frac{\partial Y}{\partial E} \to \infty$. Therefore the slope of the EKC is less than 0, that is $\frac{\partial E}{\partial Y} < 0$, or $\frac{\partial E}{\partial Y} \to 0$, meaning the EKC is decreasing in the 3rd stage. This gives us an indication that it is possible for an economy to pass through the first and second stages and proceed directly to the third stage, the sustainable growth stage, through a shortcut.

The above theory can be summarized as a new growth model for sustainable growth under the above conditions.

$$\frac{dY}{Y} = \frac{dK}{K}\omega_K + \frac{dL}{L}\omega_L + \frac{dA}{A} + \frac{dT_E}{T_E}\omega_{TE} \tag{6.39}$$

Where dA/A is technological progress for production, dT_E/T_E is environmental technological progress, and ω_{TE} is the elasticity of environmental technological progress.

We may name it as *sustainable growth model (SGM)* in which two kinds of technologies are being used: one focuses on production (e.g. energy-saving technology, sulfur deodorant technology, etc.), and the other on environment controls and emission reduction (e.g. energy-saving technology, carbon emission reduction policy, etc.). Figure 6.3 is a demonstration of the new EKC based on this model.

The SGM shows that by introducing environmental technology, the growth path of per capita GDP can move forward gradually while pollutant emissions increase slowly at first, then tend to decrease and finally arrive at the third stage in which a sustainable growth can be obtained and environmental pollution can be contained at a reasonable level. However, the model implies that per capita GDP should grow at a reasonable speed with effective pollution control but not at a high speed that excessively burdens the environment.

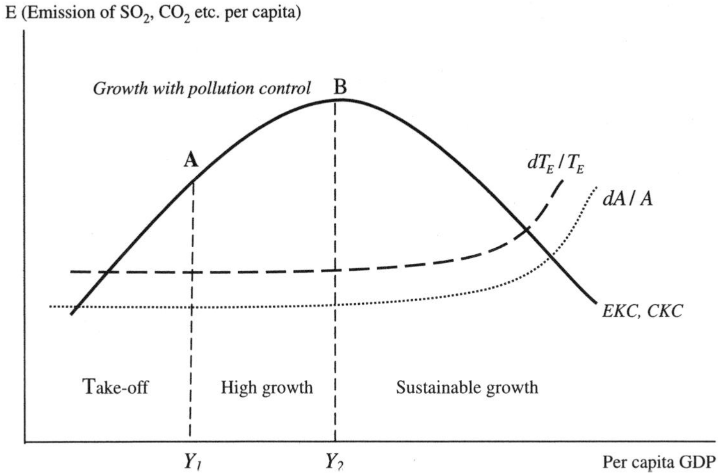

Fig. 6.3. A sustainable growth model.

6.4. A Carbon Kuznets Curve

6.4.1. *The flying geese model and the catch-up effect*

According to historical data, most developed countries spent 50–200 years overcoming the trade-off of economic growth and environmental pollution. For example, to get to their peak year of CO_2 emissions, Britain spent 185 years and the USA spent 135 years. Even the East Asian miracle economy of Japan spent more than 100 years from the Meiji Era to get to its peak year in 2005,[2] while suffering from the painful experiences of the "Four Big Pollution Diseases" (Xue *et al.*, 2002). Meanwhile, some empirical studies show most countries passing their carbon emission turning point at income levels of 10,000–23,000 U.S. dollars (at 1990 prices).

However, there are some exceptions, such as South Korea. South Korea took only 40 years from 1946–1997 to get to their turning point of CO_2 emissions at 4,114 U.S. dollars per capita GDP.[3] This phenomenon can be explained by the *flying geese paradigm theory*, an Asian version of the *theory of catch-up*.

[2]See *Nature*, No. 5, 2005.

[3]The Sustainable Development Project of the Chinese Academy of Science (2009). *China Report on Sustainable Development Strategy 2009*. (in Chinese) China: Science Press.

The Flying Geese Paradigm is a view of Japanese scholars about the technological development in Asia, taking Japan as a leading economic power in Asia. In this theory, Akamatsu (1962) postulated that Asian nations will catch-up with the West as a part of a regional hierarchy where the production of commoditized goods would continuously move from developed countries to developing ones. The underdeveloped nations in the region are considered to be aligned successively behind the advanced industrial nations in the order of their different stages of growth, in a wild-geese-flying pattern. Meanwhile, as developed countries transfer their industries to developing countries, the technology advanced in investor countries will be diffused to investing countries through investment and goods in which technologies are embodied. In this way, developing countries can catch-up with developed countries in a shorter time; namely catch-up. This theory provides a valuable explanation for the rapid economic growth of the Asian Four Dragons and current Chinese growth.

6.4.2. *The tunnel effect and a Carbon Kuznets Curve*

At present, the low-carbon economy is thought of as a shortcut of carbon reduction in order to deal with global warming issues. One of the methodologies is the so-called tunnel effect (see Fig. 6.4) which is thought of as a combination of the Flying Geese Model and the Environmental Kuznets Curve (Xue, 2011).

Figure 6.4 shows that an economy can go from the first stage to the third stage by a shortcut through the tunnel (i.e. shaped as a tunnel that drills through two sides of the hill of the inverted-U curve). The tunnel effect can be described as the results of technology transformation and the technological catch-up effect via international carbon financing, energy–saving technology, pollutant emission reduction actions, the low-carbon economy and government policies targeting climate change and green development. However, there are conditions required to realize the tunnel effect:

(1) The amount and speed of technology transfer from developed country to developing country.
(2) Absorbtion capability of receiving country, including human resources, management, institutional change, technology application and so on.

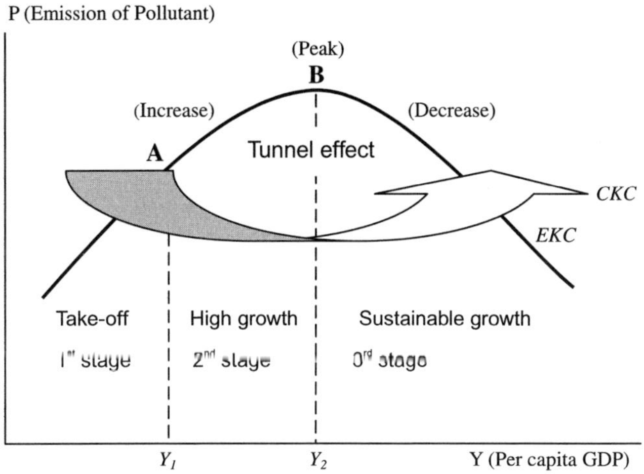

Fig. 6.4. A Carbon Kuznets Curve.

Source: Drawn by Jinjun Xue.

(3) Technology innovation, as a matching part of utilizing efficiency of technology application in the developing country.
(4) Policy orientation (one example is the low-carbon economic development path proposed by a country).

6.5. An Empirical Study of the EKC and CKC

In this section, we conduct a case study of the new EKC hypothesis using a dataset of China.

6.5.1. *China's EKC and CKC*

China has been experiencing high growth with a correspondingly high cost of environmental degradation. Some estimates put the cost of environmental degradation as high as 15% of GDP.[4] In the following we review the

[4]Chinese Academy of Science (2009). *China Sustainable Development Strategy Report 2009*. Science Press, p. 45. According to Liu and Diamond (2005), the cost of environmental damage on the economy can be accounted as 8–15% of China's GDP. Meanwhile, an official research report estimated 51 billion yuan in cost from environmental damage in 2004, 3.05% of total GDP.

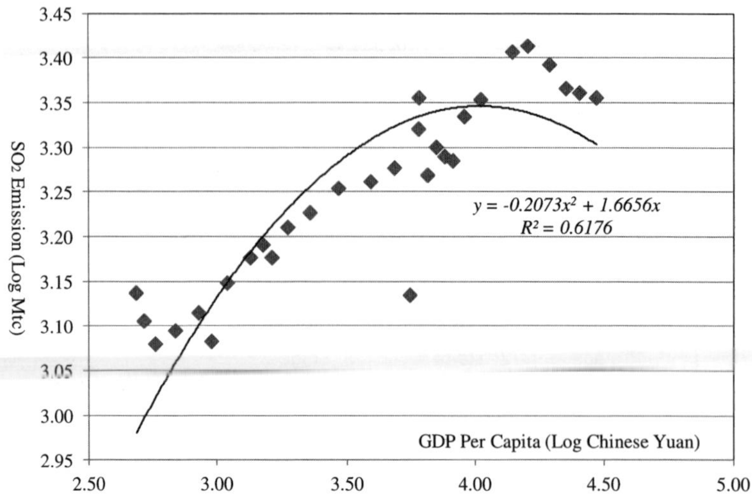

Fig. 6.5. China's EKC (1980–2010).

Source: Xue (2011).

evidence on environmental pollution issues in China during its high growth period.

We shape China's EKC and CKC using CO_2 data from the International Energy Agency (IEA), GDP data from *The World Development Indicators 2012* and SO_2 data from the *Statistics Yearbook of China 2011*.

Figures 6.5 and 6.6 are China's EKC and CKC. The most meaningful finding is that both SO_2 emission and CO_2 emission have very high correlations with GDP growth. More importantly, we note that China's EKCs begin to decrease slightly from 2007 but its CKC is still increasing and appearing in a high position in the second stage. We cannot be sure whether China passed the turning point of the EKC because it needs to be verified by long term time series data and the current decreasing might be caused by temporary government policies.

On the other hand, we found that China's CKC has not arrived at its turning point yet. This means that carbon emissions in China continue to increase and it may take a long time to get to the turning point if the Chinese government cannot impose stricter controls on pollutants.

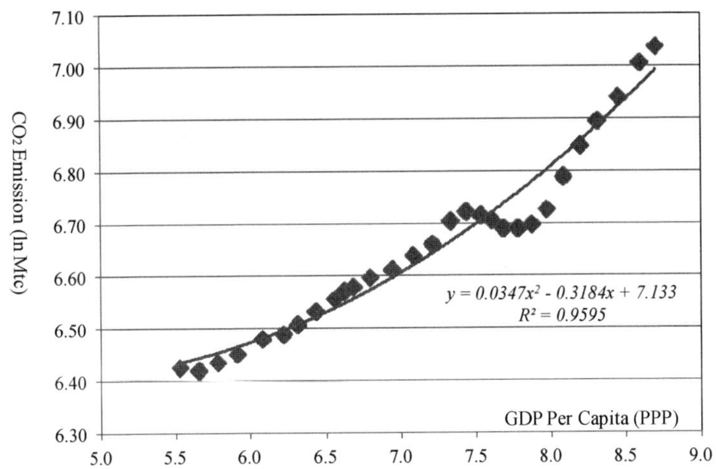

Fig. 6.6. China's CKC (1980–2010).

Source: Xue (2011).

6.5.2. *Can China apply the CKC?*

We showed in Section 6.4 that it is possible for developing countries to catch up with developed countries in environmental improvement by applying the Carbon Kuznets Curve. The question here is whether China can catch up.

To answer the question, we analyze the trend of unit CO_2 emissions by an international comparison.

Figure 6.7 shows that, compared to other developed countries, China's CO_2 emissions in total is at a very high level and will continue to increase if China continues its rapid economic growth as most economists have predicted. Figure 6.8 shows the same outcome but a decreasing trend in CO_2 emissions per 100 U.S. dollars GDP production. The decreasing trend can be explained by the achievements of energy conservation, technological progress in production and stricter control policies on industrial structure change (Zhu and Xue, 2012). This demonstrates a possibility for China to take a shortcut in reducing pollution and improving the environment at an accelerated speed.

In searching for the reasons for the decline of CO_2 emissions per unit GDP, both environmental technological progress and investment in

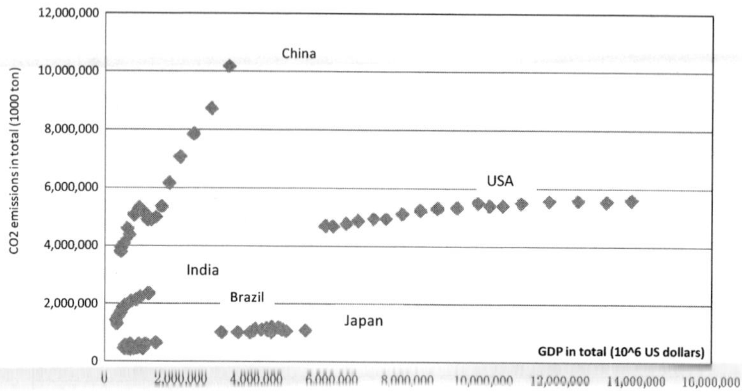

Fig. 6.7. CO_2 emissions per 100 U.S. dollars GDP (by countries).

Source: IEA (2012).

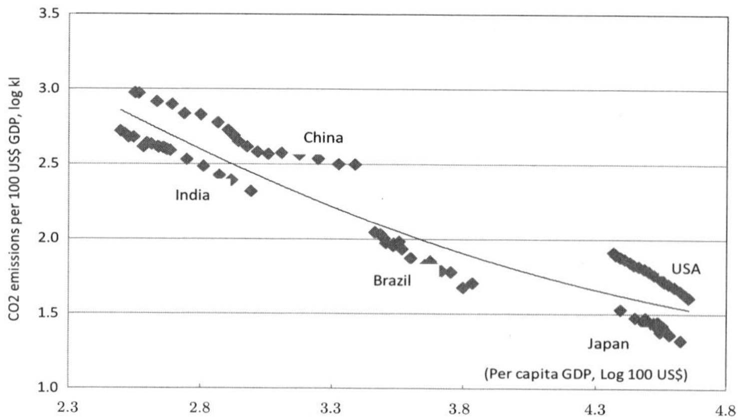

Fig. 6.8. An international comparison of carbon dioxide intensity.

Source: IEA (2012). World Bank, World Development Indicators 2009.

pollution control are considered viable options. Ninomiya (2005) estimated that environmental technological progress contributed one-third of SO_2 reductions in the period 1980–2005.

However, we believe that the China EKC has less to do with the free-market economic mechanism than it does with strong government policy on environmental protection. In other words, the apparent path of the EKC and the CKC, increasing to the turning point before decreasing, shows that government policy is active and effective in the area of environmental improvement. For example, the Chinese government added restrictions or bans of highly polluting cars and taxis during this period.[5]

As we know, the Chinese government has been promulgating a substantial number of regulations on pollution control and has invested large amounts in environmental improvement schemes. However, just as there are wide regional differences in economic growth and income distribution, the same can be said of environmental control investment.

As we assumed in the conditions of the EKC, E is a negative value and is therefore being counted as a cost in relation to GDP. Similarly the environmental control investment or the environmental improvement investment are a huge costs to the Chinese economy. The same explanation can be conducted with CO_2 emissions.

There are many reasons for environmental pollution in China. Examples include the priority afforded to heavy industry as part of the state development strategy, the shortage of investment in environmental control, differentiation amongst the regions and the logic behind the so-called, "pollute first, control second," approach that was used in the 1960s in Japan.[6] Here, we focus on the energy structure of China which indicates a heavy reliance on coal in energy consumption with a 70% share of total energy, with little to no change since 1978. The subsequent is the most important factor inducing large scale emissions both in sulfur emissions as well as carbon emissions.

6.6. Conclusion

This chapter discussed the EKC and showed that, during the high growth stage, environmental destruction may constrain future economic growth, namely inducing the Cost of High Growth. Our Environment-Constrained

---

[5]However, there may be other reasons for this result. One might be the data problem as summarized by the Rawski Puzzle, which states that the official statistical data, especially GDP growth, are not coordinated to the growth of energy production and consumption (Rawski, 2002). Therefore the data published in the *Statistical Yearbooks* is suspicious because we know that data collection regarding pollution is very difficult and the statistical system for pollution data is not yet mature in China.

[6]Xue *et al.* (2002).

Growth Model demonstrated that increasing pollution can induce a strong environmental constraint on economic growth and force the growth path to fall into a trap of low growth and high pollution. It was argued that the benevolent path of the theoretical EKC will not appear without the introduction of appropriate pollution-suppressing technologies.

However, our adapted growth theory assumes that the technological innovation of environmental measures can solve the environmental constraints placed upon economic growth, and, in turn, can promote sustainable development. The new EKC derived from this analysis shows that pollution-suppressing technology can push an economy onto the traditional EKC path. In other words, it promotes the growth path to arrive at the third stage, namely sustainable development. The theory also illustrates that pollutants and carbon emissions can be reduced through technological progress/transfer. Endowing this factor with increasing returns to scale in an adapted neoclassical framework allows the system to generate a sustainable mode of growth in output per capita. The importance of this theory is that it gives countries with environment issues and high carbon emission a theoretical and policy implication to solve the problem and catch up with developed countries through a shortcut, specifically the tunnel effect.

The Chinese case illustrates that environmental pollution has been seriously worsening since the beginning of high growth. China is not only a product factory of the world but also a pollutant factory of the world. However, the per capita and unit data of CO_2 indicate that it is possible to reduce pollution and carbon emissions by using environmental and climate change policies and technological progress of environment and energy savings. Recent developments indicate that policy changes need to be made to expedite the market-based turning point for environment-friendly growth.

Consequently it will be a long march for developing countries to arrive at the third stage of the CKC and realize a sustainable development. However, as we have shown in the SGM and the CKC, it is possible to have sustainable growth with low-carbon emissions and low-pollutant emissions if developing countries transform their development mode to a low-carbon economy, and developed countries help developing countries by providing financial support and carbon and pollutant reduction technologies.

References

Akamatsu, K. (1962). A Historical Pattern of Economic Growth in Developing Countries. *Journal of Developing Economies*, 1(1), 3–25.

Harrod, R. (1948). *Towards a Dynamic Economics: Some Recent Developments of Economic Theory and Their Application to Policy*. London: Green Wood Group.

Kuznets, S. (1955). Economic Growth and Income Inequality. *American Economic Review*, 45, 1–28.

Liu, J. and Diamond, J. (2005). China's Environment in a Globalizing World. *Nature*, 435, June 30.

Ninomiya, Y. (2005). An Empirical Analysis of Economic Growth and SO_2 Emission in China — Examination of the Environment Kuznets Curve. *Journal of International Cooperation Studies*, 21(1), 52–62.

Rawski, T. (2002). Measuring China's Recent GDP Growth: Where Do We Stand? *China Economic Quarterly*, 2.1, 347–354.

Solow, R. (1957). A Contribution to the Theory of Economic Growth. *Quarterly Journal of Economics*, 70, 65–94.

The Sustainable Development Project of the Chinese Academy of Science (2009). *China Report on Sustainable Development Strategy 2009*. China: Science Press.

Xue, J., Arayama, Y. and Peng, J. (ed.) (2002). *Economic Development and Environment Control in China — Theory and the Actual*. Northeast University of Finance and Economics.

Xue, J. (ed.) (2011). *Annual Report on China's Low-Carbon Economic Development* 2011. Beijing: Social Science Academy Press of China.

Zhu, Y. and Xue, J. (2012). China's Effort to Energy Conservation and Carbon Reduction, in Xue, J. (ed.), *China's Green Low-Carbon Development*. Beijing: Social Science Academy Press of China.

Chapter 7

THE KAYA IDENTITY FOR CARBON EMISSION ESTIMATION

*Yoichi Kaya**

7.1. Introduction

Most people believe that the main cause of global warming is CO_2 emitted by the burning of fossil fuels. So in order to mitigate global warming, reduction in dependence on fossil fuels is required. In this sense measures for mitigating global warming are mostly those for decarbonizing energy. On the other hand mankind always looks to improve its standard of living by economic development, which in most cases may induce increases in energy demand. Since the majority of primary energy of the world consists of fossil fuels, CO_2 will inevitably increase. In this sense balancing decarbonization of energy and economic development is one of the most difficult issues for mankind to solve in the years to come.

Japan adopted in the Kyoto protocol the target of reducing greenhouse gases by 6% compared with their level in 1990. Since CO_2 accounts for 95% of greenhouse gases emitted in Japan, the key for Japan to achieve its Kyoto target is reduction in CO_2 emissions. The government of Japan introduced the plan for achieving the Kyoto target and made efforts for implementing the plan. In spite of such efforts, CO_2 increased to 14.5% higher than the level in 1990 by 2007. In a sense it was lucky that the Lehman shock in 2008 induced a serious economic crisis in the world, because that led to

*Yoichi Kaya, Professor Emeritus of University of Tokyo, Deputy Council of Institute for Global Environmental Advanced Industrial Science and Technology.

heavy reductions in CO_2 emissions in Japan. By adding emission rights bought from foreign countries, Japan may eventually achieve its Kyoto target, but this situation indicates how tight the relationship is between economic growth and CO_2 emissions.

Apart from the present Kyoto target issue, there are worldwide discussions on future targets of emissions of greenhouse gases, particularly of CO_2. At the 15th session of the Conference of the Parties to the UNFCCC (COP15) held in December 2009 the Copenhagen accord was proposed; it asked participating countries to submit their own targets on mitigation of greenhouse gases by the end of January 2010. According to the UN, 55 countries submitted their target plans by that date.

Setting these targets being only the first step, countries should clarify what efforts are needed to achieve these targets. Among various methods for assessing the targets, we look forward to having those which are simple enough for people to easily understand. The author believes the factor analysis to be exhibited in the following is the one which satisfies the above condition.

It was almost 20 years ago the author proposed this method at IPCC (Intergovernmental Panel on Climate Change). IPCC is now well known partly due to the Nobel Peace Prize it received a few years ago. It was launched in November 1988 and the author participated in the energy and industry subgroup (EIS) of Working Group 3 and he felt that most participants hardly realized the difficulties in reducing CO_2 while maintaining economic development. He then tried to present this factor analysis applied to CO_2 emission at the first meeting of EIS in May 1989. Many participants expressed that they realized the difficulties in mitigating global warming because of this analysis.

7.2. The Kaya Identity

7.2.1. *Total emissions model*

The specification of the Kaya identity takes the following form:

$$\begin{aligned} CO_2 \text{ Emissions} &= \text{Population} \times (\text{GDP}/\text{Population}) \\ &\quad \times (\text{Energy}/\text{GDP}) \times (CO_2/\text{Energy}) \end{aligned}$$

Here,

$$\begin{aligned} C &= (C/E) \times E \\ &= (C/E) \times (E/G) \times G \\ &= (C/E) \times (E/G) \times (G/P) \times P \end{aligned} \tag{7.1}$$

where,

$C = CO_2$ emissions;
$E =$ energy consumption;
$G =$ GDP;
$P =$ total population;
$C/E =$ the carbon intensity of energy, representing the structure of energy;
$E/G =$ the energy intensity of GDP, representing energy efficiency; and
$G/P =$ the per capita GDP, representing a country's development stage.

The Kaya identity expresses the CO_2 emissions as a product of these four driving factors.

7.2.2. *Method of factor analysis*

Let Z be a function of $X_i (i = 1, 2, \ldots, n)$ as in

$$Z = \Pi X_i = X_1 \cdot X_2 \cdots X_n \tag{7.2}$$

Suppose that the average rate of change of X_i in a given period is γ_i, and that of Z is γ_z. Then (7.2) is rewritten as (7.3).

$$Z_o \exp(\gamma_z t) = \Pi X_{io}(\gamma_i t) \tag{7.3}$$

Therefore,

$$\gamma_z = \gamma_1 + \gamma_2 + \cdots + \gamma_n \tag{7.4}$$

where, Z_o is the initial value of Z and X_{io} is that of X_i. Equation (7.4) clearly indicates that the rate of change of Z is the sum of these of $\{X_i\}$. The factor analysis to be used in this paper is the simple analysis using only (7.3) and (7.4).

From now on let the rate of change of X be ΔX.

First we decompose CO_2 in the following three factors.

$$CO_2 = C/E \cdot E/G \cdot G \tag{7.5}$$

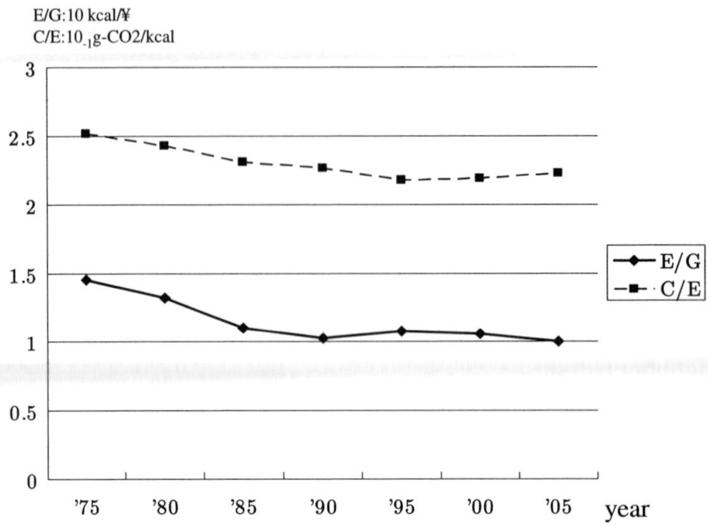

Fig. 7.1. E/G & C/E (Japan, 1975–2005).

where, *E* is primary energy, *C* is CO_2 and *G* is GDP.

The first two factors in the right side of (7.5) are interpreted as follows:

C/E: carbon intensity of energy; and
E/G: energy intensity of economy.

Shown in Fig. 7.1 are C/E and E/G of Japan after the oil shock. They decrease as time proceeds, and these trends can also be seen in other countries. Then the rates of change of these two variables can be interpreted as follows:

$\Delta C/E$: rate of decarbonization of energy (negative); and
$\Delta E/G$: rate of energy conservation (negative).
ΔG is the rate of economic growth. From (7.5)

$$\Delta CO_2 = \Delta C/E + \Delta E/G + \Delta G \tag{7.6}$$

Equation (7.6) is a simplified form of so-called Kaya identity. The first two terms cover efforts for reducing CO_2, while the last term represents economic growth, which stimulates an increase in CO_2. In this sense (7.6) is a simple but appropriate factorization when discussing the relation between CO_2 emissions and economic development.

A short comment on Fig. 7.1: shown here are 5 year average of C/E and E/G of Japan since the first oil shock. Both of them smoothly decreased for 10–15 years after the oil shock but changed little afterwards. The issue of global warming attracted public attention since around 1990, after which CO_2 emission in Japan increased almost in parallel with its economic growth. It means that the government of Japan implemented few policy measures effective in reducing CO_2 emission even after 1990. It should introduce measures more effective in reducing C/E and E/G from now on.

7.3. Example of Factor Analysis: Feasibility of 2020 Target for Japan

7.3.1. *Target of Japan*

A lot has been discussed about targets for mitigating global warming. At the G-8 in 2009 and COP15 in December 2009, a proposal was adopted that the average air temperature of the globe be no higher than 2 degrees plus its level before the industrial revolution (the author still wonders why such a proposal was made.) In past COP meetings, proposals for setting 2020 and 2050 targets were made corresponding to the 2 degrees proposal. They correspond to scenarios listed in the 4th report of the IPCC, and the target for 2020 proposed in these scenarios is a reduction in greenhouse gases by 25–40% when compared with 1990 emission levels.

The government of Japan started the investigation on the 2020 target in autumn 2008 and the former Prime Minister Taro Asou proposed the target of a 15% reduction when compared with 2005 emission levels. In September 2009 the administration changed and the new Prime Minister Hatoyama made a proposal of a 25% reduction by 2020 with 1990 as the base year. The government of Japan also responded to the request of the Copenhagen accord for a target with this proposal of a 25% reduction. This corresponds to a 30% reduction when compared with the level in 2005 and is by 15% stricter than the Asou target. What the author will try in the following is to assess this target by use of the factor analysis. It is to be noted that the target is for the total greenhouse gases but actually close to that for CO_2 as CO_2 accounts for 95% of greenhouse gases emitted in Japan.

7.3.2. *Assessment in terms of C/G*

CO_2 can be factored as in the following equation.

$$CO_2 = C/G \cdot G \tag{7.7}$$

Therefore from (7.4)

$$\Delta C = \Delta C/G + \Delta G \tag{7.8}$$

where ΔX is the rate of change of X.

Once given the 2020 target, ΔC is fixed. ΔG is the target for economic growth which the government keeps implicitly or explicitly. Since the average economic growth rate of developed countries between 2000 and 2007 has been 2.2% per year, let this value be used as ΔG. Then the corresponding $\Delta C/G$ can be calculated from (7.7). Results of major developed countries are shown in Fig. 7.2, with the future paths required for satisfying their 2020 targets.

For CO_2 the USA target is a 17% reduction when compared with its 2005 level (4% reduction with 1990 base level year), targets for UK and France

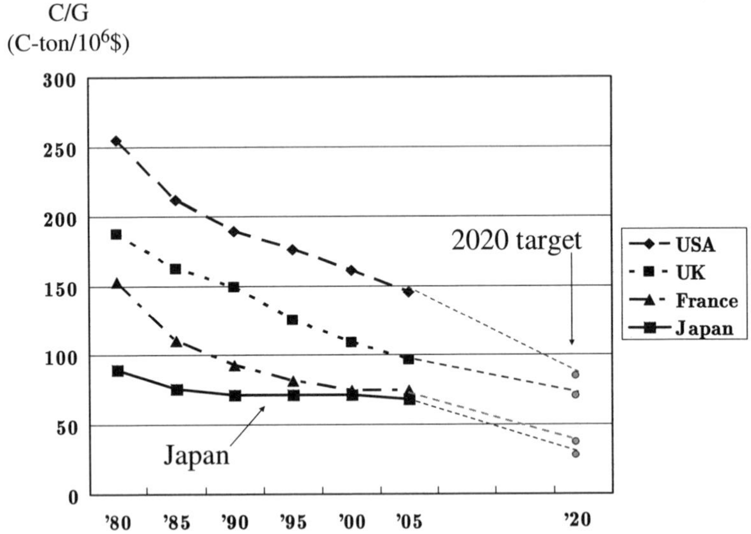

Fig. 7.2. C/G of major developed countries.
Note: Assume ΔG be 2.2% per year for all countries between 2005 and 2020.

are a 20% reduction (1990 as base year), and the target for Japan is a 25% reduction (1990 as base year). The lines for USA and UK show almost the same trends from the past to the future. In other words these countries may achieve the targets with almost the same efforts as in the past, while France experienced in the past a rapid decline in C/G, probably due to construction of nuclear power plants, but stagnation of C/G after the beginning of this century. It will not be an easy task for France to achieve the 2020 target.

In the case of Japan the past rate of change of C/G is much slower than that required for the future. It indicates that Japan has to make unprecedented efforts to achieve the 2020 target.

Figure 7.3 investigates the marginal costs of achieving the 2020 target. Several Japanese research institutes evaluated the marginal costs of reducing greenhouse gases as a function of the rate of reduction in greenhouse gases. Marginal costs for achieving a 25% reduction target lie between 40 and 90 thousand yen/ton CO_2. Taking into account that the price of emission rights in the European market is around 2–3 thousand yen/ton CO_2 you may realize how expensive the above marginal cost is. The marginal cost here is theoretically identical to the value of the carbon tax which can realize the same reduction in CO_2 emission by fossil fuel taxation. Supposing that the carbon tax to be introduced is 40 thousand yen/ton CO_2, the total amount

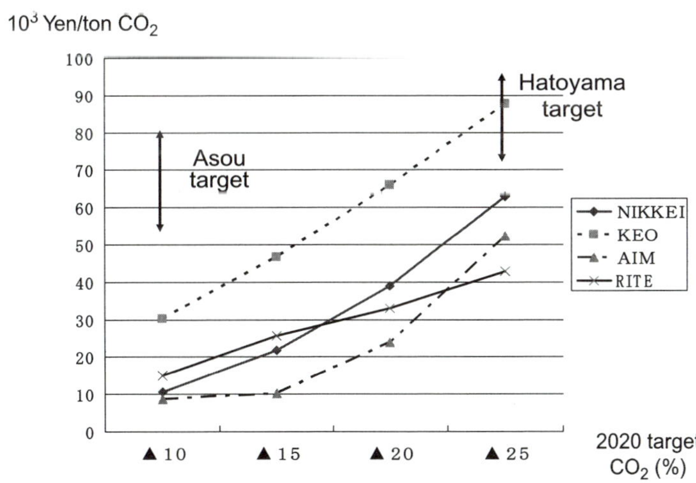

Fig. 7.3. Marginal cost of reduction in CO_2.

of revenue from the carbon tax is 1.2 Gt (total CO_2 emissions in Japan) times 40 thousand yen/ton CO_2, namely almost 50 trillion yen, which is larger than the total revenue of the government of Japan. This shows it is unrealistic to introduce such a high carbon tax in practice. If we apply this amount of carbon tax to gasoline the tax for 1 liter gasoline is about 100 yen, almost twice the present tax of gasoline.

The above calculation and analysis show that achieving a 25% reduction target is a heavy burden for Japan both in view of costs or the size of economic incentives needed. Therefore in order to realize a 25% reduction of CO_2, implementation of not only a carbon tax but also other measures such as regulation are required. Details of these measures however are not the issues to be discussed in this chapter.

7.3.3. *Assessment in terms of C/E*

For more detailed analysis on a 25% target consider the factorization as (7.9).

$$\mathrm{CO_2} = C/E \cdot E \tag{7.9}$$

Accordingly,

$$\Delta \mathrm{CO_2} = \Delta C/E + \Delta E, \tag{7.10}$$

and moreover,

$$\Delta E = \Delta E/G + \Delta G \tag{7.11}$$

Combination of (7.10) and (7.11) yields (7.6) as already described.

Among various energy demand sectors the power generation section account for about 40% of total CO_2 emissions in Japan. Then divide C/E into C/E for the non power generation sector and C/E for the power generation sector. The following equation is easily derived from this definition.

$$C/E = \frac{1}{E}\left\{\mathrm{Enp}\frac{\mathrm{Cnp}}{\mathrm{Enp}} + \mathrm{Ep}\frac{\mathrm{Cp}}{\mathrm{Ep}}\right\} = (1-\alpha)C/\mathrm{Enp} + \alpha C/\mathrm{Ep} \tag{7.12}$$

where,

α: ratio of electrification (share of electric power in primary energy);
Cnp/Enp: C/E of non power generation sector; and
Cp/Ep: C/E of power generation sector.

Equations (7.10) and (7.12) indicate that factors determining ΔCO_2 are C/E for power and non power generation sectors, α, and ΔE. Let us examine these four factors one by one.

(1) ΔE

Figure 7.4 shows ΔE of the average period of 5 or 10 years of major developed countries since 1980. It is noticeable that ΔE in most cases in most countries were positive except Germany. To examine this trend in more detail, ΔE and ΔG of Japan after 1990 are shown in Fig. 7.5, where, ΔE and ΔG are evaluated after smoothing E and G as 2 year moving averages. It is seen in Fig. 7.5 that both ΔE and ΔG change in a similar manner except 2001, and holds in almost all years. In other words, the following relation holds in most cases.

$$\Delta E > 0 \quad \text{when,} \quad \Delta G > 0, \quad \text{and} \tag{7.13}$$

$$(\Delta G) > \text{energy conservation rate}(-\Delta E/G) \tag{7.14}$$

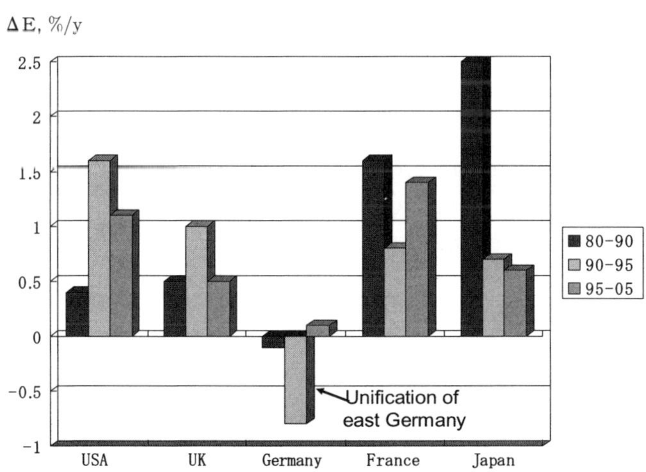

Fig. 7.4. $\Delta E = \Delta G - (-\Delta E/G)$.

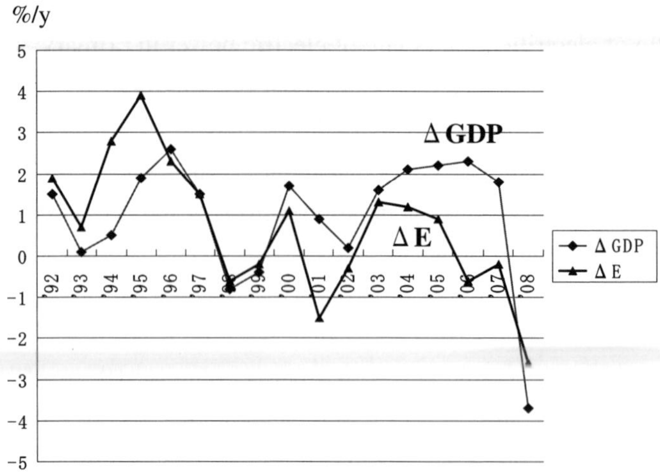

Fig. 7.5. Δ GDP, ΔE of Japan.

Note: Both GDP and E are moving average of two years.

Equation (7.14) means that energy conservation proceeds by the push of economic growth: this is easily understandable. The reason why Germany in Fig. 7.4 does not follow (7.14) may be that the unification of east Germany into west Germany accelerated the shutdown or replacement of coal power plants of low efficiency in the eastern part. Therefore we may say that in most growing economies, the lower limit of ΔE will be zero and will be hardly negative.

(2) *C*np/Enp

Energy sources in the non power generation sector are those for transportation and oil and city gas for heat demand. *C*np/Enp in Japan between 1990 and 2005 has been almost constant: around 2.27 ton CO_2/kl-oil. Little change is expected in the next 10 years about the fuel structure in this sector and so *C*np/Enp will be constant in this period.

(3) Ratio of electrification α

One of the significant changes we expect in the future of energy in Japan is advancement in electrification. Shown in Fig. 7.6 is the trend of the ratio of electrification in Japan for last 30 years. Taking into account increases in

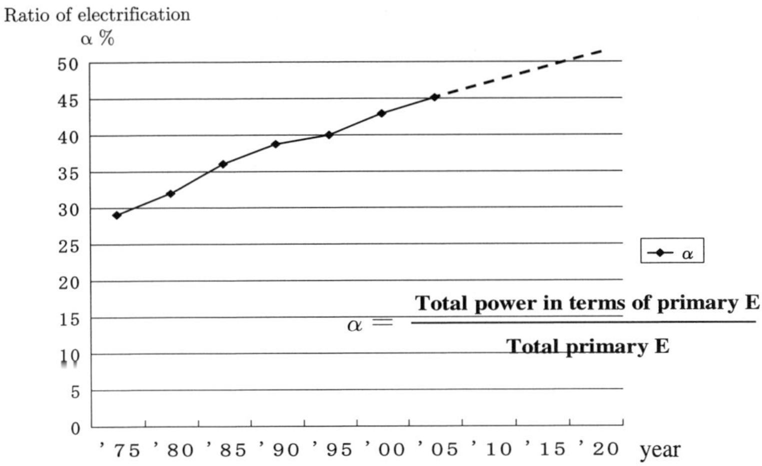

Fig. 7.6. Trends of electrification.

the kinds and numbers of electric appliances, diffusion of heat pumps and development of electric vehicles, we can expect that the trend of electrification will continue.

If we extrapolate the past trend, α will be more than 0.50 in 2020. Presently the value of *C*p/Ep is slightly less than *C*np/Enp, but in future *C*p/Ep will go down. Therefore an increase in the ratio of electrification (α) will accelerate the speed of reduction of C/E in Japan.

(4) *C*p/Ep

The carbon intensity of the electric power generation sector in Japan, *C*p/Ep continued to decrease in the past due to expansion of nuclear power and to the use of natural gas, which is the least carbon intensive among fossil fuels. Around 2020 we naturally expect changes in the power generation structure, which may reduce carbon intensity of this sector. The government of Japan has therefore been enthusiastic about revolutionary changes in the power generation structure in planning future energy systems. Shown in Table 7.1 was the government plan for 2020 with the data in 2005.

Key factors in Table 7.1 are nuclear power and photovoltaic. For nuclear power, construction of new plants and substantial improvements of plant

Table 7.1. Structure of power generation in Japan: 2005 and 2020 (plan).

	2005		2020 (plan)	
	Capacity	Share in kwh	Capacity	Share in kwh
Hydro	48 GW	8%	48 GW	8%
Nuclear (op.ratio)	49.5	26 (60%)	61.5	44 (82%)
Photo-voltaics	1.4	0	28–50	2–5
Fossil fuels	179	66	NA	46–43
Total		100		100

operation rates are planned. After the year 2000, we experienced successive troubles about nuclear power plants such as an accident at Mihama nuclear plant of Kansai Electric Co. and an earthquake at Kashiwazaki area and accordingly lowered the average operation rate of nuclear power plants in Japan, while between 1995 and 2000 the average operation rate of nuclear power plants in Japan was more than 80%. This suggests that it may be possible that the operation rate will come back to more than 80% in 2020.

The government is also eager to drastically expand photovoltaic as a symbol of renewable energy. It aims at realization of a capacity 20 to 40 times larger than the present capacity in 2020. Although the author is rather skeptical about its reliability, we assume 20 times will be realized and estimate C/E in that case.

CO_2 is emitted only from fossil fuel power plants; types of fuels and the thermal efficiency are keys in estimating the amount of CO_2. We adopt the government plan about the structure of fossil fuel plants (details are omitted here) and assume two cases in which the thermal efficiencies of fossil fuel plants are the same or by 10% better than present plants. $\Delta C/E$ are in the following.

This result indicates that CO_2 will decrease by 11 to 16% when compared with 2005 levels. The 2020 target (a 30% reduction with 2005 as the base year) is still by 15% or more far from the above results (Table 7.2).

Note: $\Delta C/E$ here is not the annual rate of change but the total change from 2005 to 2020.

Table 7.2. Reduction in C/E by 2020.

		Ratio of electrification	
		same as before	increased by 10%
Planed thermal efficiency:	same as before	▲11%	▲13%
	improved by 10%	▲14%	▲16%

Accordingly if we assume $\Delta E = 0$, the above values for $\Delta C/E$ are those for ΔCO_2.

7.4. Ways to Achieve the 2020 Target

With conventional efforts there is a gap of 15% or so between CO_2 emission and the 2020 target of a 25% reduction. In other words to achieve the Hatoyama target is really a difficult task, and then the key question is what measures exist if we are to achieve it. The government has to make a definite plan to meet this target. The followings are just suggestions the author has as measures toward achievement of the target.

(1) *Purchase of emission permits from foreign countries*

To achieve the Kyoto target, the government of Japan plans to purchase emission permits equal to 1.6% of the total emission of Japan per year from foreign countries, and has already almost achieved this plan by the purchase of CDM from developing countries and GIS (green investment scheme) from east European countries. Furthermore power companies and iron and steel companies also purchased emission permits for the purpose of achieving their voluntary targets. The total amount of their purchase is almost 5% per year of the total emissions of Japan. The sum of the government and company purchases is almost 6.6% of the total emissions per year, and the average cost is around 2 to 3 thousand Japanese yen/ton CO_2. This amount is almost 20 times smaller than the marginal costs for achieving a 25% target as described before, so these emission permits should be utilized as a measure for achieving the 2020 target.

(2) *Forest absorption*

The United Nations admitted for Japan to cover 3.8% of emissions by forest absorption in achieving the Kyoto target, and the government of Japan has made efforts for further conservation of forests in Japan. It should make similar efforts in the case of the 2020 target and keep similar amounts of CO_2 emission rights by forest absorption.

(3) *Unprecedented efforts for energy conservation*

Japan has made a lot of efforts for energy conservation, but as described before its energy conservation rate has still been lower than its economic growth rate. To clear the bar we have to introduce unprecedented measures for energy conservation. In residential sector consumers are decision makers so that it is not easy to prepare incentives effective enough for them to practice substantial energy conservation. Therefore the government should introduce not only carrots such as subsidies and tax exemptions, but also sticks such as regulation and surcharges to reduce energy consumption. For example introduce a law under which only houses satisfying certain insulation standard are permitted to be built.

The impacts of (1) and (2) are about 10% but still there is a gap of several to 10% between the emissions and a 25% target. It is then an issue to be tackled by the government of how to fill the gap by various energy conservation measures.

7.5. Concluding Remarks

The author explained how to use factor analysis, with the assessment of the 2020 target of Japan as an example. This example shows that CO_2 emissions have a close relationship to the economy. In the analysis of anti-CO_2 measures, it is indispensable to consider both the effect of implementing them on the economy and the effect of the economy on CO_2 emissions. The factor analysis utilized here in this chapter is a useful tool for doing this.

References

Fukui, T. (ed.) (2009). *Illustration of Midterm Target of Global Warming*, (in Japanese), Gyousei.

IPCC Working Group (2007). AR-4, WG-3, Chapter 13, Box 13–7.

Kaya, Y. (1989). "Impact of Carbon Dioxide Emission Control on GNP growth." IPCC/RSWG, Energy and Industry Subgroup, Geneva, May 8–9.

Chapter 8

LOW-CARBON ECONOMETRICS

*Kejun Jiang**

8.1. Introduction

Quantitative analysis on climate change and its mitigation began when climate change problems caught the attention of scientists. Climate change is a typical multidisciplinary topic of a wide range of social development issues. The United Nations Convention on Climate Change (UNFCCC), which has been the most complex international agreement, came about to cope with the complexity of climate change problems. Since its implementation in 1994, its annual Conference of Parties (CoP) has involved an increasing number and higher level of parties, of which the number of participants amounted to 30,000 in the Fifteenth CoP of UNFCCC in Copenhagen in 2009.

The human society has realized the great challenges brought by climate change and sought for possible responses. The ultimate goal of UNFCCC is to "stabilize greenhouse gas concentrations in the atmosphere at a level that would prevent dangerous anthropogenic interference with the climate system". To achieve this goal, there are many substantive issues to deal with, such as how to set the limits of appropriate target concentrations for individual countries, that emission pathways vary under different emissions pathways, how much is the cost to achieve the goals, how to allocate emission rights in a "common but differentiated" way, and how to involve all countries to equitably and effectively cooperate with each other.

*Kejun Jiang, Research Professor at Energy Research Institute of National Development and Reform Committee of China.

As a response to questions like above, the Intergovernmental Panel on Climate Change (IPCC) was launched by the United Nations, aiming to gain consensus through meetings and discussions of scientific and policy debates on climate change problems. Since the foundation of IPCC, four assessment reports have been submitted to the open public. Along with relevant researches, a large number of studies have focused extensively on quantitative analysis by presenting specific data on all aspects of climate change problems, supporting the establishment of international agreements and policies correspondingly. The recent IPCC report shows the major empirical results of a large amount of related studies.

The following sections mainly focus on the quantitative evaluation of climate change mitigation study and policymaking process, and develop a comprehensive evaluation system for climate change mitigation and the development of a low-carbon economy.

8.2. Assessment on Climate Change Mitigation

8.2.1. *An introduction to the integrated assessment of climate change*

The comprehensive evaluation system of climate change mitigation emphasizes the long-term effects of greenhouse gas emissions and aims to formulate an international or national strategy on emission reduction. In line with such formulation, responses to recent technical options and policy alternatives are also required.

There are currently various modeling methods dealing with the complexity of climate change issues, with regard to greenhouse gas emissions reduction. To address global environmental issues and global climate change in particular, scientific perspectives need to be integrated with those of natural sciences and social sciences disciplines, in order to clarify scientific insights widely and systematically to set out the basic structure of the problem and alternative solutions. Thus a large-scale multi-disciplinary model, named as "Integrated Assessment Model (IAM)" was introduced. It functions as a common platform for the academia and policymakers. However, due to the lack of investment in terms of finance and human resource, this model did not reach its climax in use until mid-1990s when the large-scale modeling group reached almost thirty. But now it drops to around ten.

The Integrated Assessment Model (IAM) system refers to energy production and consumption in the following four areas: Greenhouse Gases (GHGs) emission, recycling, climate change, biology, socio-economy.[1]

(1) Socio-economy: energy consumption and production, agroforestry, animal husbandry, urban activities, and other;
(2) Global warming effects and greenhouse gas in the global circulation system: carbon cycle, atmospheric chemistry;
(3) Climate: radiation convection, atmospheric and oceanic convection; and
(4) Hydrology and ecosystems: natural ecosystems, human ecosystems, hydrological systems, etc.

So far the integrated assessment models can be categorized into four types:

The first type is the large-scaled assessment model integrating full, detailed analyses on social economic activities and their impacts on climate change. It is the core of the assessment, and evaluates methodologies of problems and policy effects. So this type of model is called the full range integrated assessment model, such as the Asian Integrated Model (AIM) developed by Japan, the Integrated Model To Assess The Green House Effect (IMAGE) model in the Netherlands, the Global Change Assessement Model (GCAM) developed by the U.S. Pacific Northwest National Laboratory, the Massachusetts Institute of Technology (MIT) model by the Massachusetts Institute of Technology, and so on. The second is a more detailed type of models, based on the natural phenomenon of climate change, with climate change and damage mechanisms as the focus of assessment, including Model for the Assessment of Greenhouse-gas Induced Climate Change (MAGICC), Policy Analysis for the Greenhouse Effect (PAGE), etc. These models can also be seen as sub-models of the first type. The third category gives special attention to future economic development, with respect to the timetable and optimal way of measures, which cope with the climate change damages. Based on dynamic optimization, its simplified structure enables scientists to study the interactions between economic development and climate changes.

[1] IPCC (1996).

The examples include Regional Integrated model of Climate and the Economy (RICE) and A Model for Evaluating the Regional and Global Effects of GHG Reduction Policies (MERGE) models. The fourth type of models is simpler while focusing on connections with policymakers, hence an integrated assessment model of system development. The TARGETS model is a representative of this type.

The objectives of integrated assessment are:

— To identify the future development model of human society and natural system;
— To identify the main issues in assessment policy development; and
— To determine the direction of future research and to enhance the understanding of appropriate measures.

The complexity and the broad range of disciplines determine the integrated assessment method as a best means for the study of climate change problems. Deep understanding on such problems and thus appropriate actions can only be guided through a full grasp of the integrated models and assessment methods. The integrated assessment method was introduced to the international world in the 1970s, and so far has been widely used over 20 years of development. In China, however, due to its infancy of the current methods and assessment models, the large-scale integrated assessment models have not yet been developed and lagged behind developed countries.

The GHG emission model, based on energy consumption, is an integral part of the IAM model. It is relatively mature model, and has developed a lot, which can be roughly divided into two categories. The first category, named as the top-down model, is on the basis of economic models, and indexed by price and elasticity as the major economic indicators to the performance of the energy intensive in consumption and production. The second category reflects energy consumption and production in the techniques used by human activities, and then they could be described in detail. So it is also called the bottom-up model. The bottom-up model has many advantages, the biggest advantage of which is that, as human activities and technological changes are set as the basis, the prediction is very specific, and easy to explain, and convincible to policymakers.

The bottom-up model has by far more mature development, from the analysis of the impact of human activities on energy need and consumption, focused on energy supply and conversion, or targeted on energy demand

and consumption, to the current technological model with a full coverage of all energy activities. Of the first generation of the bottom-up models, the model of analyzing energy supply and conversion efficiency and introducing highly efficient techniques include the MARKAL model, developed by the International Energy Agency, and the EFOM model by France, etc. The models of analysis and computation in details, and targeted on energy needs and consumption, and the changes in energy needs and consumption caused by human activities, include the MEDEE model of France, and the LEAP model developed by the Institute of Environmental Research in Stockholm as the most well-known ones. The second generation of bottom-up models is based on energy network, which covers the entire life cycle of energy activities, from extraction of energy resources, to processing, conversion, and to energy needs. In such model, they interlink with each other, as an overall analysis on energy system. The AIM technical model developed by Japan is one example. The MARKAL model is under development and extending its end demand part. The IPAC-AIM technical model has been one of the most comprehensive energy technological models in current international and domestic spheres.

At the same time when bottom-up models achieve their progress, the top-down models develop even faster focusing on economic aspects of mitigation analysis. This type of models is mainly based on a computable general equilibrium model (CGE). Despite the weakness of CGE models, there are no complete models available, and CGE becomes a better choice to assess the impact of energy and climate change policies on economic systems. At present, the CGE model has dominated the analysis of long-term climate change policy.

8.2.2. *Determinants of IAM on climate change*

For the assessment on mitigation measures of climate change problems, some key factors need to be focused on, including mitigation cost, role of technology, and uncertainty, etc.[2]

[2]Jiang (2010).

(1) ***Mitigation cost***

The FCCC has made it clear that cost-effective is an important criteria for the implementation of climate change policy. The impact of climate policy can be defined in the introduction of the policy as changes according to the "development as usual scenario" or the "baseline scenario". In the baseline scenario, changes in general are described as "efficiency" and "cost". Efficiency typically includes the benefits in case of no climate change, and the synergies brought by the adoption of the specific climate policy. Mitigation costs and other negative results come from economic losses.

Mitigation cost depends on the choice of setting 100 or 200 years as the concentration target. The majority of models show that the cost is relatively low in the beginning, i.e. "easy to pick the fruit falls". However, most of these models also show higher cost in more reductions. The increasing marginal cost of mitigation provides the basis for the greenhouse gas emissions reduction with the introduction of an effective market mechanism.

Mitigation cost depends not only on the cost of reducing the accumulated emissions, but also on the mitigation time. Some studies suggest that the most cost-effective way is to gradually increase the emission reductions. This avoids the time changes in the immature stock of fixed assets. But other studies address the advantages of recently increasing emission reductions, by proposing that through the "learning-by-doing", the recent increase in emissions reduction achieves cost-effective, new technological development and capacity, and reduced future mitigation costs.

(2) ***Role of technology***

Due to the long time span of climate change, technological progress will be a very important factor. Development and popularization of new technology is the most conclusive of and effective way of reducing greenhouse gas emissions. Technology can be expressed in three aspects: invention, innovation and popularization. In our modeling study, we describe the hundreds of recently developed new technologies, which are able to promote energy conservation and greenhouse gas emissions reduction.

(3) *Uncertainty*

A great degree of uncertainty exists in climate change. The ambiguous relationship between greenhouse gas emissions and climate change is a constraint to understand the impact of climate change, and how it affects the value of human analysis. The uncertainty of climate change is reflected by whether it is acknowledged by mitigation policies. Some researches conclude and suggest a delay to take significant effort in dealing with climate change, mainly due to lack of understanding on climate change problems. Such views deny the possibility of a significant loss or climate change induced by the gradual accumulation of greenhouse gases. As long as the "first class" of error (an action is seen to be ultimately unnecessary) is still possible, the high-cost mitigation measures are now premature to take.

In contrast, uncertainty also brings about opposite views. Scientifically, the accumulation of greenhouse gases will eventually have negative impact on the climate, and thus substantial effect. If this is true, the "second class" of error (not to take or to take little action to reduce greenhouse gas emissions) will have enormous cost. It is possible to take actions in an earlier stage to avoid huge mitigation cost, rather than wait for decades when negative effects of climate change appear.

(4) *Factors affecting distribution and equity*

It is even more crucial for regional distribution analysis when conducting cost-benefit analysis on climate change mitigation measures. When considering the cost-effectiveness of mitigation measures at the national, regional, sector, the distribution between the generations, and their historical responsibilities for emissions, have important implication to make equitable mitigation measures. For example, carbon taxes or emission limits may result in the declines of production and profits of coal or oil industry, and their employment, as more consumers will choose natural gas. The impacts of carbon tax on renewable energy sources vary across regions, but the overall effect will be an enlarged market. Mitigation measures may lead to structural changes in the manufacturing sector. Mitigation measures will benefit the sectors supplying energy efficient equipment or low-carbon technologies. The increased energy cost of such mitigation measures has greater impact on wealthier families than low-income families. Hence, policy making should take low-income population into account and reduce the negative impact on this group of people. However, current research is dominated in developed countries while few studies exist in developing countries.

(5) *Sustainability*

In general, sustainability factors are not of major consideration by "cost-benefit analysis". The analysis of regional distribution of mitigation measures should also take into account the effect on other environmental factors. For example, the effects on sulfur, particulate matters, and land use can also be analyzed. In some scenarios with high sulfur emissions and stronger impact on health, the demand for energy will affect traditional agriculture, forests, and fallow land.

The definition of "climate policy" is not very clear. In many policy debates, mitigation policies of climate change refer to those mainly aiming at reducing greenhouse gas concentrations. They can be either direct policies to reduce greenhouse gas emissions such as carbon emission reduction, expansion of carbon sinks, and reductions of other gases emissions, or policies to promote low-carbon technology and production or processing development. If mitigation policies are these, then other policies outside the scope should be considered in the baseline scenario. However, significant impact will be caused by such activities. For example, in some areas the regional pollution control measures such as the control of hydrocarbon emissions in vehicles facilitate CO_2 emissions reduction and the production of tropospheric bromine oxide. Moreover, some policies such as poverty alleviation will ultimately have a significant effect on greenhouse gas emissions.

8.3. IPAC Model

In this section, we illustrate the policy evaluation process through the introduction of the Integrated Policy Assesssment Model for China (IPAC) which was developed by our research team as an example of comprehensive evaluation model. The IPAC model is not as a large-scale international photogenic consistent and comprehensive evaluation model as other international models, which are mainly climate models, atmospheric general circulation models, atmospheric chemistry models, and impact models. To integrate these models, we need large-scale financial support and the cooperation of different research institutions. However, in this period of model development process, we have not reached the same level of support as other developed countries, so in our ongoing development of such a large scale model, it also requires a certain degree of deep understanding.[3]

However the IPAC model has been developed as a pioneer in the forefront of the world in the analysis of the energy system. The IPAC model system includes almost all the current models of energy system analysis method, including a CGE model, a dynamic economic model, and technical models, which will be introduced as follows:

8.3.1. *The framework of IPAC model*

The IPAC model includes three parts: energy and emission models, environmental models and impact models. Energy and Emissions model is the main component of IPAC model, including a number of different types of models. Environmental model includes an atmospheric dispersion model and a simple climate model. Impact model includes a health impact model. Correlation exists among them: the results of energy and emissions model are inputted into the environment model to calculate energy activities caused by the concentration of air pollutants and possible temperature raise, and the impact is evaluated by the health effects model, and results are turned into economic value and fed back to the energy emission model. IPAC national models' main functions are described as follows:

Energy emission model:

— IPAC-SGM model: it is a CGE model, concerning the impact and relations between economic activities. The IPAC model is applied to analyze the economic impact from energy and environmental policies and also concerns energy and environment scenarios analysis.
— IPAC-material model: it is also a CGE model, yet the difference from IPAC-SGM is that it focuses more on environmental effect and policies.

---

[3] Jiang (2003).

More environmental sectors are included in the model to analyze the energy activities' impact on air pollution, water pollution, destruction of land surface, and environmental industries' impact on economy.

— IPAC-emission model: it is a partial equilibrium global model, including nine regions, and China is one of the regions. Mainly used for long-term energy and greenhouse gas emissions scenarios analysis.
— IPAC-TIMER model: it is a dynamic energy economy model, which is also a global model, including 17 regions, can be used for detailed analysis of energy import, export and investment.
— IPAC-tech model: it is a least cost optimization model for cost analysis in a certain period of time and analysis of investment policies' effect on energy under different technical progress assumptions.
— IPAC-message model: focusing on cost and market analysis of important energy technology developments.
— IPAC-AIM model: it is a detailed technical evaluation model, taking the bottom-up analysis approach. Technology policy and greenhouse gas reduction policy can be evaluated for short-term energy and greenhouse gas emissions scenarios analysis.
— IPAC-AIM/Local is a regional model which is similar to IPAC-AIM model, however it contains 31 provinces and cities, and focuses on regional energy and environmental policies assessment.

Environmental models:

— IPAC-air is a region diffusion model. Based on the results of calculation of the energy emission model, it calculates different concentrations of air pollutants, is used for input of effects models, while compared to energy development strategies and government's environmental targets.
— IPAC-climate model is a Weigley model coming from American climate center, to calculate future temperature raise.

Impact models:

— IPAC-health impact model. It is applied to analyze the health impact of air pollutions and temperature raise, then the impact on GDP.
— IPAC-water resource impact model. It is applied to analyze the impact on water resource of climate change.

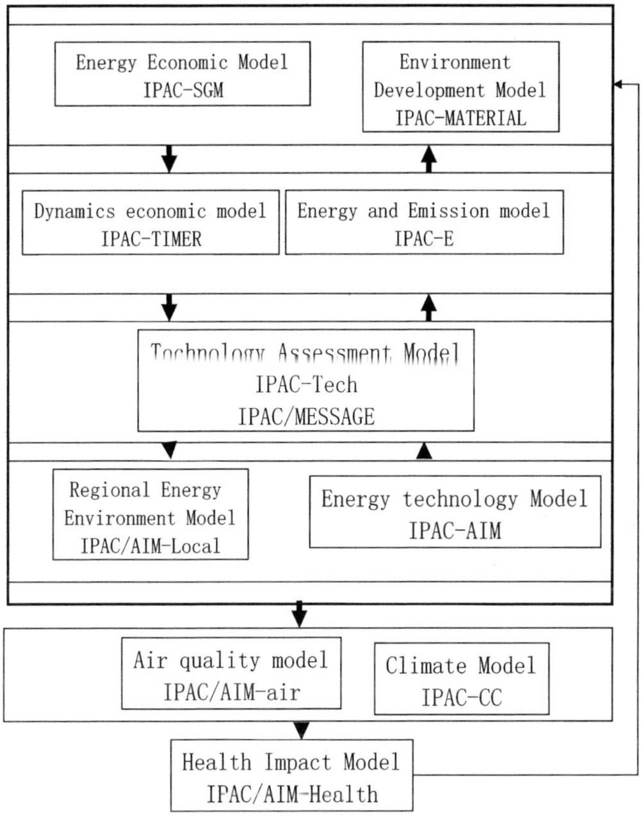

Fig. 8.1. The framework of IPAC model.

IPAC model structure is shown in Fig. 8.1. The current connection of IPAC sub-model is using soft approach, which is using the output of a model as another model's input. However, hard-connections are used between several sub-models of the IPAC-emission model.

As a model framework, IPAC application method takes the same parameter value method, or named the same development framework approach. Under this framework approach, all models use the same macro-economic development indicators, such as GDP and population, similar to the technical parameters and energy-saving measures, so the output of each model is comparable on the similar data platform, which can study different energy and environmental issues.

8.3.2. *IPAC-SGM model*

IPAC-SGM model is established based on cooperation with the Pacific Northwest Laboratory of the United States. The SGM model is developed by the U.S. Pacific Northwest Laboratory (Battelle Pacific Northwest Laboratories). It studies greenhouse gas (GHG) emissions from human economic activities based on computable general equilibrium theory.[4] SGM means Second Generation Model, is different from the first generation model which considers the development of greenhouse gas emissions model combining with the activities of all economic sectors, and establish inter-departmental interaction, reflecting technological progress's impact on economic development, studying the whole world's total emissions of greenhouse gases. Thus it is a comprehensive, systematic and comprehensive model. SGM can predict various scenarios of GHG emissions caused by human economic activities, and conduct corresponding policy analysis. Since the burning of fossil fuels is the major GHG emission sources, SGM focus on energy activities analysis.

IPAC-SGM is a dynamic simulation model, rather than an optimization model. Based on the 1990 scenario and with a step of every 5 years (referred to a Vintage), it can predict farthest of the year to 2100. The model is used for scenario analysis and policy analysis.[5]

IPAC-SGM analyzes the future through the simulation of macroeconomic scenario:

- GDP scenarios. Modeling iterative calculation and the dynamic drive investment to achieve the equilibrium of each period (or close to equilibrium), the model can estimate GDP of various periods.
- Energy demand scenarios. Getting the value of the distribution coefficient a_i through energy input/output relationship, and adjusting a_i according to energy experts estimating the future energy intensity, the model estimate the energy demand based on calculation of economic activity.

---

[4] Edmonds *et al.* (2003).
[5] Jiang (2000).

— GHG emission scenarios. Exogenously given GHG emission factors, GHG emissions can be calculated based on the level of economic activity and energy consumption.

IPAC-SGM can be used to study a variety of stable/reduction GHG emission scenarios, and conduct impact assessment and policy analysis. The policy analysis instruments provided by IPAC-SGM consists of two aspects:

— Tax on energy or carbon emissions. Control GHG emissions by energy or carbon emission tax to achieve stable or reducing emissions. The value of carbon tax can have a variety of options; the model can estimate GDP, final consumption demand, the state's tax, energy consumption, GHG emissions under different carbon tax policy, and by comparing with reference scenario, and apply the macro cost-benefit analysis;
— To promote the development of new energy technologies. Through the use of advanced energy technologies, improve energy efficiency, reduce energy consumption, to control the GHG emissions.

The constant elasticity of substitution (CES) production function in IPAC-SGM is shown as follows:

$$Xs = \alpha_0 \left(\sum_{i=1}^{N} \alpha_i X_i^{\rho} \right)^{(1/\rho)}$$

In which, X_i is the input of the ith production factors; N is the total amount of factors; a_0 is the scale coefficient; a_i is the coefficient of the ith input factor (which is a distributional parameter); ρ is the substitution parameter and the substitution elasticity is $\sigma = 1/(1 - \rho)$.

IPAC-SGM has achieved dynamic function, and investment decision is the core of dynamic mechanism of IPAC-SGM model. Parameter settings for future years starts from the allocation of investment, depends on the main pre-determined scale of investment, efficiency and current investment constraints.

Sector is the highest intensive layer of IPAC-SGM model, several sub-sectors are contained within sectors, and homogeneous products are produced by sub-sector. IPAC-SGM assumes that each sub-sectors are divided according to technology level, competition is restricted within the sub-sectors, which determine the entire sub-sector profitability. IPAC-SGM defines a logarithmic function, determines the investment share of each technology within subsectors:

$$S_{i,tech} = \frac{\left(\frac{\pi^{\sigma}_{i,tech}}{Ntech_i}\right)}{\sum_{j=1}^{NO}\left(\frac{\pi^{\sigma}_{j,tech}}{Ntech_{jNss}}\right)}$$

$$NO = \sum_{i=1}^{Nss} Ntech_i.$$

In which, $\pi_{i,tech}$ is the profit achieved by the ith subsector, N_{ss} is the number of subsector amount, $Ntech_i$ is the number of technologies used in the ith subsector, σ is the sensitive factor, reflecting the market sensitivity.

8.3.3. *IPAC-emission model*

The aim of IPAC-emission model is to evaluate the future greenhouse gas emissions policy measures may be taken of China and other regions in the global environment in long-term. Under this goal, this model is a global model, according to a major international climate change on the region, the world is divided into nine areas: the United States, Western Europe and Canada, other OECD countries, Eastern Europe, China, Middle East, other Asian countries, Africa, South America. The time interval is from 1990 to 2100 to analyze the changes and possible long-term rise of temperature.[6,7]

The greenhouse gases analyzed in the model including six gases of climate change referring to the Kyoto Protocol in support of the international negotiating process: CO_2, CH_4, CO, HFC, PHFC and SF_6, and also including SO_2, NOx emissions.

Taking into account the possible sources of greenhouse gas emissions, in order to provide a comprehensive evaluation of possible policy responses, the IPAC-e model includes all emission sources of human activities, the model is a full range of emission models, including energy activities emissions (including mining, transport and combustion emissions), land use emissions model, industrial processes and waste emission models.

---

[6]Jiang and Hu (2003).

[7]Jiang, *et al.* (2003).

The model is mainly composed of three sub-models, which are the socio-economic energy sub-model, technical and economic energy sub-model and land-use sub-model. The three models are based on the existing model and extended to connect the AIM global scenario.

Social and economic energy sub-model is developed based on the ERB model (Edmonse-Rilly-Barns model). ERB model was developed in 1982, which is used by many research groups to assess the CO_2 emissions. It is a top-down type of social and economic energy emissions model, originally developed by the Oak Ridge Laboratory of United States. The technical and economic energy model is a model is developed under the framework of AIM. AIM model is a bottom-up type model, based on the final energy technology, energy market participation cost to predict the energy demand and greenhouse gas emissions in future. It is developed by the Japanese National Institute for Environmental Studies and Kyoto University and a common structure is formed by several projects in the Asia Pacific region. AGLU (Agriculture and Land Use) model was originally the latest version of MiniCAM, which calculate the greenhouse gas emissions caused by land use. AGLU is developed by Pacific Northwest Laboratory of the U.S.

8.3.4. *IPAC-AIM/technology model*

IPAC-AIM/technology model is a major component of IPAC model, which aims for estimate energy services and equipment development in the future, and simulate the process of energy consumption. It is mainly used for calculating energy demand of various of the terminal energy sector of different future scenarios, and thus calculating of the CO_2 emissions. One of its important roles is to evaluate the impact of technology introduction and greenhouse gas emissions by different technologies policies.[8,9]

IPAC-AIM/technology model applies the minimum cost method for analysis, a variety of energy services can be selected based on the minimum cost principal. The model uses linear programming method, which can analyze the energy use under complex process from the viewpoint of process systems, rather than a single technology. In the model, various parameter settings can take different standards and methods to expand the scope

[8] Hu *et al.* (1996).
[9] Hu and Jiang (2002).

of analysis. For instance, various components of the technology operating investments compose the technical operation cost, in which investment can put into different situations, including energy, raw material input, labor input and other inputs, make the cost analysis of the technology closer to reality.

The current version of the IPAC-AIM/technology model includes 42 sectors and products, nearly 600 kinds of technology, which include existing and future technologies.

8.4. Low Carbon Scenarios in China

8.4.1. *The development of China's low carbon scenarios*

This section designs three emission scenarios based on the main factors correlated with future emission, and the IPAC's projection for 2050 (Jiang *et al.* 2008). This is in order to reflect all the possible emission pathways of greenhouse gases in China.

The first scenario, without any action to climate change problems (BaU), is designed on the basis of all possible development models, and mainly driven by economic growth. According to previous studies on scenarios, China's economic development for the coming 50 years is reviewed. Its population is projected to reach a peak of 1.47 billion in 2040–2050.

The second is Low Carbon Scenario (LC), which integrates national energy security, domestic environment, and the low-carbon way into the implementation of national policies to fulfill such a scenario. This scenario mainly considers domestic socio-economic and environmental development needs, and fulfills the expected energy and emission scenario through their own domestic efforts in strengthening the technological advances, adjusting economic structure, changing consumption patterns, and achieving low energy consumption and low greenhouse gas emission.

The third is called Enhanced Low Carbon Scenario (ELC), addressing China's further contributions with the common consensus for the mitigation of climate change. This scenario considers that, under the joint global efforts, how further technological progress contributes to a faster decline in the cost of technology, and the policies of the developed countries will be gradually expanded. Meanwhile, taking the economic strength of China after 2030 (probably the world's largest one) into account, investment in low

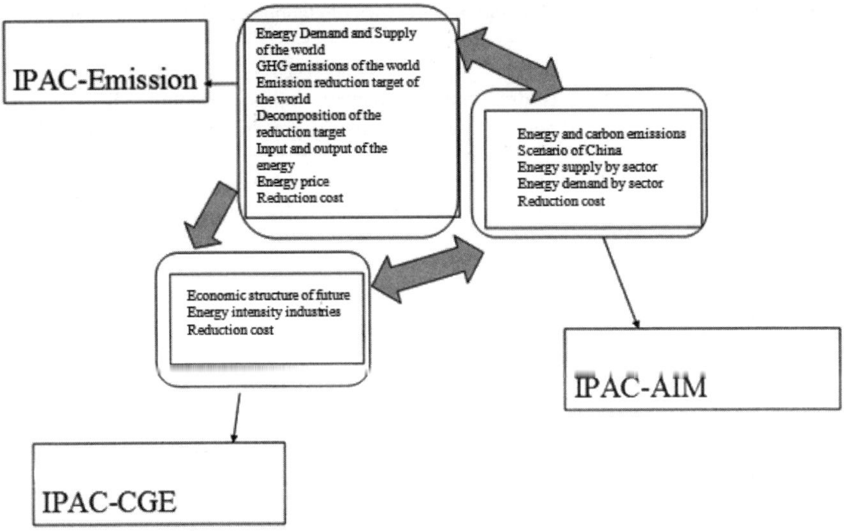

Fig. 8.2. Relationship of the IPAC-CGE model, the IPAC-emission model, and the IPAC-AIM/technology model.

carbon economy can be further increased. Hence, better use of low-carbon economy provides opportunities for economic development. At present, China is leading the technological development in clean coal technologies and carbon capture and storage (CCS), among which CCS in China has been popularized and widely applied.

In this study, three of the IPAC models are used for analysis, i.e. the IPAC-CGE model, the IPAC-emission model of global emissions, and the IPAC-AIM/technology model. Figure 8.2 depicts the relationship of the three models.

Using the above three models, according to the specific parameter designs in detail (considering the length of this study, the parameters are not described here, please refer to the relevant report if interested), the scenario of China in 2050 in terms of its energy and CO_2 emissions are projected. Tables 8.1–8.3 show the primary energy demand in each scenario, and Table 8.4 presents CO_2 emissions.

The baseline scenario estimates the primary energy demand will be increased from the 2.189 billion tons of standard coal in 2005 to 6.657 billion in 2050, of which the share of coal will be 44%, oil 27.6%, natural gas 10%,

Table 8.1. Primary energy demand, million tons of standard coal, baseline scenario.

	Coal	Oil	Natural gas	Hydro power	Nuclear power	Wind power/ solar	Bio-mass power	Alcohol gasoline	Bio-diesel	Total
2005	1536	435	60	131	20	0.8	1.9	1.8	0.6	2189
2010	2424	628	109	217	28	7	16	10	0.6	3438
2020	2991	1096	271	294	90	20	30	22	3.1	4817
2030	2932	1708	460	358	181	54	44	33	7.9	5526
2040	3001	1710	532	380	380	84	71	36	8.5	6202
2050	2925	1836	668	397	595	103	86	39	9.2	6657

Table 8.2. Primary energy demand, million tons of standard coal, low carbon scenario.

	Coal	Oil	Natural gas	Hydro power	Nuclear power	Wind power/ solar	Bio-mass power	Alcohol gasoline	Bio-diesel	Total	Coal
2005	1536	435	60	131	20	1	0	2	2	1	2189
2010	2173	528	109	207	46	12	0	9	2	1	3087
2020	2195	843	349	375	136	51	1	32	8	6	3996
2030	2091	964	529	401	301	92	4	52	28	12	4474
2040	2063	1010	628	424	471	118	9	61	36	13	4833
2050	1984	1025	745	422	760	169	20	68	44	14	5250

Table 8.3. Primary energy demand, million tons of standard coal, enhanced low carbon scenario.

	Coal	Oil	Natural gas	Hydro power	Nuclear power	Wind power/ solar	Bio-mass power	Alcohol gasoline	Bio-diesel	Total	Coal
2005	1536	448	60	131	20	1	0	3	1	0	2203
2010	2083	532	107	180	40	18	0	8	2	1	2971
2020	2144	838	330	354	145	66	1	31	8	6	3921
2030	1903	943	491	395	301	156	5	49	20	12	4275
2040	1814	993	604	429	497	214	16	59	22	13	4660
2050	1715	1032	710	420	761	239	37	63	23	14	5014

Table 8.4. CO_2 emissions of fossil fuels, million tons of carbon.

	Baseline scenario	Low carbon scenario	Enhanced low carbon scenario
2000	867.2	867.2	867.2
2005	1409.3	1409.3	1409.3
2010	2134	1943	1943
2020	2779	2262	2194
2030	3179	2345	2228
2040	3525	2398	2014
2050	3465	2406	1395

nuclear power 9%, hydropower 6%, and 3.4% from wind, biomass power and other renewable and new energy resources.

According to the low carbon scenario, in 2050 the primary energy demand will be increased to 5.082 billion tons of standard coal from 2.189 billion in 2005. Coal will account for 37.4%, oil 20.2%, natural gas 14.4%, nuclear power 14.2%, hydropower 8.4%, and 5.4% from wind, biomass power, and other renewable and new energy resources.

CO_2 emissions can be computed from the consumptions of fossil fuels, as shown in Table 8.4.

Comparing the result of low carbon scenario, the enhanced low-carbon scenario predicts that the CO_2 emissions will sharply decrease since 2030, by 48% in 2050. The enhanced low carbon scenario promotes energy saving further than the low carbon scenario, when primary energy demand will decline by 4.5%, and the share of power from renewable energy resources and nuclear will increase by 7%, up to 58%. In the meantime, coal-fired power station will be largely equipped with IGCC and CCS in 2020. CCS will be popularized in high energy-consuming industries such as iron and steel, cement, electrolytic aluminum, ammonia, oil refining, and ethylene. Building industry will commonly use renewable energy technologies such as using solar water heaters for hot water supply and heating, while wind power and photovoltaic technologies will be suitable for household uses in some areas.

8.4.2. *The way towards low carbon scenario*

According to studies on future energy and emission scenarios, from now on until the mid-21st century, China's future demand for energy will show a significant and sustained growth trend, along with its stable economic growth. To keep a low energy demand, i.e. to meet the level as in energy demand scenarios, a large amount of effort should be paid, such as optimizing the economic and energy structure, establishing and improving energy markets, achieving the 2.95% of annual average energy-saving rate through technological progress in the coming 50 years. The conclusions of these scenarios indicate that the future energy growth rate accounts for 40–65% of China's economic growth rate (i.e. the energy elasticity is 0.4–0.65), which is quite low compared to developed countries (Jiang, 2008).

In next 50 years, with the growth of energy consumption, China's greenhouse gases emissions caused by energy activities also take on an upward trend. CO_2 emission according to the 2050 baseline scenario will be close to 3.4 billion tons of carbon, while the policy scenario predicts that it is likely to reduce by 35%. The future energy demand and greenhouse gases emission will be considerably influenced by technological advancement in energy conversion, use, and supply. Economic development, demographic and income changes, and changes in the use of technologies, are the major factors correlated with the quantity and quality of energy consumption. The driving forces that guarantee the quantity and quality of energy supply are resources and conversion technologies. Technological advancement, especially those high technologies, will benefit the development and utilization of a large number of potential conventional oil, gas and unconventional oil, gas and other clean energy resources, in terms of higher efficiency and better environmental performance. Therefore, the policy making process for technological development should also incorporate economic development and adaptive measures to climate change, take advantage of our natural resources, and focus on developing clean technologies which will lead the international clean energy development, conversion and use technologies, in order to facilitate economic development and conserve the environment energy resources.

The models conclude that a low-carbon scenario approach is feasible in China. The way towards low energy demand and low-carbon economy includes: adjusting the industrial structure to be a low energy consumption and high efficiency; advancing the use of energy technology through a variety of policy measures to widely popularize advanced and efficient technologies; developing renewable energy resources and nuclear power comprehensively and rationally, and raising their importance in primary energy resources; involving public participation, changing life style, consumer behavior, and promoting low carbon consumption; developing low-carbon agriculture, and increasing forestation and forest management. Speaking of China, this is to optimize its industrial structure, control the development of high energy-consuming industries, reduce or control the export of high energy consumption products; and for the energy efficiency of industrial technologies, aims to reach the world's high level in 2025; to popularize the use of renewable technologies, such as wind power, hydropower, and to set commercial use examples for solar thermal power generation and photovoltaic power generation technologies; to fully and vigorously disseminate nuclear power, especially the third and the fourth generations of advanced nuclear power technologies; and to raise the public awareness, popularizing a low-carbon lifestyle.

According to the forecast by this study, before 2030, China will still have a certain amount of greenhouse gas emissions, even in case of the lowest energy scenario. Hence, it is very worthy of attention as all industries are facing high requirements, because different combination of polices and technologies will have various effects on emission reduction. China may not have sufficient time to develop ultra-supercritical technology, because of the need to enter the next generation of advanced power generation technology as soon as possible, i.e. the coal gasification combined cycle power generation (IGCC), in order to set the basis for the popularity of carbon storage and capture technology (CCS). The power plant equipment manufacturing enterprises in China take the IGCC technology very seriously, and have invested a large amount in R&D of this technology. This is as a result of the collaboration among several parties.

8.4.3. *Cost of low carbon development*

Low carbon economy requires more investment. In scenario research various investments and costs are analyzed, as shown in Figs. 8.3 and 8.4.

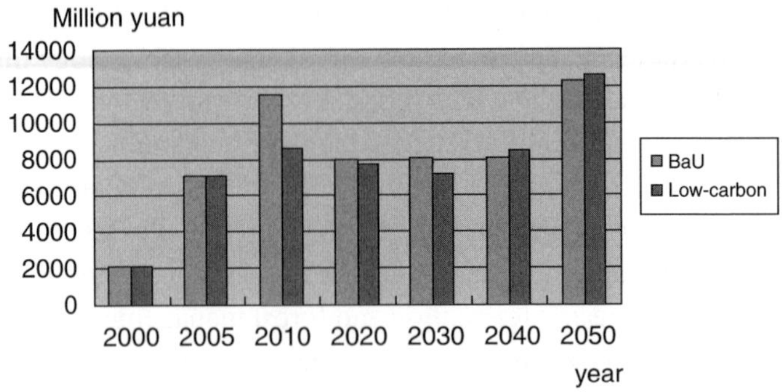

Fig. 8.3. Investment demand in energy industry.

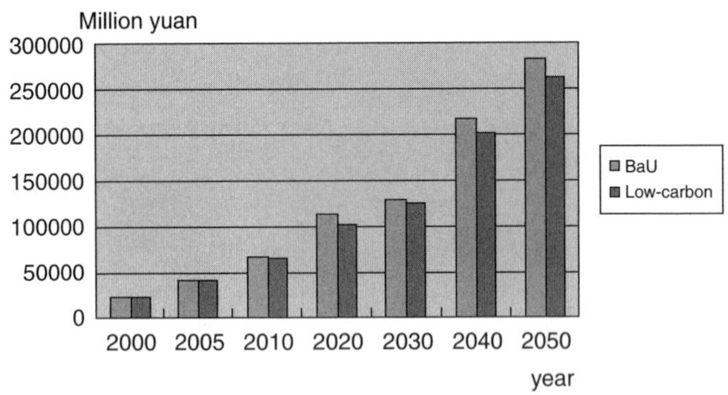

Fig. 8.4. Energy cost.

Figure 8.3 depicts the investment in energy industry. There are two factors: the first one is that because of energy efficiency (saving), energy demand in low carbon scenario is significantly less than that of the baseline scenario, leading to the result that looking at the scale of energy industry, investment will be smaller in low carbon scenario; the second one is, the higher cost of technology in low carbon scenario than the baseline one will lead to higher investment in energy industry. When considering these two factors together, the overall investment in energy industry will be slightly smaller in the low carbon scenario than in the baseline scenario, as shown in Fig. 8.3.

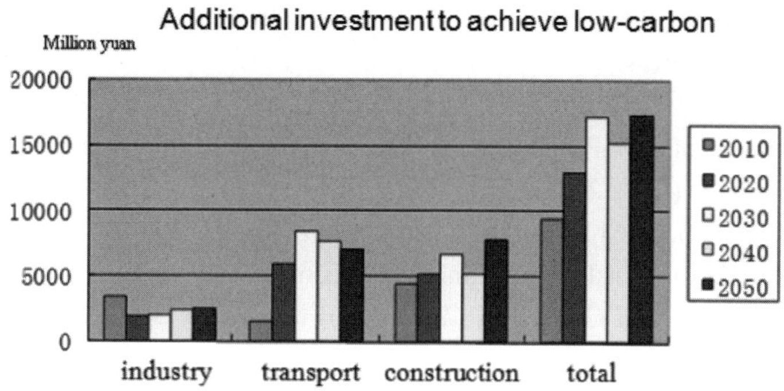

Fig. 8.5. Additional investments in the low carbon scenario.

Energy cost is another indicator of investment for a country. National energy cost refers to the amount of final energy multiplied by energy prices. On the one hand, due to energy efficiency, both the demand of energy and its cost will be declined in the low carbon scenario, but the energy price and cost will increase due to the energy tax or carbon tax. It can be seen from Fig. 8.4 that, the overall cost in the low carbon scenario will be lower than in the baseline scenario. Excluding energy tax and carbon tax, energy price in the low carbon scenario will be declining compared to the baseline scenario.

Our scenario analysis also shows the additional investment in end sectors in the low carbon scenario. Due to the difficulty in definition and in order to simplify the calculation, we only consider the main differences in investment among using technologies, which also includes inputs in public transport such as MTR (see Fig. 8.5).

References

Edmonds, J., Pitcher, H. M., Barns, D., Baron, R. and Wise, M. A. (2003). "Modeling Future Greenhouse Gas Emission: The Second Generation Model Description." Pacific Northwest Laboratory.

Hu, X., Jiang, K. and Liu, J. (1996). "Application of AIM/Emission Model in P.R. China and Preliminary Analysis on Simulated Results." AIM Interim Paper, IP-96-02, Tsukuba, Japan.

Hu, X. and Jiang, K. (2002). *Greenhouse Gas Mitigation Technology Assessment*. Beijing: China Environmental Science Press.

IPCC (1996). *Climate Change: Mitigation 1996*. Cambridge: Cambridge University Press.

Jiang, K., Masui, T., Morita, T. and Matsuoka, Y. (2009). Long-term GHG Emission Scenarios of Asia-Pacific and the World. *Technological Forecasting & Social Change*, 61(2–3), 267–287.

Jiang, K. (2000). SGM Model in China. *China Energy*, 6, 34–38.

Jiang, K. (2003). "Analyses on Emissions Scenario and the Global Greenhouse Gas Models and China." In *China Energy Issues in 2002*. Beijing: China Environmental Science Press.

Jiang, K. and Hu, X. (2003). "China and the World's Greenhouse Gas Emissions Scenario Analysis Model." In *China Energy Research 2002*. Beijing: China Environmental Science Press.

Jiang, K., Hu, X., Zhuang, X., Liu, Q. and Zhu, S. (2008). China's Energy Demand and CO_2 Emission Scenarios by 2050. *Climate Change Research*, No. 5, 17–19.

Jiang, K. (2008a). Low-Carbon Society in China: Opportunities and Challenges. *China's Reform*, 6, 59–61.

Jiang, K. (2008b). Low-Carbon Society, Our Future. *Lvye*, 10, 44–46.

Jiang, K. (2010). "The Meaning of the Costs of Climate Change." Energy Research Institute Report.

Chapter 9

LOW-CARBON INTERNATIONAL ECONOMICS

Shijin Liu and Yongsheng Zhang*

9.1. Introduction

Global warming is a major challenge that human beings are facing. It brings about rises in sea-level, frequent extreme weather, and furthermore, political, economic, and military conflicts. Greenhouse gas emissions are the main cause of climate warming. CO_2 discharged from fossil fuel burning accounts for the main part of the GHG emissions of human beings. Climate warming and GHG emission are multidisciplinary subjects. Although there is still debate about whether human activities cause global warming, once most countries agree that human activity is a direct cause of climate warming and are ready to take actions, the climate problem will extend from purely scientific aspects to become a global politic issue. Regardless of the scientific conclusions about the impact of human activities on climate in the future, countries currently have to deal with the political and economic shocks brought on by climate negotiations.

The Kyoto Protocol is the first international action plan for GHG emissions. The "cap and trade" and flexible emission reduction mechanism of the Kyoto Protocol have won high praise, however, its drawback is also criticized severely. On the one hand, it can only cover 30% of global emission; moreover, the emission that is really brought into the Emission Trade Scheme (ETS) is only 8% of the total global emission. Under the current framework of the protocol, the emissions in year 2010 were just 1.5% more than the emissions without control. Even if the protocol gets the extension,

---

*This chapter is written by a research team led by Shijin Liu, Vice Minister of Development Research Center of the State Council, China. The first draft was authored by Dr. Yongsheng Zhang and finalized by Shijin Liu. We thank Junkuo Zhang, Hongchun Zhou, Peilin Liu, Zhenyu Wu and Anbo Xiang for their valuable comments.

its role in curbing global warming is also very limited. On the other hand, under the current framework of the protocol, countries, especially developing countries, do not have incentive to take part in global emission reduction actions (Nordhaus, 2006). Therefore, a new effective protocol is urgently needed as the Kyoto Protocol expired in 2012.

At present, various options have been proposed to replace the Kyoto Protocol. Among them, some advocate abandoning the Kyoto Protocol and replacing it with a new framework, while some advocate replacing the cap-and-trade mechanism by using the carbon tax approach. In regard to GHG emissions allocation schemes, some are lacking internal consistency in their theoretical basis, while others are filled with pragmatic and subjective value judgments. These programs more or less have a common characteristic that is ignoring the interests of developing countries, intentionally or unintentionally (see Aldy *et al.*, 2003; Aldy and Stavins, 2007; Garnaut, 2008; Nordhaus, 2006, 2008; Stiglitz, 2006; Weitzman, 1974; Sagar and Kandlikar, 1997). Although the Kyoto Protocol, under the United Nations Framework Convention on Climate, institutes a "common but differentiated responsibilities" principle, which does not require developing countries to reduce emissions, the legitimate rights and interests of developing countries have not received adequate protection. In this case, it is difficult for developing countries to have incentives to participate in global emissions reduction; as a result, it is also difficult to reach global emission reduction targets. Therefore, the new international agreement must take full account of the legitimate rights and interests of developing countries, in order to make it acceptable.

This chapter covers the global GHG emission reduction theory that adequately protects the reasonable rights and interests of developing countries, and presents a global solution covering all countries, to succeed the Kyoto Protocol. In Section 9.2, we will discuss the economic attributes of GHG emissions in order to find a solution to the problem. In Section 9.3, we will propose a theoretical framework of GHG emission rights and the global solution of emission reduction. In Section 9.4, we will conduct a digital stimulation in order to explain the theoretical framework more concretely. And in Section 9.5, the real meaning of our theoretical framework will be revealed in depth, and the related major issues of principle will be clarified. The last section will be the conclusion.

9.2. The Economic Attributes of GHG Emission

Economic attributes of greenhouse gases depend on the following natural and technological factors. First, because of the mobility of the atmosphere, regardless of which country discharges greenhouse gases, all human beings on the planet are affected. Secondly, the retention and harm of greenhouse gases in the atmosphere lasts for a long period. For example, half of CO_2 decays in 100 years, but 20% remains after hundreds of years (IPCC, 2007a, 824). Currently the greenhouse gases retained in the atmosphere were mostly discharged by developed countries during the Industrial Revolution. Thirdly, the effects of greenhouse gas are not competitive, that is, each person will not suffer less by the increase of population.

Under the existing technology, production activities require direct or indirect emissions of greenhouse gases. However, the intensity of per unit output of greenhouse gas emissions differs per country. Although greenhouse gas emissions harm human beings, they do bring benefit to mankind at the same time. Greenhouse gas is the indispensable condition for the development that human beings are pursuing. However, if global greenhouse gas accumulates to a certain level, then the resulting damage will eventually exceed the benefits. Most scientists agree that greenhouse gases must be controlled within 450–550 PPM CO_2-e. Beyond this limit, the earth would suffer disastrous consequences. Although whether this result will really happen remains uncertain (IPCC, 2007b; Stern, 2007; Garnaut, 2008), once it is recognized that the resources that produce emissions of greenhouse gases are not unlimited, the amount of greenhouse gas emissions produced by these resources will decline. However, how to encourage all countries to act in a concerted effort so as to ensure the realization of global emission reduction targets is a great challenge.

The benefits of greenhouse gas emissions are occupied exclusively by the states that produce them, but the harm is undertaken by all generations on earth. As the atmosphere is a public resource, national emissions are not clearly defined, each country's best option is to discharge "the more the better". As a result, excessive emissions lead to global warming; it is the so-called "tragedy of the commons" (Hardin, 1968). Meanwhile, since the low emission countries also suffer a net external hazard caused by high emission countries, high emitters should compensate low emitters for their excess emissions. However, only when the national emissions are clearly defined, can the excess emissions of the country be calculated in accordance with actual emissions? Once national emissions are clearly defined and strictly protected, the country's emissions would be strictly bound, and the amount of total global emissions might not exceed disastrous levels. It is clear that the definition of property rights directly changes the national emissions behavior.

Carbon reduction will reduce the expected loss of GDP. At a given level of income, the expected loss reduction is equivalent to an increase in welfare, which can be thought of as the benefit of carbon reduction. Because of the mobility of greenhouse gases, emission reduction costs to individual countries should be undertaken by those countries, but the benefits are shared by countries all over the world. In other words, the benefits of emission reduction are non-competitive, i.e., they are not exclusive to the emission reduction country. In this case, in order to minimize their costs, each country will choose the "no emission reduction" or "free-rider" option, and rational behavior of global cooperation to reduce emissions would be difficult to achieve spontaneously. If the countries of emission reduction could occupy the benefits exclusively, the optimal strategy of countries would be changed from "no emission reduction" to "reduction". However, in reality, it is difficult to achieve this.

Therefore, global GHG emissions can actually be classified as a "how to define or allocate national emissions" problem. According to Coase (1960), as long as the initial emission rights of countries have been clearly defined, the international emissions trading system (IETS) will allow countries with the lowest emission reduction cost to reduce emissions, the costs of global emission reduction will be minimized, and eventually an optimal global structure for total emissions will form. Coase emphasized the role of defined property rights to minimize social costs. But how to define initial property rights is not taken into account. Therefore, there is no mechanism to define initial emissions rights. Grossman and Hart (1986) and Hart and Moore (1990) stressed that, under the present transaction costs, different ownership structures have different levels of efficiency, so there will be a kind of ownership structure that optimizes overall efficiency.

So, how do we define the country's greenhouse gas emissions rights, namely property rights? The atmosphere is a typical global public resource.

The allocation of public resources more often obeys the rule of "first-come first-serve" or relies on the result of bids depending on the country's power. The definition of greenhouse gas emission is related to historical accumulated emissions, in which developed countries emitted much more than developing countries. Therefore, we need to start from generally accepted principles, according to the unique natural attributes and economic attributes of greenhouse gas emission, in order to find a fair and simple theoretical basis and solution for defining and distributing initial national emissions rights.

9.3. A Theoretical Framework and Global Solution

9.3.1. *A theoretical framework of global GHG emission reduction*

The core of the theoretical framework we explored is to solve the problem of allocation of national emissions so that the global emission reduction targets can be endogenous. The corresponding global greenhouse gas emission reduction solution consists of three steps (see Fig. 9.1). The first step is calculating the historical accumulated emissions of each country. According to the current contribution of greenhouse gases retained in the atmosphere of each country, we can determine whether the accumulated emissions of each country during period T0–T1 exceed actual emissions. If those levels exceed actual emissions, the current (time point T1) emission rights balance will show a deficit; if not, a "national emission rights account" can be established for each country. The second step is increasing the global total emissions quota from T1 to T2, and allocating the quota to each country. The increased quota plus the emissions balance of each country is the total emissions quota of each country on time point T2. If the global total emission quota is set, the global emission reduction quota can be determined

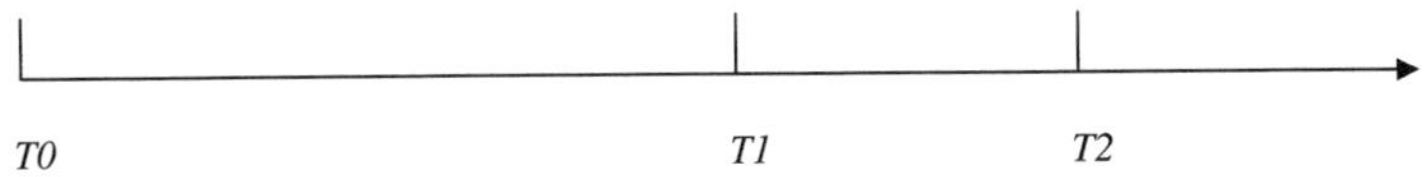

Fig. 9.1. Time dimension of International GHG emissions rights distribution.

Note: Suppose T0 stands for Industrial Revolution period; T1 stands for current year (e.g. 2008); T2 stands for future levels (e.g. 2050).

simultaneously. The third step is proposing each country's reduction road map under their respective quotas, determining the actual emission each year during T1–T2, and conducting international trading of greenhouse gas emission rights. However, each country will have to eliminate their emission rights deficit before the target year T2.

In the solution mentioned above, the core problem is how to define and distribute the national emission rights of each country during the period T0–T1 and T1–T2. We only define the limit of state emissions, as further definition of emission rights within each country is not in the scope of this article. As greenhouse gas is liquid, inert and non competitive in the harm it causes, we assume that all people suffer equal harm from greenhouse gases. Each country's greenhouse gas emissions bring harm not only to their own citizens, but also to all other nations. There will be a net external hazard to each individual country when the harm caused by all countries takes effect on a global scale. The definition of property rights is reciprocal. Coase was concerned about social costs, and did not care about the distribution of property rights, whether in regard to the emitting party or the injured party.

Because allocating property rights to the emitting country or the injured country is of great importance to the fairness of emission rights distribution, we set the following principle: "no country has the right to impose net external hazard on any other country without payment". If a country's net emissions generated net external harm to other countries, the victims should be compensated. In this case, a baseline for emissions should be established to prevent one country from causing harm to others. We define the baseline emissions as the initial emission rights which should be allocated to each country. From the principle and definition mentioned above, we can prove the following proposition:

Proposition: When the per capita emission of all countries are equal the respective populations suffer the same degree of harm caused by greenhouse gas, and the emissions of each country will not bring net external harm to others. Emissions under these conditions will be determined by the initial emission rights allocated to each country. If the actual emissions of a particular country are higher than the initial emission limitation, this country shall compensate the victim for the excess emissions.

Demonstration: Because we will not discuss the emission allocation within the individual country, we suppose the actual emission of each country reflects an even distribution of per capita emissions. Per capita actual emission of a certain country is e_{ij}, the population of that country is N_i, world population is $N = \sum N_i, i = 1, 2, \ldots, n$, i stands for country, $j = 1, 2$ represents the periods T0–T1 and T1–T2 respectively. The global per capita accumulated retained emission during period j is $\overline{e_j} = \frac{\sum e_{ij} \cdot N_i}{N}$. Due to the fact that the emission harm caused by each person is undertaken by all people on Earth, the external harm caused by each person to all the others is $e_{ij}(N-1)$. The degree of emission harm that each person suffers is $\sum_{i=1}^{n} e_{ij} \cdot N_i = \overline{e_j} \cdot N$. The degree of harm caused by themselves is the actual emission of each person e_{ij}. The degree of external harm caused by other people to each person is $\sum_{i=1}^{n} e_{ij} \cdot N_i - e_{ij}$. Due to the assumption that the emissions of each person are equal within one country, the harm caused by all people can balance out one another within one country. Thus, all the external harm caused by each person is the net harm to people in other countries. Providing no compensation mechanism, each person's emissions do not produce net external hazards to other countries. If and only if:

$$e_{ij}(N-1) - \left(\sum_{i=1}^{n} e_{ij} \cdot N_i - e_{ij}\right) = 0 \tag{9.1}$$

And, the emissions of each country do not bring external harm to other countries, If and only if:

$$[e_{ij}(N-1)] \cdot N_i - \left[\left(\sum_{i=1}^{n} e_{ij} \cdot N_i - e_{ij}\right)\right] \cdot N_i = 0 \tag{9.2}$$

Putting $\sum_{i=1}^{n} e_{ij} \cdot N_i = \overline{e_j} \cdot N$ into formula (2), then $e_{ij} = \overline{e_j}$, we can get the emission volume under the condition of each country does not bring external harm to other countries, namely e_{if}^{r}, the initial per capita emission rights that should be allocated to each country as we have already defined. In this case the initial per capita emission rights of each country will be,

$$e_{ij}^{r} = \overline{e_j} \tag{9.3}$$

Therefore, the initial per capita emission rights of each country should be equal, that is to say $e_{ij}^{r} = \overline{e_j}$. If the actual emissions of a certain country are more than the per capita initial emission, namely $e_{ij} > e_{ij}^{r}$, or more than the average world level, the emission of this country has already caused net external harm to other countries, and this country shall give compensation for the excess emissions. If we use e_{ij}^t to stand for the per capita excess emission, then, $e_{ij}^t = e_{ij} - e_{ij}^r$. Suppose p is the equilibrium price of per unit emission rights under IETS, or the shadow price without IETS, then, the compensation amount is $y = p \cdot e_{ij}^t$.

In this case, distribution of global per capita emissions of countries in different periods is the matrix $[e_i^r]$. According to the principle that "no country has the right to impose net external hazard on any other country without payment", we can solve the emission rights distribution problem, and then the allocation of emission rights of each country is based on an objective rather than subjective criterion. After the definition of property rights has been established, the optimal emission structure can be formed through IETS, namely the matrix $[(e_i^r + e_i^t)]$. The optimal emission structure is just the optimal property right structure of minimized social cost as described by Coase (1960). However, because the historical emissions of the period T0–T1 occurred under the circumstances in which greenhouse gas emission rights had not been defined, as long as we consider the cost of the emission, it does not matter how we define the historical emission rights of each country because the global actual emissions structure (namely, matrix $[e_{i1}]$) during T0–T1 is not optimal. This non-efficiency condition also has to do with $e_{i1}^r = \overline{e_1}$. The global actual emission structure of period T1–T2 is Pareto Optimal in that it can realize the target of performance and fairness on emission. In reality, as measuring the actual emissions of all countries completely and accurately involves large transaction costs, there will be a dilemma of conflict among improving measurement accuracy, the degree of implementation of property rights and reducing transaction costs. Thus, there will always be an error between the emissions measured and allotted to a given country and the true emissions of that country. In this case, $(e_{ij} - e_{ij}^t) - e_{ij}^r = \varepsilon_{ij} \neq 0$ may not always happen. The value ε_{ij} of the dilemma of conflict mentioned above represents the efficient externalities described by Wuchang Zhang (Cheung, 1970).

9.3.2. *Solution of global GHG emission reduction*

Based on the formula $e_{ij}^r = \overline{e_j}$, we can determine each country's emission rights, and calculate the emission account balance for each country according to the actual emissions and international emissions trading and establish accounts accordingly.

Total emission entitlements of each country:

$$E_{ij}^{r} = e_{ij}^{r} \cdot N_i = \overline{e_j} \cdot N_i = \frac{E_j}{N} N_i \tag{9.4}$$

E_j stands for the future accumulated total emission credits. If the actual emission E_{ij} of a certain country is higher than its total emission limitation E_{ij}^{r}, the country should buy the emission index E_{ij}^{t} from other countries. We have:

$$\overline{e_j} \cdot N_i + E_{ij}^{t} = E_{ij} \tag{9.5}$$

Account balance of national emissions rights of each country at a different time point: B_{ij}:

$$B_{ij} = \overline{e_j} \cdot N_i + E_{ij}^{t} - E_{ij} \tag{9.6}$$

We apply the consistent principle in (9.4) and (9.6) to period T0–T1 and T1–T2, and all countries.[1]

This means that the historical accumulated per capita initial emission rights of different countries should be the same, and the future initial emission rights of different countries should also be the same. Substituting $j = 1, 2$ into (9.4) and (9.6), we then get the emission rights of each country during period T0–T1:

$$E_{i1}^{r} = \overline{e_1} \cdot N_i = \frac{E_1}{N} N_i \tag{9.7}$$

E_1 represents the accumulated global emissions retained in the atmosphere, and $\overline{e_1}$ represents the accumulated per capita emissions retained. This formula means the historical emission rights of each country should be distributed based on the principle that the per capita historical accumulated emissions of countries should be the same.

Due to countries not setting emissions limits, and the international emissions trading system having not existed during the period T0–T1, actual

[1]The lack of internal consistency includes: some agreement on the equal per capita principle, but disagreement over the implementation of that principle in terms of a target year (e.g. 2050); some deal with the historical emission responsibility of period T0–T1 based on a different principle; some relate the emission growth rate of a given country to its GDP growth rate; some deal with the problem based on different principles according to different countries (see Garnaut, 2008).

trading E^t_{i1} is zero. $E_{i1} - E^r_{i1}$ is the excess emission. Since the emission cost is not included in the prices of goods and services trading, the net embedded emissions of international trade should be reckoned in the actual emissions right E_{i1}.

In this case, the national emissions account balance for each country on time point T1 calculated based on the historical emission of period T0–T1 is as follows:

$$B_{i1} = \overline{e_1} \cdot N_i - E_{i1}, \qquad \sum_{\text{且 } i=1}^{n} B_{i1} = 0 \tag{9.8}$$

The new increased emission rights for each country on period T1–T2 is as follows:

$$E^r_{i2} = \overline{e_2} \cdot N_i = \frac{E_2}{N} N_i, \qquad \sum_{\text{且 } 1}^{n} E_{i2} = E_2 \tag{9.9}$$

In the formula above, E_2 represents the increase in total global emissions of global greenhouse gas, and $\overline{e_2}$ represents the increase in per capita global emissions. This formula means that the future initial emission rights of each country should be calculated according to the equal per capita principle. Once E_2 is determined, the emission quota of global greenhouse gas can be determined at the same time. This emission quota is the global GHG emission quota without reduction E^b (namely, business as usual) subtracts E_2. $(E^b - E_2)/E^b$ is the reduction rate of global greenhouse gas emissions.

The new increased balance of each country on period T1–T2.

$$B_{i2} = \overline{e_2} \cdot N_i + E^t_{i2} - E_{i2} \tag{9.10}$$

Due to the fact that each country can increase or reduce its emission quota by emission index E^t_{i2} trading through IETS, the emission quota of each country is always equal to its actual emissions. Furthermore, because emission rights are allocated by person, and the emission cost has already been included in the price of international goods and services in the period T1–T2, there is no longer any need to consider embedded emissions when determining the actual emissions of the country.

In this case, the accumulated balance of each country considering the two periods, T0–T1 and T1–T2 is as follows:

$$B_i = B_{i1} + B_{i2} = (\overline{e_1} \cdot N_i - E_{i1}) + (\overline{e_2} \cdot N_i + E^t_{i2} - E_{i2}) \qquad (9.11)$$

The balance of each country is zero on the target year T2, namely $\sum B_i = 0$. This means that not only can the distortion between emission rights and actual emissions of each country on period T0–T1 be modified, but the increased global GHG emission quota, determined scientifically, can also be used up. Thus the global reduction target of greenhouse gas emission can be realized.

9.3.3. *An example of measurement*

We use a virtual example to illustrate how to set up a country's national emission account according to the above theory and program (the actual program will need to be estimated based on real data). Assuming that the world is composed of A-type developed countries and B-type developing countries. The data of population, current emissions, historical cumulative emissions of the countries and the future global emission quota are shown in Table 9.1.

In Fig. 9.1, T0–T1 represents the period of the Industrial Revolution to the present, and T1–T2 represents the present to a certain future point

Table 9.1. Virtual greenhouse gas emissions of countries. (Person, Unit)

Country	Country A		Country B		Global	
	Total	Per capita	Total	Per capita	Total	Per capita
Population	200	—	400	—	600	—
Net emission of current year (T1)	400	2	400	1	800	1.33
Historical cumulative emissions during period T0–T1	28,000	70	14,000	35	42,000	70
Future global emission quota during period T1–T2	—	—	—	—	30,000	50

Note: Historical cumulative emissions are the net retention of greenhouse gas after considering the attenuation of greenhouse gases over time.

(e.g. 2050). Assuming that we are at the point T1, currently emissions of greenhouse gases retained in the atmosphere is 42,000 units. According to the scientifically determined global emission reduction requirements from point T1 to T2, the total amount of GHG emissions allowed is 30,000 units.

To make our theory simple and clear, we assume that the population size of every country will not change in the future. We first need to define the national emissions at point T1 according to the historical emissions during the period T0–T1, then allocate the greenhouse gas emission rights of period T1–T2 among countries. The amount of total emissions at point T1 of a certain a country is the surplus rights at point T1 plus the new increased emission rights of period T1–T2.

First step: Definition of emission rights of countries at point T1 and the surplus rights (Note: Because the current welfare level of the population of a country is based on the historical emission, the current population is the beneficiary or the victim).

The world's historical per capita cumulative emission is 70 units on the period T0–T1. For simplicity, embedded emission hasn't be taken into consideration and is assumed to be zero when calculating the actual emission of countries. The result of Table 9.2 means that country A has emitted 14,000 units of greenhouse gas over the limit in the past, while country B has emitted 14,000 units less than the limit. In other words, country A has a 14,000 unit emissions rights deficit at point T1, while country B has a 14,000 unit emissions rights surplus.

Second step: Setting the total global emissions targets during the period T1–T2, and calculating the national emissions based on the principle of equal per capita.

Table 9.2. Emission rights account of countries (T0–T1).

	Country A	Country B	Global
Emission rights	70 × 200 = 14,000	70 × 400 = 28,000	42,000
Actual emissions	28,000	14,000	42,000
International trade credit	0	0	0
Emissions account balance	14,000 − 28,000 = −14,000	28,000 − 14,000 = 14,000	0

Suppose the total amount of global emissions is 30,000 units during the period T1 to T2, and per capita global emission is 50 units (the growth rate is less than the past). At point T2, the total accumulation target of greenhouse gas in the atmosphere is 72,000 units. Suppose the population from T1 to T2 will not change. According to the principal of equal per capita allocation, country A can get a 10,000 unit quota, and country B can get a 20,000 unit quota. Taking the emission rights consumption of the past period (T0 to T1) into account, country A still has a 4000 unit emission rights deficit, while country B has a 34,000 unit surplus which can be used.

Third step: The establishment of mechanisms to facilitate international cooperation for emission reduction.

After the definition of national emission rights, the international emissions trading system (IETS), Joint Implementation (JI), Clean Development Mechanism (CDM), and other flexible compliance mechanisms in the Kyoto Protocol can continue to be used. If there is no market-based mechanism for international cooperation, it would be difficult for countries with an emission rights deficit to eliminate the deficit, and reduction resources could not be optimally allocated worldwide. Each country could adopt a cap-and-trade model, a carbon tax, or a mixed mode (McKibbin and Wilcoxen, 2008). Furthermore, each state may decide the emission credits of period T1–T2 in their reduction road map, as long as it does not exceed its total emission credits (attenuation factors in the short term can be ignored).

The international emissions trading called TBD in Table 9.3 can change the emission account surplus of the country, while the emission trading within the country cannot change the emission account surplus. Domestic emission trading can only provide a more efficient market-based compliance

Table 9.3. Emission rights account of countries (T1–T2).

	Country A	Country B	Global
Emission rights	$50 \times 200 = 10{,}000$	$50 \times 400 = 20{,}000$	30,000
Actual emissions	TBD	TBD	TBD
International trading credit	TBD	TBD	TBD
Cumulative emissions account balance (T1 point)	$-14{,}000 + 10{,}000 = -4{,}000$	$14{,}000 + 20{,}000 = 34{,}000$	30,000

mechanism. If country A buys an emission rights credit from country B, then country A's emissions account balance increases while country B's emissions account balance is correspondingly reduced. As to international transactions completed through the CDM mechanism, it would be required to increase the balance of country A, and correspondingly reduce the balance of the country B. Thus, the additionality of CDM can be resolved (European Commission, 2008).

From the above virtual example we can see the following points:

First, a simple comparison of current total emissions between two countries is misleading. In the example, the current annual emissions of country A and country B are both 400 units, however, per capita emissions are significantly different. The per capita global emissions are 1.33 units. The per capita emission of country B is only 1 unit, which is much less than 2 units of country A. This means country A has violated 0.67 units of country B's per capita emission, while country B still has a 0.33 unit emission quota.

Secondly, only comparing the current per capita emissions of the two countries is also misleading. Suppose we adjust the current emission amount of country B in Table 9.1, changing 400 to 800, making the per capita emission of country A and country B equal. Even in this case, due to high historical emissions, country A has still violated the emission rights of country B. The emission deficit of country A would then be 14,000 units, while country B would still have a 14,000 unit surplus of emission rights.

Thirdly, once the emission rights of countries are defined and can be traded, the actual emissions of all countries will no longer be considered as a measure of their contribution to global emissions reduction. For example, all the manufacturing and other emission industry are concentrated in country B, and country A can sell all its emission index to country B, then country A can focus on their service industry without emissions and import manufactured goods from country B. Although all the emissions come from country B, it does not mean that country A contributes more to the global emission reduction. So, country B is not deserving of criticism. As long as the structure of emission rights distribution (pattern of global division of labor) is determined by the market, then the structure is the optimal allocation of global greenhouse gas emissions.

Fourthly, the reason why country A and country B have different emission reduction responsibilities is not because country A is rich and country B is poor, as people usually believe, or that country B has a right to develop, but because country B undertakes the harm of emission caused by country A (country A has 14,000 units emission deficit). This means that although the result of our research is also that the rich country should bear more emission responsibility, the theory that our solution is based on is totally different from others.[2]

9.4. Discussion

The global solution proposed above not only retains the advantages of the Kyoto Protocol, but also deals with its shortcomings, including small coverage, lack of incentive of developing countries and poor emission reduction, etc. First, developing countries have incentives to add new protocols since they have surplus emission reductions and the more emissions they reduce the more credits they can sell. Secondly, the distribution of the national emissions under total global emission allotment means that the global emission reduction targets are fully guaranteed. Thirdly, as the national emission limits are clearly defined and all countries are covered by the agreement, the classification of developing country and developed country, and the classification of mandatory and voluntary emission reduction, are no longer meaningful. At this point, voluntary emission reduction has become a self-serving behavior.

In the following section, the real meaning of our theoretical framework will be revealed more deeply, and the related major issues of principle will be clarified.

9.4.1. *How can "Contraction and Convergence" affect the interests of different countries?*

"Contraction and Convergence" to which some developed countries responded, was initially proposed by India. "Convergence" means each country's current emissions level is determined by the current actual emissions, but per capita emissions of high emitters will decrease, while the per capita emissions of low-emitters countries will increase, and the emission rights of countries will become equal in some particular year (e.g. 2050).

[2]For example, UNFCCC (2007) emphasizes emission capacity. Some suggest relating the emission target of each country to their income level (e.g. Pew Center, 2005; Frankle, 2007). This solution is unjust in theory.

"Contraction" means the current total global emission is gradually compressed in the convergence process (see Garnaut, 2008, p. 203). The distribution of each country's emission limit, based on the current actual emission level in the "Contraction and Convergence" plan, has no justification, and it makes allowance for the reality of high per capita emissions and vested interests of developed countries (namely country A). It also gives room for country B to increase its per capita emission levels, which gives the appearance of accommodating the interests of the developing B countries, but it actually impinges on the legitimate rights and interests of country B. Irrespective of whether or not it considers the convergence of 14,000 units of country A's historical responsibility, it postpones the per capita emission limit of T1 to T2, which means country A can continue to occupy a large part of the emission rights of country B during period T1 to T2 without payment. The emission rights of country A will continue to bring net external hazard to country B.

9.4.2. *How can "common but differentiated responsibilities" affect the interests of different countries?*

The United Nations Framework Convention on Climate stipulates that developed and developing countries should undertake "common but differentiated responsibilities" in emissions reduction. It establishes a principle for protecting the legitimate rights and interests of developing countries. According to our plan, after the emission rights of countries are clearly defined, the balance in the national emissions account are its legitimate interests. In the demonstration, country A's accumulated emissions is as high as −4,000 units, while country B's accumulated emissions is up to 34,000 units. These deficits and surpluses, clearly represent the rights and responsibilities of global greenhouse gas emissions reduction of different countries. However, the rights of developing counties may be obscured if the "common but differentiated responsibilities" of each country cannot be clearly defined. Developed countries will be required to gradually reduce emissions, based on the current emissions framework of the Kyoto Protocol, while developing countries are not required to reduce emissions compulsively. It seems that only developed countries would undertake the reduction responsibility; however, country A still occupies country B's emission reduction quota, since this free use is concealed.

9.4.3. *Is it effective to set up a "Global Emission Reduction Public Fund"?*

The transfer of technology and funds from developed countries to developing countries has been an important issue in international climate negotiations. The reason why the developed countries are required to do so, is simply because they are expected to deal with their responsibility of large historical emissions. At present, the proposal of the establishment of a "Global Emission Reduction Public Fund", is being widely discussed (Bhagwati, 2006; Garnaut, 2008, Ch. 10). If the fund is established, it may mean that the historical issues on emissions of developed countries will have been "satisfactorily resolved", and all countries will stand at the same starting point on emission reduction issues. Developed countries requiring developing countries to reduce emissions compulsively will be justifiable. However, according to the previous experiences, the "Global Emission Reduction Public Fund" cannot really be set up and operated effectively. It is difficult for developing countries to be granted rights and interests corresponding to their "emission surplus" even if the fund can be set up, because it is controlled by certain international organizations. Finally, developing countries may fall into such a situation: on the one hand they had to bear the responsibility for mandatory emissions reduction, on the other hand they will not be able to procure benefits from the fund. According to the present suggestion, the fund size is over 100 billion U.S. dollars annually. It is hard to imagine that such a bureaucratic management of public funds can really promote technical invention. New technology is able to appear because inventors and entrepreneurs can make a profit in the market through invention (Romer, 1990). The global solution we advocate is to use a more efficient decentralized market mechanism to replace the establishment of "Global Emissions Reduction Public Fund".

Due to an emission right deficit, lack of emissions quota, or high opportunity costs of emission reduction, developed countries have to purchase emission quota from developing countries, or to use new technologies to help developing countries to reduce emissions in order to exchange emission quota. Thus, capital and technology transfer has become a simple exchange relationship of interests.

9.4.4. *How can embedded emissions affect the welfare of different countries?*

Embedded emission is an important issue that has been widely discussed in international climate negotiations. In our study program, embedded emission is only considered in the period T0–T1 when the historical emission responsibility of states are recognized. This is because the cost of emission is not included in the prices of international goods and services. In period T1–T2, the cost of emission has already be included in the price of goods and services, and embedded emission cannot influence the welfare level of the exporters. However, if there are countries that are not included in the international agreements and not subject to emission limits, the original exporter of high-emission products is likely to be replaced by exports from countries outside the agreement, which will impact their welfare level, while the welfare of the importing country is not affected. Consequently, only when all the major countries sign the agreement of global greenhouse gas emissions reduction, the signed countries will not attempt to transfer their high emission industries to those countries without signing the agreement in order to avoid emission control (namely, carbon leakage). Furthermore, the emission reduction action will not bring great impact on the existing patterns of global industry division.

9.4.5. *Global warming: market failure or government absence*

Global GHG emission is considered an example of market failure by many people (e.g., Stern, 2007). This argument is misleading. Reduction of emission is essentially a problem of definition of rights. The property rights system is the basis that the market works on. Actually, defining, implementing limitations and monitoring greenhouse gas emissions are just the functions of government. Greenhouse gas emissions are a global cross-border issue, which needs a global "government" that can provide the public service about distribution, implementation and supervision of greenhouse gas emission rights. The market function has no way to play a role without the definition and protection of emission rights of countries caused by "global government" failure. In that case, classifying the global climate issue as a market failure has no basis, and it is an example of a government failure. It indicates that, in response to global climate change, accelerating to form an effective global governance structure is needed.

9.5. Conclusion

In this chapter, based on the principle that no country has the right to impose net external hazard on any other country without payment, a theoretical framework of historical and future GHG emission rights distribution has been established. The analysis in this chapter has shown that the net external hazard caused by GHG emission will not take place, only if historical and future GHG emission rights are allocated based on an equal per capita principle. The global solution can be found following the theoretical framework: "distribution of emission rights→international emission rights trading→optimal allocation of global emission reduction resources". Global emission reduction resources can be allocated optimally if the trading of emission rights is conducted based on the original emission rights of countries.

When verifying the actual emissions during T0 to T1, the transfer of national emissions should be considered, but the transfer of emissions no longer needs to be considered during T1 to T2. Our solution aimed at covering all countries, but the major countries can be covered in advance in order to improve operability. Other countries may join at any time as long as the conditions are allowed.

GHG emissions relate to multiple disciplines, especially those related to complex scientific problems. The program is still a matter of principle, in which many issues still need further study through interdisciplinary cooperation. For instance, how to measure various historical emissions reasonably; how to decide the new increased total emissions of certain future periods; how to promote the international and domestic institutional requirements by IETS; how soon the historical emission rights deficit of developed countries can be eliminated; how to design international transactions, monitoring and enforcement mechanisms, etc. The final determination of the international emission reduction program depends on the game of complex international political power; and it is not a mere academic question. In any event, in international climate negotiations, there should be a fair, reasonable and widely accepted benchmark for the evaluation of the interests of gains and losses. This is particularly important to developing countries, including China.

At present, it is generally assumed that there is a conflict dilemma between emissions reduction and economic development in the discussion of GHG emissions. With technological advances, especially the accelerating innovation of new energy, a low-carbon economic development mode may gradually replace the traditional development model of high-emissions, and currently shows great potential. The reason for the emergence of new technologies is that the market provides profit for this innovation mechanism. If the emissions rights of countries can be distributed clearly and protected strictly, moreover having a market trading mechanism established, emission reduction will be a profitable behavior and provide strong impetus to the low-carbon technologies and the development of a low-carbon economy. In the long run, greenhouse gas emissions reduction would make human society more prosperous and provide sustainable development.

References

Aldy, J. E. Barrett, S. and Stavins, R. N. (eds.). (2003). "13 + 1: A Comparision of Global Climate Change Policy Architectures." Discussion Paper, Resources for the Future.

Aldy, J. E. and Stavins, R. N. (eds.). (2007). *Architectures for Agreement: Addressing Global Climate Change in the Post-Kyoto World*. New York: Cambridge University Press.

Bhagwati, J. (2006). "A Global Warming Fund Could Succeed Where Kyoto Failed." *Financial Times*, 16 August.

Cheung, S. (1970). The Structure of a Contract and the Theory of a Non-exclusive Resource. *Journal of Law and Economics*, 13, 49–70.

Coase, R. (1960). The Problem of Social Cost. *Journal of Law and Economics*, 3, 1–44.

European Commission (2008). "Questions and Answers on the Commission's Proposal to Revise the EU Emissions Trading System." MEMO/08/35, Brussels, 23 January.

Frankle, J. (2007). "Formulas for Quantitative Emissions Targets." In Aldy, J. E. and Stavins, R. N. (eds.). (2007). *Architectures for Agreement: Addressing Global Climate Change in the Post-Kyoto World*. New York: Cambridge University Press.

Garnaut, R. (2008). *The Garnaut Climate Change Review*. New York: Cambridge University Press.

Grossman, S. and Hart, O. (1986). The Costs and Benefits of Ownership: A Theory of Vertical and Lateral Integration. *Journal of Political Economy*, 94, 691–719.

Hardin, G. (1968). The Tragedy of the Commons. *Science*, 162, 1243–1248.

Hart, O. and Moore, B. (1990). Property Rights and the Nature of the Firm. *Journal of Political Economy*, 98, 1119–1158.

IPCC (Intergovernmental Panel on Climate Change) (2007a). *Climate Change 2007: The Physical Science Basis. Contribution of Working Group I to the Fourth Assessment Report of the IPCC*. S. Solomon, D. Qin, M. Manning, Z. Chen, M. Marquis, K. B. Averyt, M. Tignor and H. L. Miller (eds.). Cambridge and New York: Cambridge University Press.

IPCC (2007b). *Climate Change 2007: Mitigation of Climate Change. Contribution of Working Group III to the Fourth Assessment Report of the IPCC*. B. Metz, O. R. Davidson, P. R. Bosch, R. Dave and L. A. Meyer (eds.). Cambridge: Cambridge University Press.

McKibbin, W. J. and Wilcoxen, P. J. (2008). "Building on Kyoto: Towards a Realistic Global Climate Agreement". Working Paper for Brookings High Level Workshop on Climate Change, Tokyo, May 30.

Nordhaus, W. (2006). After Kyoto: Alternative Mechanism to Control Global Warming. *American Economic Review*, May, 2006.

Nordhaus, W. (2008). *The Challenge of Global Warming: Economic Models and Environmental Policy*. New Haven, Connecticut: Yale University.

Pew Center on Climate Change (2005). *International Climate Efforts Beyond 2012: Report of Climate Dialogue at Pocantico*. Arlington, Virginia.

Romer, P. (1990). Endogenous Technological Change. *Journal of Political Economy*, 98, 71–102.

Stern, N. (2007). *The Economics of Climate Change: The Stern Review*. Cambridge: Cambridge University Press.

Stiglitz, J. (2006). "A New Agenda for Global Warming." *Economist's Voice*, www.bepress.com/cv.

UNFCCC (2007). "Report of the Conference of the Parties on its Thirteenth Session, Held in Bali from 3 to 15 December 2007, Addendum, Part Two: Action Taken by the Conference of the Parties at its Thirteenth Session." FCCC/CP/2007/6/Add.1.

Weitzman, M. L. (1974). Prices Vs. Quantities. *Review of Economic Studies*, 41(4), 477–491.

Chapter 10

LOW-CARBON ENVIRONMENTAL ECONOMICS

*Hidefumi Imura**

10.1. Common but Differentiated Responsibilities

The protection of global assets such as the climatic system, the oceans, the ozone layer, and biodiversities is the collective responsibility of all human beings. However, there are many views on the responsibilities of individual nations, communities, and economic sectors. Conflicts exist between industrialized and developing countries. The former enjoy material affluence and cause larger environmental impact, while the latter have a large population but each of their people uses a smaller amount of environmental resources. The Rio Declaration on Environment and Development therefore proclaims that states have "common but differentiated responsibilities" (United Nations, 1992).

10.2. Environment and Climate Change

Human life and economic activity are based on the use of natural resources and the resulting environmental impacts — the decrease or depletion of limited resource stocks on Earth, the production of polluting substances, and impacts on natural ecosystems. For sustainable development, the use of exhaustible resources must be reduced by expanding the use of renewable resources, and the emissions of polluting substances such as greenhouse gases that may cause irreversible change to the global environment must be limited. The future of humanity depends on two major challenges. One is the improvement of "environmental efficiency" (or "eco-efficiency") in order

*Hidefumi Imura, Professor at Yokohama City University, Japan.

to achieve the highest possible levels of wealth and welfare with the least amount of resource consumption and environmental impact. The other is the solution of "environmental equity", which embraces two questions: how equally we can distribute among people the wealth and welfare produced, and how we can share the responsibility to reduce environmental impacts among various actors in society.

Environmental efficiency is realized by technological advancement and appropriate incentive systems such as setting environmental standards and regulations and using market forces. In general, environmental and economic efficiency show a positive correlation if proper incentive systems are designed and put into force. Conversely, because there can be such a positive correlation between the two, a big challenge in environmental policy is to determine what incentive systems should be introduced to balance the economy and the environment. On the other hand, environmental equity is a matter of income allocation, which involves two perspectives. The first relates to how we can equally enjoy the wealth and welfare realized at the cost of environmental impacts, and the second relates to how we can equally share the burden, required to reduce environmental impacts, among various actors including nations, regions, corporations, and citizens.

What is difficult in the discussion of environmental equity is how to define "equity". One of the strong norms regarding environmental equity is the thought that all humans are equal, and therefore the amount of natural resources one person can use (for instance, daily per capita food intake) and the amount of permissible environmental impacts (such as per capita carbon dioxide emissions) should be equal. In reality, however, significant disparities, or "inequities," exist among countries and regions in the distribution of various natural resources — for example, mineral resources like oil and ores, fertile farmland, water, forests, and so forth. Meanwhile, large differences exist in the possession of the technologies that produce high added value utilizing these natural resources. Considering the current world situation, while some countries like Japan are producing advanced industrial products with highly educated human resources and technologies, despite a lack of natural resources, there are many developing countries that still cannot fully utilize their own rich natural resources because of various social and political reasons.

According to the Hecksher-Ohlin model of international trade, a country's economic and trade structure is determined by given conditions on factors such as natural resources, labor and capital. By utilizing comparative advantages regarding the given factors, each country can produce particular goods and services at advantageous conditions compared to others, and this therefore creates an international division of labor. One of the important factors in comparative advantage is differences in the level of environmental regulation and standards. Goods and services can be produced cheaply if there are loose regulations on environmental pollution and if people have low concern about environmental issues. As a result, it is an issue for discussion on international trade: how to limit exports of products produced cheaply by exploiting loose environmental regulations. Under economic systems that exist today, regulations and price incentives to internalize the costs of environmental impacts are utterly inadequate, making it difficult for one to conclude that the environment is being properly considered in international trade rules.

From these perspectives, Imura, Nakamura and Morisugi (2005) presented a study on the environmental equity between ten Asia-Pacific countries including Japan, the United States and eight Asian nations from the actual status of resource utilization and environmental impacts in each country, by examining the international interdependency on natural resources and environmental impacts arising through international trade. For this discussion, they employed national per capita carbon dioxide emissions, energy for agricultural use, the area of farmland, and the area of forest cover as indicators to evaluate environmental impacts. The "direct impacts" and "indirect impacts" are quantified using a standard methodology of input-output analyses (Miller and Blair, 1985), using the Asia International Input-Output Tables for 1985, 1990 and 1995. First, the environmental impact that one unit of final demand in an industry in a given country induces directly and indirectly, or the embodied environmental impact, is estimated. Second, the export-import structure of the embodied environmental impact accompanying trade between countries is analyzed. Thus we analyze environmental impacts induced by a country's industrial sector when it conducts productive activity, whether occurring directly within the country or indirectly in other countries through trade. In particular, we consider environmental equity among the countries within the Asia-Pacific region by comparing per capita direct and indirect environmental impacts in each country, with carbon dioxide emissions as the indicator. We also consider the environmental efficiency and equity concerning the production and trade of agro-products, with a focus on the agricultural energy and farmland area necessary for agricultural production, particularly in China, Japan, and the United States.

10.3. Resources, Trade and Environmental Impacts

The world population was about 2.5 billion in 1950 and reached 6 billion in the year 2000. Meanwhile, the world's total GDP expanded by about 5.5 times during the same period. On the other hand, the Earth's resources, such as farmland, energy, and forests, are limited, and there is only a limited possibility to increase these stocks in a short period of time. As a result, the amount of resources that each person can consume is declining. As with carbon dioxide emissions, in some cases emissions are increasing beyond the level of emissions per capita which is allowable to sustain the earth's climatic system. A number of indicators have been proposed to discuss natural resource consumption and environmental impacts on the global scale. The simplest indicator is a country's resource consumption per capita. For mineral resources such as iron, steel, and aluminum, industrial countries' resource consumption reaches a hundred times the level of developing countries'. Figure 10.1 shows the comparison of pollutant emissions per person and per unit of GDP in the U.S., Japan and China, demonstrating significant differences among countries.

1999	US	Japan	China
GDP (1995 Billion U.S. dollars)	8,582	5,356	964
Population (Million)	278	127	1,254
GDP/Capita	30,845	42,318	769
CO2 (Mt-C)	1,542	320	834
SO2 (Mt)	(16)	(0.7)	18.6
CO2/GDP (ton-C/M U.S. dollars)	180	59.7	865
SO2/GDP	1.86	0.13	19.3
CO2/Capita (ton-C)	5.54	2.53	0.665
SO2/Capita (kg)	57.6	5.51	14.8

Fig. 10.1. Comparison of resource consumption and pollutant emissions in U.S., Japan and China.

In line with the Framework Convention, many industrialized nations presented national action plans for stabilizing the emissions of CO_2 and other greenhouse gases in order to protect the climatic system of the earth (IEA/OECD, 1992) . Questions, however, may be raised as to the definition of national stabilization targets. First, most of the targets are determined with reference to the baseline CO_2 emissions in the year 1990. These baseline CO_2 emission per capita exhibit remarkable disparity between North and South regions as well as among countries: for example, 6.15 tC in the United States, 2.69 tC in Japan, and 0.60 tC in China (Energy Data and Modeling Center, 1994). From the viewpoint of "equity", the reduction of such disparities should be a matter for international discussion. The second question concerns the definition of CO_2 emissions: if a country moves its production facilities to other countries and imports finished products as much as possible, it can reduce its apparent CO_2 emissions. What should be considered here is the "embodied energy" or the total energy input for the production of goods and services traded between countries.

The amounts of direct natural resource consumption in each country alone are not sufficient to discuss environmental equity. National economies are mutually and intimately linked through international trade, and the trans-boundary movements of goods and services are ever-increasing. Technological innovation in communications, transportation and information since the 1980s has expanded the technological gaps between countries. At the same time, a shift is evident in which industries with intensive consumption of natural resources and heavy environment impacts are moving to less developed countries. That is to say, the goods and services that cause heavy environmental impact during production are not being produced domestically, but rather imported after being produced in other countries. This transfer matches the Hecksher-Ohlin model of international trade, but importing goods and services from other countries indirectly increases resource consumption and causes a heavier environmental burden in that country. Our lives depend on international trade, meaning that environmental burdens must inevitably occur somewhere on Earth. Therefore, when we consider ways to reduce the impacts on and damage to the global environment caused by human activities, it is necessary to consider indirect environmental impacts and resource consumption arising through international trade.

The concept of the "ecological footprint" (EF) is a typical example of an indicator that can be used to evaluate the impacts by which, through the economic system, the activities of one country affect either the environment in another country, or the global environment. According to one estimate, the land area that one Japanese person needs to live is 5.94 ha, far larger than the 2.18 ha per capita that actually can be provided globally. Since Japan's food self-sufficiency has dropped to about 40 percent on a caloric basis, a large area of arable land overseas is being used for food production for the Japanese. Besides arable land, large amounts of water, fertilizer, pesticides and fuel are required to produce this food.

The situation is similar with energy consumption, which is inseparably connected with every human activity. As many of the goods and services that are consumed in a given country are imported from other countries, much energy is being used as an input to produce those goods and services, leading to global-warming-causing carbon dioxide emissions in those countries. Considered this way, it is not enough to simply compare the amounts of carbon dioxide directly emitted within a country. It is also important to look at carbon dioxide emissions induced indirectly through trade. In an extreme case, in a country that specializes in service industries, it would be possible to rely completely upon imports and have no domestic industrial production with higher environmental impacts at all. As one can see, the total sum of a country's carbon dioxide emissions, as well as the direct and indirect components, significantly reflect its economic wealth, industrial structure and lifestyle. A larger indirect component means a country is more dependent on other countries through trade, or transferring into another country the emissions that would otherwise have been emitted domestically. On the other hand, part of the direct portion of one country's emissions to produce goods and services is incurred to export them to other countries.

10.4. International Interdependency of Resources and Environmental Impacts

10.4.1. *Embodied environmental impact*

Various natural resources are used as inputs for the production of the goods and services we consume, and these result in various environmental impacts.

Said another way, we can consider the goods and services we consume to embody various environmental impacts that occurred during their production processes. This concept we call the "embodied environmental impact," which includes environmental impacts that occur in every phase until final provision of a good or service to the user, including resource extraction, processing, transportation, distribution and so forth.

There are two kinds of resources necessary to produce final products in a given industry sector. One is the resources directly put into the sector itself, and the other is the resources that are put in during the production phases for intermediate products that are provided as inputs into the sector from other sectors. Thus, there are two types of environmental impact that occur to produce a finish product: the direct environmental impact that occurs in the sector that produces the product, and the indirect environmental impact that occurs during the production phases of the intermediate products. The Input-Output Model is suitable to systematically analyze such cross-sector resource flows and the structure of the environmental impacts. Under the concept of input-output analysis, when final demand arises for the goods and services provided by an industry in a given country, the industry's production is induced to fulfill it, and at the same time demand propagates to other industries that produce inputs for that industry's production. As a result, demand arises for industries in that and other countries, requiring various resource inputs and leading to environmental impacts.

10.4.2. *Embodied environmental impacts of products in the Asia-Pacific countries*

The United States and China show high levels of embodied carbon dioxide, reflecting the fact that these two countries are among the top carbon dioxide emitters in the world and also in the Asia-Pacific region. The emission intensity of a country depends upon the country's industrial structure (e.g., the relative importance of the unprocessed, semi-processed or finished goods sectors in the total national economy), and the efficiency of production technology, especially energy-related technology. In addition, however, it depends upon the industrial structure and production technology of other economies from which each sector receives goods and services for its own output. Therefore, there is an interdependence between sectors within an economy and between different sectors of other economies with which trade in goods and services takes place.

China has the highest level for embodied energy in agricultural products, followed by the United States. Japan's trend is increasing, approaching U.S. levels. The United States and China, with rich land resources, also have high levels for embodied farmland, but the levels in the other countries are not so high. These patterns appear because agriculture is characteristically highly dependent on domestic land resources, and these two countries have established their status as agricultural producers based on their rich land resources. On the other hand, Japan, with fewer land resources, has a low level of embodied farmland, but a high level of embodied energy in agricultural products.

For embodied forest land, the United Sates has a high level, followed by China, Indonesia and Japan. Japan's high level of embodied forest land, despite its limited domestic forest area per capita, reveals a glimpse of Japan's high timber consumption based on imports.

10.4.3. *Trade of embodied environmental impacts*

As mentioned earlier, if we are to examine environmental equity while taking into account the globalization of economic activities, it is necessary to quantify the status of international interdependency in environmental resources and environmental burdens arising through trade. For this purpose, we calculate the amount of carbon dioxide emissions induced indirectly in other countries by industry in a given country, and then evaluate the balance of embodied environmental impacts among the countries, i.e., the inflows and outflows of carbon dioxide and environmental resources that accompany international trade. The results for 1995 are shown in Fig. 10.2.

First, on the carbon dioxide balance, one notes that the carbon dioxide inflows and outflows of the United States are significant. This outcome arises because the monetary trade between each country and the United States is large, and indicates the strong influence of the United States on the burden of carbon dioxide emissions arising through international trade. The carbon dioxide outflows from China are most significant, and inflows to Japan and the United States are particularly conspicuous. This is because, although the amounts of trade are small in monetary terms, in their trade with China these two countries are mostly importing goods and services from China,

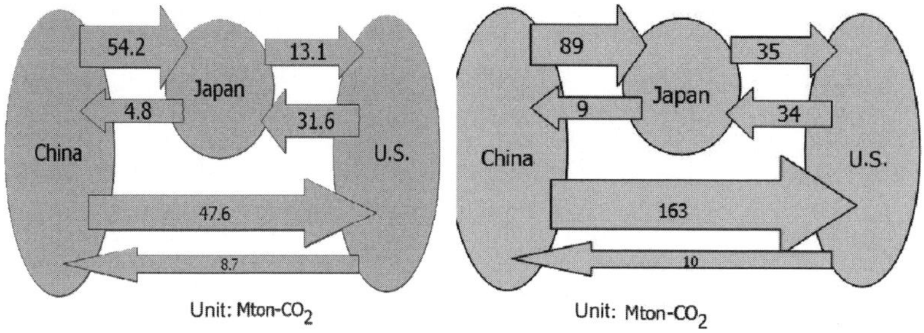

(a) Year 1995 (b) Year 2000

Fig. 10.2. Flow of embodied CO_2 associated with the trade in Asia and the Pacific region.

which has low production efficiencies. Regarding total outflows, the carbon dioxide outflow from China is the highest in our study, and it can be said that, in a way, China is taking over responsibility for the carbon dioxide emissions of other countries.

On the balances of energy of agricultural products and farmland, the inflows into Japan are significant. This indicates that because Japan's food self-sufficiency is low, with its low amount of land per capita Japan is indirectly consuming large amounts of energy in agricultural products, as well as farmland, from outside the country. One should note from the inflows from the United States and China that Japan depends heavily on those two countries for energy consumption and the use of farmland for agricultural production.

Figure 10.3 shows the relationship between per capita GDP and embodied carbon dioxide per capita in each country. As a general trend, total (direct and indirect) CO_2 per capita increases with increasing income per capita and a rising standard of living. In terms of CO_2 emissions, countries are categorized into four groups, i.e., the United States, Japan, Singapore and other nations, revealing disparities among the countries of the Asia-Pacific region. Embodied carbon dioxide per capita can be considered to be the carbon dioxide emissions produced by people who live at different economic levels and lifestyles. The outcome shows that the carbon dioxide emission burdens of Singaporeans are increasing with economic growth to U.S. levels. Meanwhile, the levels of Japan's embodied carbon dioxide per capita are almost unchanged at the three points of time shown (1985, 1990

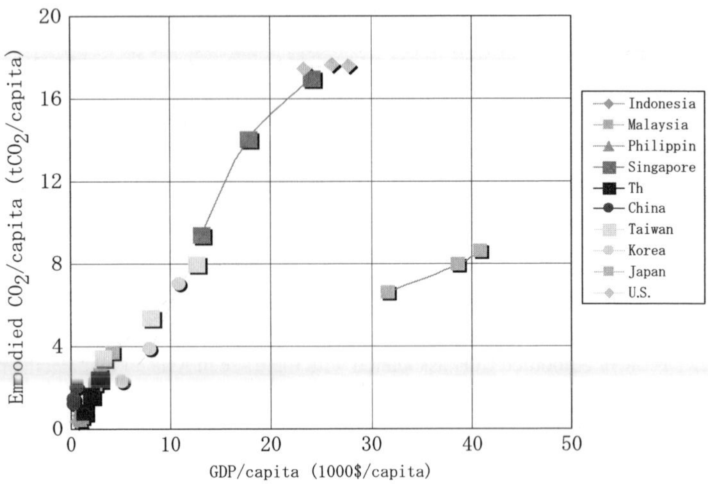

Fig. 10.3. Economic growth and embodied CO_2 per capita.

and 1995), and other countries (excluding the United States and Singapore) are catching up. Amid concern about increasing carbon dioxide emissions associated with the rapid economic growth in East Asian nations expected in the future, if we are to realize environmental equity, there is an urgent need to ensure that embodied carbon dioxide per capita in those countries levels-off at Japan's levels.

Based on these findings, environmental equity is examined based on carbon dioxide emissions. Figure 10.1 already suggested that large disparities in environmental burden exist among countries in the Asia-Pacific region. The United States, for instance, shows high carbon dioxide emissions for domestic production, and conversely, Singapore shows large carbon dioxide inflows from outside the country. Meanwhile, Japan's domestic carbon dioxide emissions per capita are about half those of the United States, but the situation differs when considering inflows and outflows. This is because, although the environmental efficiency of production in Japan has attained the highest levels, the importing of goods and services from overseas increases the country's total environmental burden. Likewise, Taiwan and South Korea are also receiving net inflows, and the net inflows of these three countries are due to their high proportion of trade with China, with its low environmental efficiency.

Increased environmental efficiency and the realization of environmental equity are both necessary conditions for sustainable development. One effective approach is, therefore, to improve environmental efficiency on the global scale by transferring technology, for instance, from a country like Japan, with its environmentally-efficient technologies, to China, which is producing with less-efficient technologies in some areas. At the same time, Japan is a country with net inflows of carbon dioxide emissions, and given that the primary source of those inflows is from China, such technology transfers effectively help satisfy conditions of environmental equity. Thus, it can be said that transferring highly environmentally-efficient technologies from industrial countries to their trading partners that produce using less-advanced technologies could be an effective method to induce the improvement of environmental efficiency, and at the same time, to realize environmental equity.

The embodied energy in agricultural products per capita and embodied farmland per capita are another matter of concern. According to our estimates, in terms of embodied energy in agricultural products, countries can be categorized into two groups: the United States, Japan and South Korea, all with high levels of embodied energy in agricultural products per capita, and other countries, with low levels. In terms of embodied farmland, countries can be also categorized into two groups: the United States, with a large area of embodied farmland per capita, and other countries, with small values.

10.5. Environmental Implications of Foreign Direct Investment

The world economy is steadily moving toward globalization based on free trade. In East Asia, it is also clear that rapid economic development, called the "East Asian Miracle (World Bank, 1993)"[1] was largely due to the benefits of active international flows of capital, goods and services. These flows made it possible for China and the Asian NIES, including the ASEAN countries, to succeed in export-driven industrialization and achieve dramatic economic growth. Moreover, these countries share a common feature in their high dependence on foreign capital. Therefore, this approach is termed

[1] World Bank (1993).

"foreign-capital dependent export-driven industrialization" (Wu, 2005). In China, the positive introduction of foreign direct investment is a major force in economic growth.

Driven by elevated domestic labor costs and foreign exchange fluctuations, Japanese industry has been moving overseas, with the result that considerable industrial capacity is being lost in Japan herself. In the 1990s, more and more Japanese affiliated firms began switching to local production to enter the Asian market. This dynamic change in industrial location in recent years bears great environment importance.

The transfer of industry from Japan to other countries is nothing but the transfer of production equipment, and its accompanying environmental load, to other Asian countries. For the industrializing countries which receive this transferred technology it means that they take over the production of goods with large environmental loading from the industrialized countries, and achieve economic growth by exporting the products to those industrialized countries. The industrializing countries receive production facilities and equipment from the industrialized countries, acquiring advanced technology and the opportunity to shift to cleaner production technology with lower environmental loading. This change in the international relocation of industry is directly related to industrial restructuring in both Japan and recipient countries, and also to the structural change of trade. Moreover, countries with extensive land area, such as China, may end up with disparate levels of development between different internal regions due to differences in technology transfer from other countries.

International division of labor and industrial location in the region has deep implications for East Asia in terms of achieving regionally balanced and environmentally sound economic development. In order to have clearer views about this, the study focuses on China, which has achieved rapid economic growth in recent years.

10.6. Discussion and Conclusion

In today's globalized world economy, countries closely depend upon each other for transactions of goods and services. Through trade, economic activities and their private lives, even citizens of a small local community are linked to the rest of the world. Responsibility for causing environmental impacts should be discussed taking into account such economic interdependence between nations, regions, and economic sectors.

A simple example of the problems of trade and the environment is where hazardous materials are exported to other countries and cause damage to the environment of the importing countries. In general, however, the relationship between trade and the environment is more indistinct and complicated. Traded goods and services give rise to environmental impact throughout the process of their production, transportation, consumption and disposal. The imported goods may not even be consumed in the country, but be changed into other forms and exported to other countries. Hidden within the trade there may exist disparities between countries in the cost and the required level of environmental countermeasures as in the price of products; these may reflect differing policy priorities, regulatory systems and standards.

This study quantified the international interdependency on environmental resources and environmental burdens through trade, using industrial input-output analysis with a focus on the Asia-Pacific region. Embodied environmental impacts per capita were calculated, and these were employed and assessed as indicators to estimate environmental equity, environmental resource consumption and environmental burdens in the region. The following findings were obtained as a result.

(i) In terms of both carbon dioxide emissions and environmental resource consumption, the United States and China have high levels of embodied environmental burdens, and net outflows from the United States and China are consistently large. These findings indicate that both countries account for a large portion of the carbon dioxide emissions and environmental resource consumption in the Asia Pacific region, and are significant in the context of resources and environmental issues in the region. In particular, the influence of China, which has been establishing its role as a factory for the world in recent years, can be considered as an important factor in environmental issues and food-related matters in the Asia-Pacific region.

(ii) The imbalances in embodied environmental burdens are a result of international trade based on comparative advantage, and they cannot be instantly judged as being either good or bad. Trade liberalization, however, imposes the same level playing field on all countries, without adequate consideration of each country's industrial production efficiency and international interdependency on environmental resources. Questions of the appropriateness of approaches to deal with these issues from the perspectives of resources, the environment, and environmental equity remain unresolved.

(iii) In terms of embodied environmental impacts per capita in each country, the United States has high levels in terms of both carbon dioxide emissions and environmental resource consumption. There are disparities between the United States and the other countries in the study. Regarding carbon dioxide emissions, the amount of embodied carbon dioxide per capita is rapidly increasing along with economic growth in East Asian nations. It will be important to limit increases in the burden of carbon dioxide emissions through appropriate environmental policies and technology transfers in these countries.

(iv) If the domestic portion of embodied carbon dioxide per capita alone is considered, Japan is only at about half the level of the United Sates, but conversely, in terms of net flows Japan has the largest net carbon dioxide inflows. This is because the environmental efficiency of industry in Japan has reached the highest levels in the Asia-Pacific region, and at the same time Japan is importing goods and services from other countries that have relatively lower levels of environmental technology. Thus, technology transfers to China, which apparently has lower levels of environmental technology and is a primary trading partner, would contribute to both the realization of environmental equity and the improvement of environmental efficiency in the region.

Although this chapter focused on the embodied environmental burdens of energy, there are many other important categories of environmental resources, such as water, for example, that form the basis for human survival. In addition, when discussing agriculture, the international balances of environmental burdens and environmental resources are likely to differ depending on the agricultural product in question. Furthermore, when considering national differences in industrial structure, it will be necessary to discuss the environmental efficiency in each industrial sector. In particular, regarding carbon dioxide emissions, it will be important to consider disparities in technological levels of the same industrial sector among countries, with greater attention to industrial structures. These issues remain as topics for future discussion.

References

Akira, H. M. (2009). *Economic Development and Environmental Policy in East Asia*. (in Japanese) Minerva.

Global Footprint Network (2009). *Ecological Footprint Atlas 2009*. Global Footprint Network.

IEA and OECD (1992). *Trade Issues in the Transfer of Clean Technologies*, OECD, Paris, www.oecd.org.

Imura, H. (ed.) (1998). "Environmental Issues in East Asia: Regional Structure and Its Future Generation of Environmental Impact when Viewed From Industrial Distribution and Trade." (in Japanese). Viewpoint to East Asia, International Centre for the Study of East Asian Development.

Imura, H. *et al.* (2000). "Urbanization and Environmental Issues in East Asia: The Applicability and Effectiveness of the Kitakyushu Model (Part 2)." (in Japanese). Viewpoint to East Asia, International Centre for the Study of East Asian Development.

Imura, H. (2002). Environment and Development, in Miyamoto, K. and Yoshida, F. (eds.), *Environmental Issues in Asia* (in Japanese). Tokyo: Iwanami Shoten.

Imura, H. (2004). Developing Countries and Global Environmental Issues. In Imura, H. and Matsuoka, S. (eds.), *Environment and Development* (in Japanese). Tokyo: Iwanami Shoten, pp. 120–127.

Imura, H., Nakamura, C. and Morisugi, A. (2005). Trade Structure and Environment Load of Japan, USA and Asia, Japan Society of Civil Engineers Proceedings, No. 790, pp. 11–23.

Imura, H. (2007). *China's Environmental Issue: What's Happening Now?* (in Japanese). Tokyo: Kagaku Dojin.

Katsuhara, K. (2001). *Development and Environmental Issues in East Asia* (in Japanese). Tokyo: Keiso Shobo.

Miller, R. and Blaire, P. (1985). *Input-output Analysis: Foundations and extension*, NJ USA: Prentice-Hall.

The Energy Data and Modeling Center (1994). Energy and Environment Statistics 94, Energy Conservaton Center Tokyo.

The United Nations Conference on Environment and Development (UNCED) (1992). Rio Declaration on Environment and Development, 1992. U.N. Doc. AICONF, 1515/Rev. 1.

World Bank (1993). *East Asian Miracle*. Oxford: Oxford University Press.

Wu, Jinglian (2005). *Understanding and Interpreting Chinese Economic Reform*, Oakland: Texere.

Chapter 11

LOW-CARBON DEVELOPMENT STRATEGY*

Mukul Sanwal†

11.1. Introduction

India combines features of high vulnerability to climate change due to persistent poverty and a high share of agriculture in GDP, with large and increasing emissions, ranking fourth after China, the US and Japan. Thus India sees climate change as a co-benefit of sustainable development, in the context of three interacting concerns. First, it considers itself as a victim of historical (and current) emissions by industrialized countries. Second, it is acutely aware that social and economic development both depends on access to modern forms of energy. Third, as India is also coming up as an economic power it seeks to contribute substantially to a containment of future global emissions. Therefore, in the context of an addition of around 6%–8% to India's average annual GDP over the next 40 years, there is a debate within the country on the pathway to a sustainable development, or low carbon strategy, which puts emphasis on poverty first or on growth itself. This has been expressed as an energy policy that is DEFENDUS (development focused, end-use oriented, service directed) or GROSSCON (growth-oriented, supply-sided, consumption directed). The emerging approach

*Based on documents of and the author's discussions with officials of the Ministry of External Affairs, the Ministry of Environment and Forests, and the Ministry of Non-Conventional Energy Sources, as well as the Planning Commission, Government of India, and statements made at the meetings of the Climate Convention.

†Mukul Sanwal, Visiting Professor at the University of International Business and Economics, Beijing. He has held senior policy positions in the Government of India and the United Nations He was closely involved with the Rio Conference, 1992, and the World Summit on Sustainable Development in 2002, and is associated with the South Centre, Geneva, and with think tanks in India. These are his personal views.

stresses the social rather than the environmental dimension, with the shift in focus changing the nature of the solutions from those adopted in industrialized countries.

11.2. National Context

The first in-depth national (and international) discussion of the Indian position on global climate policy was triggered by the publication of country-specific greenhouse gas emission data by the Washington-based World Resources Institute in 1990, which focused on the current rather than the accumulated level of emissions, and pointed at India as a major culprit due to the country's high methane emissions from rice cultivation. While the estimate of U.S. emissions reached 3.67 billion tCO_2 equivalent, India's value was 0.84 billion, of which 43% were attributed to methane. Overall, the report held developing countries responsible for 45% of the global problem at hand. The WRI numbers led to an angry rebuttal by the "Centre for Science and Environment" The rebuttal not only pointed at the questionable assumptions underlying the calculations, but also argued for a differentiation between "luxury emissions" in the North and "survival emissions" in the South. Moreover, it underscored that emissions should be accounted for on a per capita basis and considers historical responsibility. Consequently, the Indian position hardened and called even more strongly for differentiated responsibilities and far-reaching technology transfer. For India, the issue addressed fundamental questions of equity, and from this perspective, its positions reflected a moral imperative rather than a simple policy stance. By hosting COP 8 in Delhi in 2002, India showed its full commitment to the international process, but remained firm in its rejection of emissions commitments for developing countries. The National Action Plan on Climate Change (NAPCC) released in June 2008 shortly before the G8 summit further demonstrated India's willingness to participate in the global mitigation effort as long as this effort was self-determined and did not hamper economic growth.

However, at the same time, at Copenhagen in December 2009, India announced a self-imposed intensity goal of reducing carbon emissions per unit of GDP by 20–25% from the 2005 levels until 2020. While this may not sound very ambitious, notably when compared to the Chinese objective of reducing emission intensity by 40–45% (equally announced before COP15 in Copenhagen), it should be taken into account that Indian emission intensity is already much lower, even lower than in the typical OECD country. Therefore Indian energy intensity is more difficult to reduce further as compared to China where the potential is immense across all types of industry.

Energy efficiency has also become an economic imperative, independently of environmental or climate policy concerns. In fact, even today, energy intensity (measured as kg of oil equivalent, kgoe, per USD of GDP) is remarkably low for an emerging economy at the current stage of development. It reached its peak as early as in 1985 and is declining since. With 0.16 kgoe per USD (PPP) it was below the OECD mean or countries like Germany in the mid 2000s. Yet, as compared to OECD countries, there is substantial scope for a further increase in energy efficiency since the range of technologies used in Indian industry is extremely large. Along with a modern production sector, many small industries with more or less medieval technologies still continue to exist in sectors like pulp and paper, iron and steel, and engineering. Overall, energy security concerns implicitly strengthen the case for national climate policy in India. They generate economically motivated incentives not only for energy efficient production, but, also, at the policy level, for the search for and promotion of alternative renewable energy resources (solar, water and wind).

11.3. International Context

Climate policy has now been elevated to a policy field of strategic importance. Linking up with China, Brazil and South Africa in the BASIC group, India now tends to be put in the same category as China, an economically much more powerful country. As compared to all other BASIC countries, India is still a relatively poor country. At 3,296 U.S. dollars, it's GDP per capita (PPP) is less than half of the corresponding figure for China, and not even one third of the corresponding figures for Brazil and South Africa.

While India forms part of the BASIC group (i.e. Brazil, South Africa, India and China), its socio-economic parameters are much weaker than those of the three other countries in the group.

- India's total GDP in 2008 was about 1.2 trillion U.S. dollars, which ranks it in twelfth place in the World compared with China's third place with 4.3 trillion U.S. dollars.
- India's per capita GDP (in terms of purchasing power parity or PPP) in 2007 was 2,800 U.S. dollars compared with 5,400 U.S. dollars for China, 9,800 U.S. dollars for South Africa and 9,600 U.S. dollars for While India forms part of the BASIC group (i.e. Brazil, South Africa, India and China), its socio-economic parameters are much weaker than those of the three other countries in the group.
- India's total GDP in 2008 was about 1.2 trillion U.S. dollars, which ranks it in twelfth place in the World compared with China's third place with 4.3 trillion U.S. dollars.
- India's per capita GDP (in terms of purchasing power parity or PPP) in 2007 was 2,800 U.S. dollars compared with 5,400 U.S. dollars for China and around 9,800 U.S. dollars for South Africa and around 9,600 U.S. dollars for Brazil.
- India is the fifth largest emitter of greenhouse gases (GHGs) in the world, accounting for approximately 4.7% of total global emissions, In 2005, India's total GHG emissions were 1,866 $MtCO_2e$, about a quarter of China's total emissions at 7,234 $MtCO_2e$.
- India's per capita GHG emissions were 1.7 metric tons of CO_2e, less than a third of China's per capita emissions at 5.5 metric tons of CO_2e and the world average at 5.8 metric tons of CO_2e.3 India has declared that its per capita emissions will never exceed those of the developed countries. India's economy is less GHG or energy intensive than other major developing economies.
- In 2005, the GHG intensity of the Indian economy was about 760 tCO_2e per million international dollars in PPP terms, exceeding the world average of 670 tCO_2e, but outperforming China (1,350 tCO_2e) and South Africa (1,060 tCO_2e).
- India's energy intensity in 2006 was about a half of China's, lower than the US and only slightly higher than the EU's Brazil.
- India is the fifth largest emitter of greenhouse gases (GHGs) in the world, accounting for approximately 4.7% of total global emissions, In 2005, India's total GHG emissions were 1,866 $MtCO_2e$, about a quarter of

China's total emissions at 7,234 MtCO_2e, while China has become the largest emitter with 23% of total GHG emissions.

- India's per capita GHG emissions were 1.7 metric tons of CO_2e, less than a third of China's per capita emissions at 5.5 metric tons of CO_2e and the world average at 5.8 metric tons of CO_2e.3 India has declared that its per capita emissions will never exceed those of the developed countries. India's economy is less GHG or energy-intensive than other major developing economies.
- In 2005, the GHG intensity of the Indian economy was about 760 tCO_2e per million international dollars in PPP terms, exceeding the world average of 670 tCO_2e, but outperforming China (1,350 tCO_2e) and South Africa (1,060 tCO_2e).
- India's energy intensity in 2006 was about a half of China's, lower than the US and only slightly higher than the EU's.

India is often compared with China, despite the former having lower levels of emissions, higher poverty levels and lower economic capacity relative to China. India's per capita emissions of 1.5 tones CO_2 eq are far below the world average. With 17% of the world's population, India contributes only 4.6% of the world's GHG. However, India is the fourth largest GHG emitter in absolute terms. Since 1990, emissions have grown by 65% and they are projected to increase by 70% by 2020. Rapid economic growth and making provision for access to energy to those previously without access account for much of this growth. Some 250 million people live on less than a dollar per day, and 400 million people in India do not have access to electricity. Providing electricity to these people while moving to low-carbon electricity generation is a social imperative. Consequently, Indian electricity supply and demand are projected to increase fivefold to six fold between now and 2050. This development will require massive investment, but it also creates unique opportunities to transform the country's carbon intensity, even as India's per capita emissions are far below the world average.

India is highly vulnerable to climate-related events including extreme weather events. The country has particular concerns about the anticipated impact of the receding of the Himalayan glaciers on water runoff from these glaciers, especially in the northern part of the Indian subcontinent. Densely populated urban centre and coastal areas are greatly exposed to heavy rainfall, storms (e.g. monsoons and hurricanes) or consequential floods.

Particularly sensitive areas are at risk of submergence due to sea-level rise. An increasing frequency and intensity of storms and flooding or droughts resulting from changes in patterns of precipitation and temperature are expected to have substantial effects on climate-sensitive economic sectors such as agriculture, which supports the livelihood of two out of every three Indians. A resulting decline in crop yields, degraded lands, water shortages, and health problems would further undermine the capabilities of India's rural poor, widening the economic and social disparities. Loss of habitats and the means for production would lead to migration from rural communities to urban slums. These slums may further attract refugees from neighboring countries (e.g. Bangladesh) that are even more prone to the repercussions of climate change.

India's climate agenda as defined and shaped by its national circumstances — development, vulnerability, mitigation and energy challenges.

11.4. Policy Development

India's action on integration of renewable energy and energy efficiency can be traced back to the early 1980s, when the country started to focus on harnessing alternative sources of energy and — more specifically — locally available resources. These gave genesis to large national level programs on renewable energy (biogas, biomass, solar PV and thermal) and energy efficiency (e.g. improved energy efficient wood fuel stoves, energy efficiency in industries). With the opening of the economy and a greater role for the private sector, throughout the 1990s and beyond, India diversified and scaled up its ambition on renewable energy and energy efficiency. These plans are spelled out in the many existing policies that are aimed at reducing GHG emissions.

India's economy has progressively de-carbonized since the early 1990s. Emission intensity has declined by more than 30% during the period 1994–2007. Recent studies have projected that the emission intensity will reduce to 0.15 $KGCO_2e$ by 2030 from a level of 0.35 $KGCO_2e$ in 2003–2004. One of the key reasons for this continuous decline in emission intensity is the increase in share of services sector in the economy. The government has also implemented a renewable energy and energy efficiency program, and put strong incentives in place for their further promotion.

For a long time India has pursued domestic regulatory and funding programmes focused on energy conservation and the deployment of renewable energy technologies. These actions have been supported by legislation, regulation and tariff arrangements, e.g. the Energy Conservation Act in 2001, the New and Renewable Energy Policy in 2005 and the Integrated Energy Policy in 2006. The promotion of renewable energy can be traced back to the early 1990s. India has a number of publicly funded programmes to address both the direct impacts of climate change (e.g. through scientific research) and the reduction of its vulnerability to climate-related risks (e.g. through poverty eradication and rural development programmes). In 2006–2007, the Indian government spent no less than 12% of its annual budget or 2.63% of GDP on these programmes. The current expenditure of the Indian government on adaptation exceeds 2.6% of GDP, with an emphasis on agriculture, water resources, health and sanitation, forests, coastal zone infrastructure and extreme weather events. Among other things, the government intends to increase not only the quantity but also the quality of forest cover, thereby maintaining (at least in the next decade) the current level of carbon sequestration by forestry, which equates to 10% of annual GHG emissions.

A more recent development is the National Action Plan on Climate Change (NAPCC) — an effort by the government to focus — among other things — on harnessing renewable energy and providing a framework for following an ecologically sustainable development path, ensuring energy security and social and economic development, while yielding co-benefits of climate change impacts. This plan also describes other ongoing initiatives — including energy efficiency, renewable energy and decarbonising power generation — and outlines eight priority missions that will promote sustainable development and mitigate the impact of climate change. The Action Plan was released by the Prime Minister in June 2008 and consists of eight Missions:

- **National Solar Mission** seeks to deploy 20,000 MW of solar electricity capacity in the country by 2020. The first phase (2010–2012) is currently underway during which 1,000 MW are planned to be installed.
- **National Mission for Enhanced Energy Efficiency** creates new institutional mechanisms to enable the development and strengthening of

energy efficiency markets. Various programmes have been initiated, including the Perform, Achieve and Trade (PAT) mechanism to promote efficiency in large industries, and the Super-Efficient Equipment Programme (SEEP) to accelerate the introduction of deployment of super-efficient appliances.

- **National Mission on Sustainable Habitat** promotes the introduction of sustainable transport, energy-efficient buildings, and sustainable waste management in cities.
- **National Water Mission** promotes the integrated management of water resources and increase water use efficiency by 20%.
- **National Mission for Sustaining the Himalayan Ecosystem** establishes an observational and monitoring network for the Himalayan environment so as to assess climate impacts on the Himalayan glacier and promote community-based management of these ecosystems.
- **National Mission for a "Green India"** seeks to afforest an additional 10 million hectare of forest lands, waste lands and community lands.
- **National Mission for Sustainable Agriculture** focuses on enhancing productivity and resilience of agriculture so as to reduce vulnerability to extremes of weather, long dry spells, flooding, and variable moisture availability.
- **National Mission on Strategic Knowledge for Climate Change** identifies challenges arising from climate change, promotes the development and diffusion of knowledge on responses to these challenges in the areas of health, demography, migration and livelihood of coastal communities.

Solar Energy

India announced in its solar mission a target of generating 22 GW (20 GW grid-connected, 2 GW off-grid) of electricity from solar energy by 2022 (MNRE, 2008–2009). Solar capacity is still very small compared to other countries. India had 10.2 MW grid connected photovoltaic systems and 2.5 MW stand-alone systems in March 2010. On top of that there were about 592,000 solar street and home lighting systems, and 7,300 agricultural pumps driven by PV. Some estimates are modest in terms of PV growth, to around 100 MW in 2022. Others are more ambitious and project a potential for 1.8 GW solar PV in the next five to six years. India's solar mission is structured in three phases: the first aiming for a deployment of 1 GW of grid-connected solar by 2013, the second 4 GW by 2017 and the final one to reach 20 GW by 2022.

Thanks to a major initiative by Prime Minister Indira Gandhi, a solar PV R&D programme was started by Central Electronics, Ltd. (CEL), as early as 1976. In 2010, India had a solar cell production capacity of 490 MW per year, from nine manufacturers. The aggregate annual module production capacity rose from less than 60 MW in 2005 to more than 1.27 GW in 2010 with 15 assemblers having an annual production capacity larger than 30 MW. Indian companies do not currently cover the entire PV value chain from silicon supply, through ingot and wafer production to cell and module manufacturing. They focus on the more labour-intensive and less technology-intensive part of cell and module production. Therefore, India's PV industry depends largely on the import of silicon wavers as input for its PV cell production. By 2020, India's solar mission aims for an annual PV production capacity equivalent to 10 GW, including dedicated manufacturing capacities for poly silicon material corresponding to 2 GW capacity of solar cells. Solar PV module costs in India are around INR 180,000/kW (3,850/kW U.S. dollars).

India also has good concentrated solar power (CSP) potential, notably in Rajasthan, with a solar insulation of 2,400 kWh/m^2 and high proportion of sunny days. The land requirement for 100 GW is 3,600 km^2. Rajasthan has more than 175,000 km^2 of desert land. So far, no CSP plant exists in India, but several projects with a combined capacity of 381 MW are in the planning or construction phase in Gujarat and Rajasthan. Within its solar mission, India strives to reach a CSP capacity of 10 GW by 2022. Three ongoing CSP projects in Rajasthan have been included as so-called migration projects with special tariff conditions within the solar mission, each having a capacity of 10 MW, but based on different technology concepts (tower systems, parabolic dish, parabolic trough with storage). Costs vary from INR 150,000/kW (3,208/kW U.S. dollars) for the tower system to INR 400,000/kW (8,555/kW U.S. dollars) for the trough project. For its base tariff, India assumes calculation investment costs of INR 153,000/kW (3,330/kW U.S. dollars), which can be considered very low compared with commissioned plants worldwide.

Within the first phase of the solar mission, 446 MW of PV and 470 MW of CSP capacity are being allocated in tender processes by 2012, in which project developers are bidding for a discount on base tariffs. Independent of these national support measures, several Indian states are promoting solar power through feed-in tariffs. India plans to achieve the 1 GW target of the first phase of the solar mission, including 54 MW of PV and 30 MW of CSP capacities already under development before the launch of the solar mission in November 2009 (so-called migration projects). The migration projects benefit from special tariff conditions, since they do not have to undergo the bidding process of the regular projects in the solar mission.

11.5. Low-Carbon Strategies

Government of India has also set up an Expert Group on low-carbon Strategies for Inclusive Growth, which is inter-sectoral and based in the Planning Commission. The Expert Group submitted its interim report in May, 2011. The Expert Group's Final Report is likely to contain greater details of technology, policy and finance options that are necessary to pursue a low emission, inclusive growth trajectory in the future. Some of these recommendations will be built into the growth strategy for the 12th Five-Year Plan. The Expert Group has emphasized action on the following main fronts:

(i) Power: In the power sector, the Expert Group has suggested action both on the supply and demand side. On the supply side, they have suggested the need to adopt super-critical technologies in coal based thermal power generation as quickly as possible. Gas being in limited supply; its best use is not in base load power, but in combined heat and power systems in large establishments. They have also stressed the need to invest in renewable technologies, particularly solar, wind and second generation bio-fuels. Development of hydro-power, in a sustainable manner, is also considered critical to maximising generation of renewable energy, which poses additional load curve problems on the supply side.

On the Demand side, they have stressed the need to accelerate adoption of super-efficient electrical appliances through a combination of market and regulatory mechanisms, and also the need to enhance efficiency of agricultural pump-sets and industrial equipment using power by facilitating adoption of best available technology.

Last but not the least, they have stressed the need to modernize the transmission and distribution systems to bring technical and commercial losses down to the world average levels, while at the same time universalizing access to electricity for the poor. This would not only require acceleration of power reforms, but also adoption of new and frontier technologies like the smart grids.

(ii) Transport: On the transport front, they have stressed the need to increase the share of rail in overall freight transport. This is not possible unless the efficiency of rail freight transport is drastically improved, and also made price competitive by bringing down the levels of cross subsidisation between freight and passenger transport. Completion of dedicated rail corridors is to be taken up on top priority. The need is to improve both share and efficiency of the public transport system; and also further improve the fuel efficiency of vehicles through both market based and regulatory mechanisms.

(iii) Industry: The Expert Group has identified major sources of industrial emissions and made specific recommendations for sectors like Iron & Steel and Cement, which account for over 60% of industrial process emissions. It has been stressed that is important that green-field plants in these sectors can adopt the best available technology; existing plants, particularly, small and medium ones, modernize and adopt green technology at an accelerated phase. For this to happen, financing mechanisms will need to be tied up in an equitable and transparent manner.

(iv) Buildings: While efficient appliances can reduce demand for power to some extent, change in the design and structure of the building itself can act as a multiplier in reducing overall energy demand. India is fortunate that most of the commercial buildings that will be extant in 2030, are yet to be built. There is a need to both evolve and institutionalize Green Building Codes at all levels of Government: Centre, State and Urban Local Bodies.

(v) Forestry: Up to the Eleventh Plan the focus was on increasing area under forest and tree cover. Given the scarce land availability and the trade-offs involved, achievement in this front has been limited. However, there is tremendous scopc for increasing the stock and quality of existing forests.

"Green India Mission" is being designed to regenerate at least 4 million hectares of degraded forest; increase density of cover on 2 million hectares of moderately dense forest; and overall, increase the density of forest and tree cover on 10 million hectares of forest land, waste lands and community lands.

Finally, there is the broader issue of adapting agricultural practices to serious alterations in climatic conditions, and to manage our water resources in a more comprehensive and efficient fashion. There is an urgent need for developing agro-climatic zone specific water harvesting and management technology to enable rural communities to withstand the effects of climate change. Similarly, genetic improvement of agricultural crops to develop a flexible portfolio of plant varieties that can thrive in drier or wetter environments, flash floods, pest attacks due to increase and decrease in humidity, etc. is an important area of research that needs to be pursued in the Twelfth Plan.

Initiatives on State Climate Action Plans

All levels of Government are required to act together to combat the challenge of Climate Change. The State Governments need to develop a State Action Plan for Climate Change that can be dovetailed to the National Action Plan for Climate Change, by identifying vulnerable areas and communities, and by developing a State specific action programme for the above mentioned areas that will facilitate mitigation and adaptation action against the challenge of climate change. Some States have already made a beginning; other States need to take similar action during the Twelfth Plan period.

A major initiative in the form of developing individual State Climate Action Plans have been initiated within the country in collaboration with the Ministry of Environment and Forests, State Governments and international organizations. These plans will be guided by the overall principles and aims of the NAPCC. There would be a common framework of strategies to be developed in due course, under which the each sate will be identifying their respective areas of priority and scale of intervention. Overall, India is currently undertaking a multilayered approach to its climate actions. This involves developing inter-region network for study of change impacts in South Asia with neighboring countries, approval and implementation of several NAPCC Missions within the country, while also initiating the development of state level climate change action plans. Besides as part of the commitment to reduce the use of fossil fuel by increasing the share of renewable energy in the energy mix, several Indian states are developing their own areas and scale for climate actions that fit within the NAPCC framework, for evolving state specific mitigation and adaptation strategies.

11.6. Challenges

Poverty eradication through economic development is a national priority. Since India has a large population in which the majority is very poor, development is regarded as the means to move the entire population out of absolute poverty. India's development agenda is far from completion and remains focused on poverty alleviation. Economic development is also seen as the principal way to foster adaptation to climate change, since adaptive capacity is a function of the level of economic development of the rural and urban poor in India.

Energy is essential for development. Primary energy demand in India rose by 68% over the period 1990–2005 at an annual growth rate of about 3.5%. Nearly half of the commercial, primary energy demand is met by coal but the share of coal in India's total primary energy is about 38%. That is because non-commercial primary energy, such as biomass and cow dung, which is mainly used for household energy, accounts for about 28% of the total primary energy consumption in India.

At the end of 2009, coal accounted for 53.3% of the total installed utility-based capacity, followed by that of hydro at 24.7%, gas at 10.5%, nuclear at 2.9% and oil at 0.9%. Renewable energy sources (primarily wind) make up 7.7% of the total installed capacity. Although the installed renewable capacity has recorded double-digit growth in recent years, its contribution to the primary energy mix remains at about 1%.

The principal energy-related challenge is access to energy, which has two distinct facets: ensuring energy supply to meet the growing demand of fuelling economic growth; and providing access to clean commercial energy for the poor, chiefly electrification and cooking gas/kerosene.

To sustain a growth rate of 8–9%, it is inevitable to some extent that investments in large infrastructure, of which the most important is power-generation capacity, will continue. The Economic Advisory Council to the

Prime Minister (2009) proposes a number of policy interventions in the electricity sector. Among the proposed options are facilitating more private investment in power generation, including recently from China, developing an "active" plan for the creation of power capacity over the next 15 years, diversifying fuel sources and establishing natural gas- and nuclear energy-based power plants.

Not only power generating capacity but also facilities for international and domestic transport of energy sources hold the key to success. India has never been a big importer of thermal coal for power generation. Imported coal accounted for about 10% of the total demand for raw coal in 2008–2009. India imports coking coal mostly from Australia for use by the steel sector for quality considerations, and non-coking coal mostly from Indonesia and South Africa for use by the power and cement sectors (also for cost and quality considerations). For example, an expert committee on integrated energy policy suggests that on the western and southern coasts imported coal is more cost-competitive compared with domestic coal, and even more so compared with imported gas at these locations, and that such a cost is an advantage of imported coal over imported gas. To date coal remains the most realistic option for power generation in the short to medium term, even though an increase in the supply of coal is constrained by the ability to raise domestic production beyond the current level, and the absence of both port and rail capacity for the import and movement of coal. The second concern is that the majority of the Indian population lacks access to energy. More than 400 million persons do not have access to electricity and more than 700 million depend on noncommercial biomass for cooking. The disparity in access to energy between urban and rural areas and across income groups has stayed the same or even widened. Lastly, deficiencies in India's network for the transmission and distribution of electricity undermine people's access to energy.

India is set on a low-carbon path with declining energy intensity, i.e. with the objective of decoupling economic growth and rises in GHG emissions. Its success in decoupling growth in energy use and resulting emissions from GDP growth has been acknowledged. India's energy-use efficiency has improved over the years as reflected in a continuous decline in energy intensity (i.e. energy consumption per unit of GDP). India's energy intensity showed an economy-wide decline in PPP terms by nearly 50% between

1980 and 2006 and a reduction in the industrial sector by more than 40% between 1990 and 2005. Consequently, the country's emissions intensity also declined by 17.6% between 1990 and 2005 and its CO_2 intensity declined by about 20% over the same period. Despite this trend, India's future absolute emissions will grow.

Economic modelling has shown that defining business-as-usual (BAU) emission trajectories for a dynamic economy like India's is difficult if not impossible. Thus, it will be a challenge to operationalise the suggestion that developing countries as a group should limit the growth of their emissions to 15–30% below BAU by 2020. Existing model results show a wide variance in the range of future increases in GHG emissions, depending primarily on the different assumptions regarding GDP growth, energy and carbon intensities. For example, the Climate Modelling Forum summarises the initial results of five studies:

- NCAER-CGE, a computable general equilibrium (CGE) modelling study by India's National Council of Applied Economic Research (NCAER);
- TERI-MoEF, a market allocation (MARKAL) modelling study by the Energy & Resources Institute (TERI);
- IRADe-AA, an activity analysis modelling study by Integrated Research and Action for Development (IRADe);
- TERI-Poznan, another MARKAL model-based study by TERI, presented at the 14th Conference of Parties (COP) on Climate Change at Poznan; and the McKinsey study, a detailed sector-by-sector analysis of GHG emissions by McKinsey and Company.

These five independent studies show the result that in absolute terms India's annual GHG emissions in 2031 could reach between 4.0 billion tons of CO_2e and 7.3 billion tons (4.0–7.3 Gt) under the BAU scenarios. These projections rest on assumptions of a high, annual GDP growth rate averaged over the period up to 2030, ranging from 7.51% (McKinsey) to 8.84% (NCAER) per annum.

The following conclusions may be drawn from this analysis:

(i) India's total GHG emissions will grow from the current level but will likely remain under 4 Gt by 2030, based on the assumptions of a GDP growth rate in the range of 6–7% per annum.

(ii) Both energy intensity and carbon intensity is likely to continue falling. It is possible to reduce India's energy intensity by up to 25% from the current level and to reduce the GHG intensity of the economy by as much as one-third.

(iii) Power-generation sector alone would account for more than 50% of GHG emissions; based on the assumption that two-third of generation capacity will remain coal-based, and would more than triple between 2005 and 2030.

(iv) The investment in India needs to meet the demand would reach 1.25 trillion U.S. dollars (in year 2006 dollars) in energy infrastructure, three-quarters of it in the power sector.

(v) There would be a clear shift to demand-side management and renewable sources of energy.

11.7. Conclusion

The National Action Plan on Climate Change provides a framework for an ecologically sustainable low carbon development path. So far, the plan is the most comprehensive document crafted to address climate change within the context of sustainable development for the country. The NAPCC uncovers India's readiness to commit to a gradual shift from economic activity based on fossil fuels to one based on renewable sources of energy. India's policy action on climate change is broader than carbon emissions reductions and incorporates a vision of following an ecologically sustainable development pathway. It aims at enhancement of the country's energy security while meeting its social and economic development objectives, and yielding co-benefits of addressing climate change effects at the same time. Within the South-Asian region, and within the country, India has launched proactive support for developing science based evidence of climate impacts in key physiographic areas.

Reference

Ministry of New and Reviewable Energy of India (2008–2009). Annual Report, Delhi.

Chapter 12

LOW-CARBON TARIFF THEORY

*Libo Wu and Weiqi Tang**

12.1. Introduction

The Copenhagen Summit, i.e. the 2009 United Nations Climate Change Conference, closed on December 18, 2009 without reaching any concrete results. The only fruit of the highly anticipated international conference was, sadly, a legally insignificant agreement made by five countries including the U.S., China, India and Brazil, which wasn't officially approved by all the contracting countries. The fierce controversy at the summit revealed the wide divergence among all the interest groups, such as the non-EU developed countries led by the U.S., the EU countries, the emerging economies like China and India, G77, and the Alliance of Small Island States. There seems to be no way, at least in the short term, to bridge the gaps between these interest groups. Given the urgency and complexity in climate change issues, better regimes and stronger policy supports are required to build a new international emission reduction cooperation framework, which needs to be more active and effective than the current Kyoto Protocol. However, regardless of the calls for multilateral collaboration, many developed countries are planning to impose their will on emerging economies and force them to take emission reduction efforts through unilateral activities. One of the most effective and commonly used methods is carbon tariffs — tariffs levied on commodities exported from developing countries according to their carbon intensity, unless the exporters no longer subsidize energy utilization and implement a carbon tax domestically.

*Libo Wu, Professor and Executive Director, Center for Energy Economics and Strategy Studies at Fudan University; Weiqi Tang, Ph.D. Candidate of Institute of World Economy, Fudan University, China.

China, known as the biggest carbon emitter in the world, now faces great pressure for emission reduction. The Chinese government has made a commitment to reduce CO_2 intensity by 40%–45% by 2020, compared to 2005. Basically, this was unacceptable for most developed countries. Many of them, including the U.S., U.K., France, and Japan, claimed that they would levy carbon tariffs on commodities imported from China, so as to force China to pay for CO_2 reduction efforts implemented worldwide. So far it is still uncertain whether China would be trapped by a new round of carbon tariff sanctions. However, given the fact that developed countries are struggling to recover from the financial crisis, trying to preserve domestic employment by protecting traditional industries with a carbon tariff could also serve as a convenient way for developed countries to legalize their trade protectionist behavior.

In this chapter, we will discuss the theoretical mechanism and the policy environment of carbon tariffs from three perspectives: (1) possibility of carbon tariff on Chinese exports, (2) theoretical meaning of carbon tariffs, as well as their rationality and legality, (3) effect of carbon tariffs on China's international trade, and feasible policy for China to mitigate their impact.

12.2. Possibility of Carbon Tariffs on Chinese Exports

12.2.1. *The implications of China's emission reduction for global climate change*

The International Energy Agency made a projection in 2008 (IEA, 2008), saying that if all economies keep growing at their current rates and under present policy scenarios, total CO_2 emissions will increase by 45% by 2030, and China will take up about 46.9% of the total increment. At present, China's CO_2 emissions account for 20.3% of the world total, and they will increase to 28.9% in 2030. In contrast, the U.S. currently takes up almost the same proportion of CO_2 emission as China does, but it will decrease to 14.3% in 2030. The remarkable increase of CO_2 emissions predicted in China made her the target for censure at the Copenhagen summit. Although the emissions per capita of China roughly equal the world average, and is only 24.7% of U.S. emissions per capita, or 57.3% of the European average, when it comes to 2030 China's emissions per capita will exceed Europe and equal 50.6% of the U.S. per capita level and 163% of the world average. Surely, we can expect significant change in China's industrial structure during this period, which will result in decreases in CO_2 intensity. The decrease is estimated to be 3.1% each year. However, such a reduction rate is still not enough to fully eliminate the emissions expansion created by fast economic growth. The commitment of 40–45% reduction in CO_2 intensity is equivalent to the target of reducing CO_2 emissions per unit GDP by 2.6%–3% annually, which will make China's emissions reach the level predicted by IEA, given an economic growth rate of 6%. This is obviously beyond developed countries' tolerance.

12.2.2. *The meaning of emission reduction burden allocation to China*

There are wide gaps between the commitment made by the Chinese government and the widely accepted global warming threshold. The IPCC 4th Assessment Report claimed that the CO_2 concentration must be kept under 450 ppm before it exceeds the 2°C global warming threshold compared to pre-industrial level and causes irremediable results. Unfortunately, if the world develops under existing policies and economic growth patterns, by 2020 emissions from developing countries alone will push the CO_2 concentration higher than 450 ppm. This indicates that developed countries could not achieve the emission reduction target if developing countries do not reduce their emission substantially. In order to reach the emission reduction target, the world's total CO_2 emissions must decrease by at least 10% by 2020. How to allocate this 10% among economies is the key to international climate change issues. Considering the diverse social, economic and political situations of developed and developing countries, there could be different principles of allocation, each with different economic and social costs.

Table 12.1 shows the comparison of possible allocation criteria proposed by the International Energy Agency (IEA). According to different allocation criteria, the emission reduction target of each economy, calculated as the required reduction ratio compared to the benchmark scenario, is also different (see each column in Table 12.1). The analysis shows that the required reduction for China in 2020 would be much higher than the benchmark,

Table 12.1. Comparison of criterions for emission reduction burden allocation (required emission reduction rate compared to baseline scenario).

Country/region \ Criterion	Equalized reduction	Emission per capita	Current emission	Current GDP
U.S.	−10%	−75%	15%	19%
E.U.	−10%	−47%	17%	77%
Japan	−10%	−55%	23%	85%
Other OECDs	−10%	−49%	−5%	23%
China	−10%	−41%	−34%	−67%
India	−10%	153%	−33%	−36%
Africa	−10%	378%	−8%	9%
Russia	−10%	−71%	−4%	−49%
Middle East	−10%	−51%	−27%	−50%
Latin America	−10%	76%	−10%	0%
World Total	−10%	−10%	−10%	−10%

Source: IEA, 2008.

regardless of which criterion is taken. It's also noteworthy that the baseline scenario has already taken the energy conservation target of the "11th Five-Year Plan" into account, which requires energy consumption per unit GDP to be reduced by 20% from 2005 to 2010. Reduction in energy intensity will surely lead to reduction in carbon intensity, and the commitment of reducing CO_2 intensity by 40%–45% by 2020 is doubtlessly based on the former policies. Under such policy scenarios, China will have to reduce her emissions by 34%–67% by 2020. During the Copenhagen summit, some Chinese scholars and media claimed that "per capita emission" is a fairer criterion that should be used as a benchmark indicator in allocation of the emission reduction burden. However, from the analysis here, we can see that even if per capita emissions were accepted as the criterion, China still has to reduce CO_2 emission by a further 41% before 2020. It's quite clear that the commitment made by the Chinese government in the Copenhagen summit was insufficient for tackling global climate change.

Therefore, we can see that although China has taken positive actions in energy conservation during the 11th Five-Year Plan period by requiring energy intensity to be reduced by 20%, it's still far from the minimum requirement for tackling global warming because the explosive expansion of the Chinese economy in the baseline scenario would overwhelm the decrease in carbon intensity. This is the main reason for all the accusations against China in the Copenhagen summit. If China does not change its development strategy and make further commitments, it is very likely that developed countries would levy carbon tariffs against China.

From the perspective of feasibility, whether Chinese exports would be subject to tariffs depends on three factors: (1) does China have a lot of energy-intensive products exported to countries where carbon emissions or energy consumption are taxed; (2) whether those products are subsidized by the Chinese government for the energy consumed during production; and (3) if there are any appropriate emission restriction policies for China that can effectively reduce CO_2 emissions but are being ignored.

Figure 12.1 shows that Chinese exports of steel, cement and some other energy-intensive products have significantly increased since the beginning of this century; for example, China's steel exports in 2008 accounted for 12.1% of the world's total. One of the most important reasons is the relatively low energy cost for Chinese producers compared to foreign ones. Since the start of the 21st century, there has been a sharp rise in international energy prices. However, energy prices in the Chinese domestic markets were only slightly related to international price movements, thanks to prevailing price regulations and energy subsidies. Stable and low energy prices provided Chinese energy-intensive products with significant advantages in international trade, and thus, boosted exports. But energy

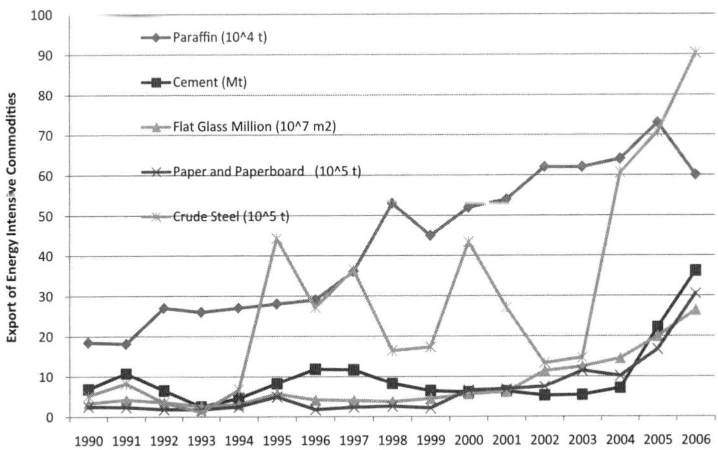

Fig. 12.1. Export of energy intensive commodities.

Source: *China Energy Statistical Yearbook* (2008).

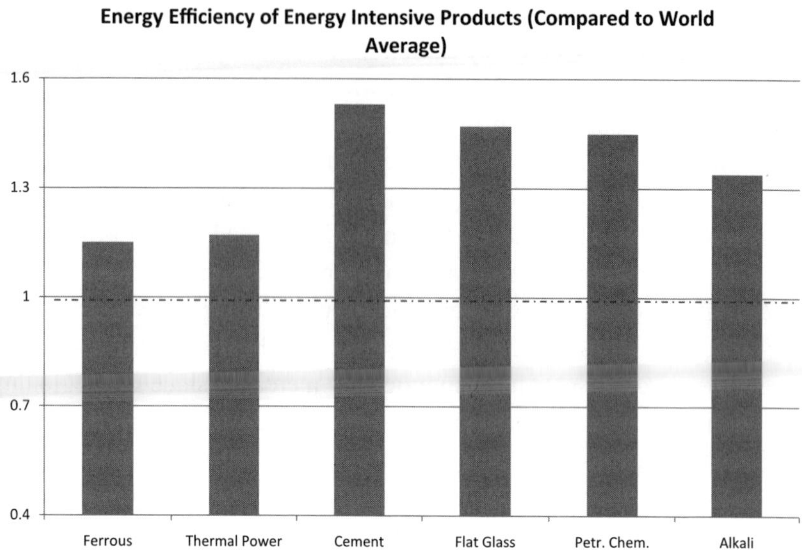

Fig. 12.2. Comparison of energy efficiency of China and world average (energy consumption per unit of product, Chinese/world average).

Source: Author calculated.

subsidies also discouraged energy efficiency improvements and distorted the energy market mechanism. The cross country comparison shown in Fig. 12.2 implies that energy efficiency in most energy-intensive industries in China is significantly lower than the world average. Since carbon tariffs are based on CO_2 emission related to production, products exported from China would thus face higher carbon tariff rates, and that would to a remarkable degree dampen the competitiveness of Chinese exports.

12.3. The Rationality, Legality and Feasibility of Carbon Tariffs

Since 2001, when the U.S. unilaterally withdrew from the Kyoto Protocol, European countries started seeking market-oriented means to force countries that denied emission commitments back to global climate cooperation. The idea of "carbon tariffs" was born in that background. Although the U.S. House of Representatives approved the bill of "Clean Energy and Security Act", which authorized the U.S. government to levy carbon tariff on commodities imported from China and India, it's noteworthy that carbon tariffs were originally designed by European countries against the U.S. Therefore, carbon tariffs are unfavorable to both emerging economies like China and India and developed economies including the U.S., Canada, and Australia — it's actually a double-edged sword. The U.S. advocated carbon tariffs mainly because of the threat of explosive expansion of energy-intensive product exports from China. The growth rate in this century of Chinese energy-intensive product exports is astonishing: steel products exported from China accounted for 5.12% of the world's total in 2004, but after 5 years, in 2008, the figure more than doubled to 12.09%; in the same period, the share of chemical products exported from China expanded from 2.68% to 4.65%; and that of textile products grew from 16.98% to 26.08%. Domestic producers in the U.S. are severely threatened by Chinese exports, and carbon tariffs make a good excuse for trade protectionism, especially when the aftereffect of the financial crisis still haunting the U.S. However, there're a lot of controversies over carbon tariffs within the U.S. as well. If the U.S. government wants to levy carbon tariffs on goods imported from countries that don't have reasonable emission reduction targets for 2020, the corresponding domestic industries of the U.S. would have to be charged a carbon tax first, which is strongly opposed by producers and industry association from iron and steel, cement, chemical industry. Despite the bill of "Clean Energy and Security Act" gaining approval by the House of Representatives, whether it would be approved by the Senate, where conservative forces are powerful and entrenched, is still uncertain. Aside from political thoughts, there is still a lot of uncertainty in the future for carbon tariffs from economic and technical perspectives.

12.3.1. *Legality*

Some developed countries, including Britain and France, have already started or plan to tax domestic carbon emissions, and energy costs for producers in energy-intensive industries in those countries would be higher. Therefore, carbon tariffs are needed to adjust producing costs at the border, so as to maintain fair competition in international trade and prevent "free riders" in global climate cooperation. If carbon tariffs are accepted, all those countries that don't have a carbon tax would be involved.

So far, there is a huge controversy about whether carbon tariffs are trade-protectionist. The Chinese government has insisted that carbon tariffs are discriminatory and violate WTO principles. According to WTO principles, inputs and processes of production are free from international trade policies. Importing countries can only set regulation on imported commodities; they are not allowed to regulate how producers in other countries produce certain products, nor can they regulate what inputs foreign producers can or cannot use.

According to this principle, energy consumed and carbon emitted during the production process cannot be regulated by trade policies, let alone subject to tariffs. But supporters for carbon tariffs — mainly developed countries, including the U.S. — invoke the Shrimp/Sea Turtle Case to defend the legality of carbon tariffs. They argued that exhaustible resources are protected by WTO principles, since they are public resources and don't belong to any single country. In the Shrimp/Sea Turtle Case, the U.S. restricted imports of shrimp from Indonesia and Malaysia because they didn't take measures to protect sea turtles during fishing for shrimp. The U.S. protested that it would harm the wellbeing of sea turtles, which is a common heritage of mankind. The U.S. won this controversial case. According to this case, preventing global warming is the requirement of well-being for all human kind, which can be a perfect excuse for levying tariffs under the legal framework of WTO.

12.3.2. *Rationality*

China's becoming the "world's factory" and a "giant emitter" is, to a large extent, the result of international specialization in the trend of globalization. The optimization of resource distribution requires factors to flow into places with the highest factor productivity. An abundance of labor and low energy prices in China attracted a lot of energy intensive industries. More than 1/3 of the CO_2 emitted in China is related to exports. Exports from China are mainly primary products, including commodities and industrial raw and processed materials, which usually have high price elasticity and low income elasticity, while products produced in developed countries are the reverse. Thus, it's easy to understand that carbon tariffs for Chinese exports would affect Chinese producers mostly, while carbon taxations in developed countries mainly affect domestic consumers. Although Prof. Paul Krugman — the Nobel Prize winner — denied the discriminatory aspect of carbon tariffs given corresponding carbon taxation in developed countries, the facts say the opposite. Krugman protested that if developed countries tax domestic CO_2 emissions at a corresponding level, carbon tariffs would not interrupt the ordinary trade flow among economies. However, carbon tariffs mainly affect producers in developing countries, but are incapable of adjusting consumers' activity, and that may cause essential discrimination.

One of the most important purposes of carbon tariffs, as well as other tariffs and taxes, is to redistribute social wealth and adjust activities of agents; thus, increasing governments' income is not the ends, but the means — the key is how to use the tax revenue. In developed countries, carbon tax revenues are transferred to low income residents who have low energy consumption, to subsidize research and development of clean energy and energy conservation technologies and to support national energy security strategy. In order to mitigate the impact of carbon taxes and to reduce objections, some countries even reduce some other taxes to maintain the overall tax burden to residents and producers. Unfortunately, in the context of international trade, carbon tariffs would transfer wealth from developing countries to developed countries. If there were not any reverse wealth transfer mechanism, like tax feedback, adjustment and compensation mechanisms, carbon tariffs would not be able to support carbon emission reduction policies in developing countries and, even worse, diminished export demand would cause unemployment and lower income in developing countries. Carbon tariffs without a transfer payment mechanism run in the opposite direction of the fair development of mankind, and depress the rationality of climate change policies proposed by developed countries.

12.3.3. *Feasibility*

Feasibility is definitely the biggest problem for applying carbon tariffs. Carbon tariffs require carbon accounting, which involves monitoring GHG emissions of all the production processes of the whole industrial chain of corresponding products. It's an unimaginable amount of work to set standards and measure the technological features, including energy and production efficiency, of the production processes of thousands of commodities from all the exporting countries. Even the data on energy consumption and carbon emissions for each industry, which is required for rough calculation, is hardly available in most countries given the totally different energy statistical system for each country. None of the energy research and statistics organizations, including IEA, EIA and other agencies, is able to provide energy consumption on an industry level, let alone on a product or commodity level. The Clean Development Mechanism (CDM), under the framework of the Kyoto Protocol, is based on projects, and all the emission reductions achieved need to go through a strict examination and approval process which is carried out by a chartered third party and takes a long period of time before it can be certified. Carbon tariffs would need a similar procedure, and its connection to the existing international trade system needs further research.

12.4. The Economic Impact of Carbon Tariffs on Chinese Exports of Energy Intensive Products — Estimation Based on a Carbon Tax Standard

Carbon taxes have a long history in Europe, and recently there is also an active discussion on details about carbon tax systems in the U.S., including tax rates, tax base and tax revenue compensation mechanisms. Some states have also set up pilot areas to test the real effect of carbon taxes. Tables 12.2 and 12.3 list items of energy and carbon taxes in the U.S. and EU countries.

As can be seen from Table 12.3, there are big differences between energy/carbon tax schemes in different countries. Generally speaking, Denmark and Sweden are high carbon tax countries; Germany is a middle carbon tax country; and carbon taxes in Britain and the U.S. are relatively low.

Given the above mentioned tax schemes in the U.S. and EU countries, we can approximately assess the probable cost of carbon tariffs to Chinese exportation. We followed an assessment procedure as follows:

(1) Calculate the overall energy consumption and CO_2 emissions related to energy-intensive export industries.

Table 12.2. Comparison of carbon tax schemes in the U.S.

Carbon tax scheme	Tax base	Tax rate	Subject
Option I:			
Clinton's 1993 BTU energy tax	Fossil fuels, hydropower, nuclear power	25.7 cents/mill. BTU 34.2 cents/mill. BTU (for oil)	Petroleum refiners; Coal, electricity end-users & importers
Option II:			
Save the Climate Action, HR2069 2007	Coal, petroleum and natural gas	\$10/tCO_2	Manufacturing importer
Option III:			
HR3416 2007 energy security funding bill	Coal, petroleum and natural gas	\$15/tCO_2, 10% higher each year	Manufacturing importer
Option IV:			
BC climate protection tax plan	Electric Power	0.49 cents/kwh (household) 0.09 cents/kwh (commercial) 0.03 cents/kwh (industry)	Electricity end-users
Option V: San Francisco Air Quality Management	CO_2	4.4 cents/tCO_2	Industrial sector Business sector

(2) Calculate the emissions embodied in exports, according to total CO_2 emissions and the proportion of exports to overall production in the corresponding industries.

(3) Calculate the corresponding carbon/energy tariff according to each carbon tax scheme of the U.S. and EU countries, assuming that developed countries set carbon tariff schemes in accordance with their domestic carbon taxes.

In this paper, we analyzed four energy intensive industries with the highest growth of exports from China: ferrous metal smelting and processing, non-metallic mineral products, paper and paper products, and non-ferrous

Table 12.3. Comparison of carbon and energy tax schemes in EU countries.

	Danmark plan	Germany plan	Germany manufacturing industry plan	Sweden	Sweden manufacturing industry plan	UK plan
Light fuel oil EURO/kl	281.9	45	61.4	60.5	369	135.8
Natural Gas EURO/thousand m^3	300.6	35.8	59.6	45.3	241.8	24.4
Power Euro/MWh	89.4	12.3	20.5 Household sector	0.5	28.6 Household sector	6.4

Table 12.4. Estimated carbon tariffs for energy intensive products exported from China (according to U.S. schemes).

Mill. U.S. dollars	Option I	Option II	Option III	Option IV	Option V
Ferrous smelting and processing	184.17	750.38	1,125.57	5.19	3.30
Non-metallic mineral products	39.43	139.64	209.46	1.77	0.61
Paper and paper products	16.42	61.48	92.22	1.12	0.27
Non-ferrous smelting and processing	119.82	550.83	826.24	2.07	2.42
Total	**359.84**	**1502.33**	**2253.49**	**8.08**	**6.6**

metal smelting and processing. The calculated results based on 2007 data are listed in Tables 12.4 and 12.5.

As we can see from Tables 12.4 and 12.5, the four energy intensive industries would have to pay over 2 billion U.S. dollors in carbon tariffs in the most demanding U.S. scheme (option III), and 7.68 billion U.S. dollars according to the Denmark scheme. The impacts of carbon tariffs in each industry vary widely due to the different proportion of exports in total production. Figures 12.3 and 12.4 shows the ratio of carbon tariffs to total exports. From the figures, we can see that the non-metallic mineral product and non-ferrous processing industries would be the most affected. In "high tax scenario (Denmark scheme)", the ratio of the carbon tariff to

Table 12.5. Estimated carbon tariffs for energy intensive products exported from China (according to EU schemes).

		Germany		Sweden		
Mill. U.S. dollars	Denmark	manf.	other	manf.	other	UK
Ferrous smelting and processing	2154.49	492.85	296.61	733.07	22.46	167.25
Non-metallic mineral products	643.99	146.03	89.35	280.05	20.57	70.90
Paper and paper products	298.84	68.39	41.24	103.42	3.39	24.32
Non-ferrous smelting and processing	4578.63	1048.09	630.63	1542.63	43.60	353.22
Total	**7675.95**	**1755.36**	**1057.83**	**2659.17**	**90.02**	**615.69**

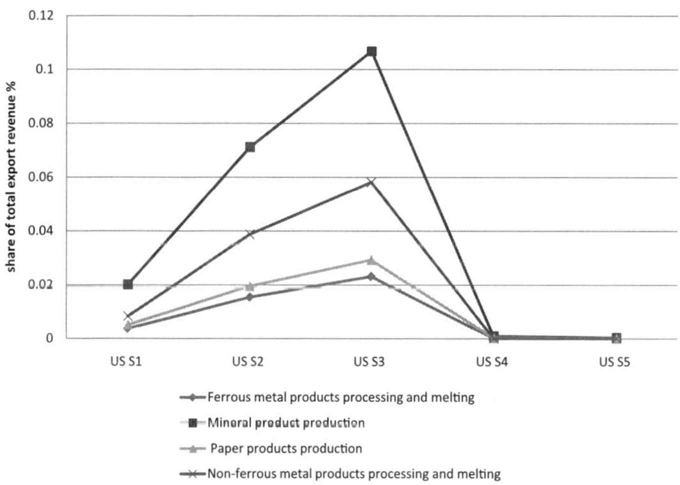

Fig. 12.3. Proportion of carbon tariff in total export value (according to U.S. scenarios).

total export value would be as high as 32.8% for non-metallic products, such as cement, flat glass, etc.; carbon taxes and tariffs would sum up to 13.4% of the total cost of doing business. That is indeed unacceptable for Chinese producers, since their original margins are very low. In the "middle tax scenario (German scheme)", carbon tariffs would account for up to

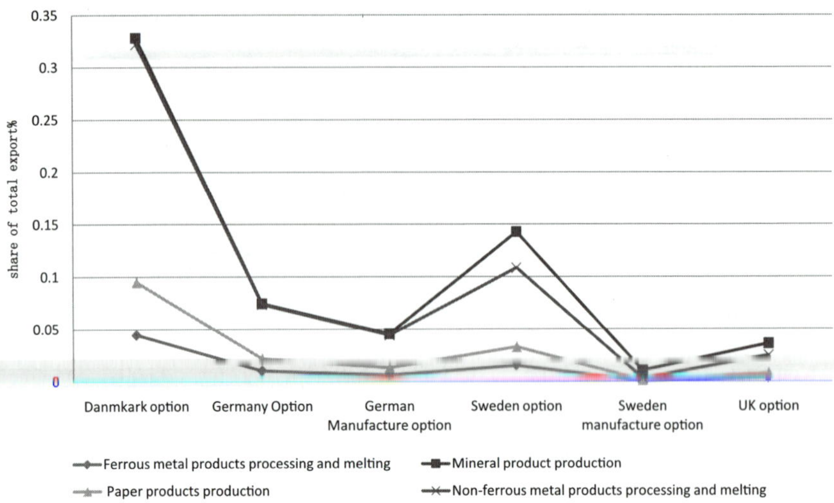

Fig. 12.4. Proportion of carbon tariff in total export value (according to EU countries scenarios).

10% of total export value, and 3–7% in "low tax scenario (British and US schemes)". In 2007 the gross margin of the non-ferrous processing industry in China was 12.1%, so that even in the "middle tax scenario", producers could hardly earn any profit, not to mention the "high tax scenario".

From this study, we can get one basic conclusion: for Chinese exporters, carbon tariffs will definitely become a heavy burden and a real threat. In order to mitigate the impact of carbon tariffs, effective means are to improve energy efficiency, reduce carbon intensity, and export less energy-intensive products. As we mentioned before, once the developed countries impose carbon tariffs on Chinese exports, it would serve as a direct wealth transfer from China to developed countries. It could neither compensate for the cost of domestic energy conservation, nor adjust domestic income distribution. Thus, it's more rational to respond more actively by reforming the domestic energy market: cancel unnecessary energy subsidies, rationalize the pricing system, and tax energy consumption or carbon emissions, so as to protect Chinese exports from carbon tariffs. At the same time, we can redistribute the tax revenue to subsidize low income groups and to financially support energy conservation and the development of new energy.

References

China Energy Statistical Yearbook (2008). China National Bureau of Statistics.

China Statistical Yearbook (2008). China National Bureau of Statistics.

Chapter 13

ENERGY TAXATION AND LOW-CARBON ECONOMY

Soo-cheol Lee and Kazuhiro Ueta†*

13.1. Introduction

Japan currently has three types of energy-related taxes (referred to generally as "energy taxes" in what follows): (1) taxes that are levied when energy sources are imported or extracted, such as the Petroleum and Coal Tax, (2) taxes that are levied on the use of transportation fuels, such as the Gasoline Tax, and (3) the tax levied on the energy conversion sector, namely, the Tax for Promoting the Development of Electric Power Facilities.

These energy taxes are not directly intended to control external diseconomies resulting from energy use, such as carbon dioxide emissions. While it is true that these taxes may have contributed to reductions in environmental stress and can thus be regarded as second-best environmental taxes, Japan's energy taxes are special-purpose revenue sources earmarked for designated uses including subsidies for the coal and petroleum industry, the construction and maintenance of roads, and the development of electric power facilities centering on fossil-fuel or nuclear power generation. In other words, it can be said that, in terms of the usage of tax revenues, Japan's energy taxes have actually contributed to increases in external diseconomies resulting from energy uses by stimulating energy consumption.

*Professor, Faculty of Economics, Meijo University, Japan.
†Professor, Graduate School of Economics, Kyoto University.

Recent years have seen various reforms of energy taxes in Japan. Although traditional special accounts that received energy-related, special-purpose tax revenues have been abolished or combined, similar special accounts have been newly created at the same time, indicating that the traditional basic framework, in which revenues from energy taxes have been treated as special-purpose revenue sources, has not changed.

In this context, this chapter clarifies the characteristics of energy taxes, and provides a detailed examination of the effects of Japan's energy taxes on carbon dioxide emissions, focusing on how the taxes are levied and used. This chapter also discusses how the energy taxation system should be reformed in order to promote energy sustainability and the transition to a low-carbon economy.

13.2. The Current State and Characteristics of Japan's Energy Taxes

13.2.1. *Energy taxation in Japan*

Japan's energy taxes are, as mentioned earlier, introduced for the purpose of collecting revenues. Many other countries also have energy taxes in order to raise revenue. Energy resources have become major targets of taxation in raising government revenues, and, in fact, energy taxes play a significant fiscal role in various countries. For example, the share of energy taxes in total tax revenues is 7% to 30% (1% to 3.5% relative to GDP) in developing countries and 10% to 15% on average (2.0% to 2.5% relative to GDP) in the OECD countries (OECD, 2010).

Figure 13.1 is an illustration of the Japanese energy tax system. It shows that energy resources are revenue sources with a high tax-bearing capacity; that is, the government can gain stable tax revenues with low administrative costs. Generally, compared to other goods, the price elasticity of demand for energy is low, and the level of demand is not significantly affected by business cycles. Also, in the case of upstream taxation, the administrative cost associated with levying and collecting energy taxes is low compared to other goods because the number of businesses engaging in the extraction, importation, and processing of energy resources is small.

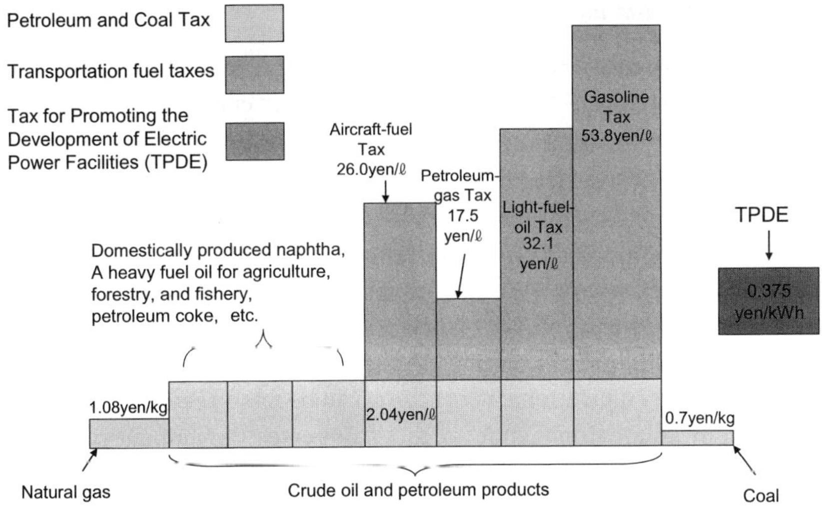

Fig. 13.1. Japan's energy taxation (as of August 31, 2010).

Note: The Petroleum and Coal Tax levied on domestically produced naphtha, Class A heavy fuel oil for agriculture, forestry, and fishery, petroleum coke, etc., is fully refunded.

Source: The graph is created by the authors based on data from the Petroleum Association of Japan (every year) and the Ministry of Finance (every year).

Furthermore, with energy taxes, the relationship between benefits and costs associated with energy use can be made relatively clear. This has provided a major justification for the taxes related to transportation fuel, including the Gasoline Tax. As examined in detail in the next section, energy taxes have the characteristics that the road users bear the cost of the construction and improvement of roads, which is compatible with the beneficiary pays principle. Therefore, these taxes are regarded as taxes that are easily accepted by taxpayers and thus face relatively weak opposition.

From these observations, it can be said that Japan's energy tax system is, after all, a development-oriented tax system emphasizing government policy objectives such as support for the energy industry, road improvement and construction, industry promotion including the development of power plants, energy security, and the promotion of local economic and industrial development. Consequently, the system has increased energy consumption and associated externalities.

13.2.2. *Energy tax and energy consumption*

The discussions thus far can be summarized theoretically as shown in Fig. 13.2. S_1 is the private marginal cost curve for energy suppliers. S_0 is the energy supply curve: the current energy tax t_0 is added to the private marginal cost curve S_1. D_0 is the demand curve for energy.

If the energy tax t_0 was not levied, the energy supply curve would coincide with the private marginal cost curve S_1, and energy consumption and carbon dioxide emissions would increase by $Q_1 - Q_0$ from the current level. It is assumed that the quantity of energy consumption is proportional to the magnitude of pollution (represented by the amount of carbon dioxide emissions). Therefore, the current energy tax t_0 can be interpreted as performing a function of an environmental tax that suppresses the amount of carbon dioxide emissions by $Q_1 - Q_0$.

However, a careful examination of the characteristics of the energy tax system considered thus far reveals a dubious basis for the above discussion. The private marginal cost curve for energy supply is close to S_0 in the first place, instead of S_1. This is because with the current energy tax t_0 and reasons such as energy security, the government collectively covers costs that should be included in the private cost of energy-related industries. For

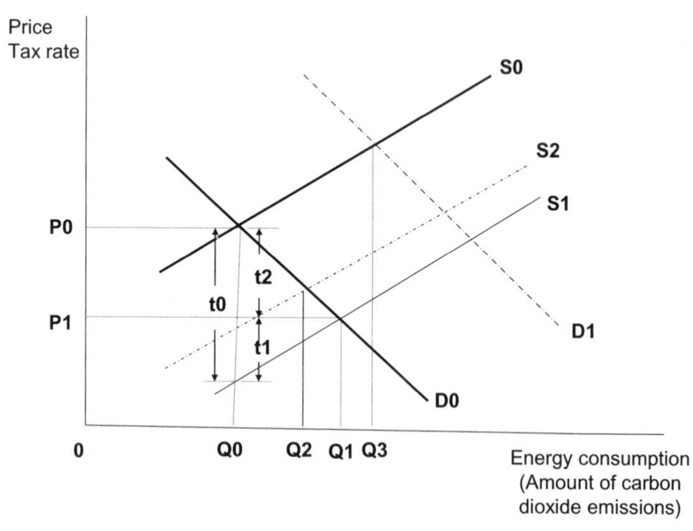

Fig. 13.2. Energy tax and energy consumption.

example, it can be considered that, as energy taxes, the government collects (1) the Petroleum and Coal Tax, (2) the Tax for Promoting the Development of Electric Power Facilities, and (3) the taxes related to transportation fuel; these cover, respectively, (1) costs mainly associated with petroleum-related development and transportation which should be borne by the petroleum industry, (2) costs mainly associated with the establishment and development of power plants which should be borne by the electric power industry, and (3) costs mainly associated with road construction which should be borne by road users.

The traditional energy taxes, therefore, have the characteristic that revenues from them are used to cover the costs that should be paid by energy-related industries or road users as private costs. Given such a characteristic, it is difficult to consider the energy taxes as performing a function of restraining energy consumption and internalizing externalities. Hypothetically, if the cost of energy security and road improvement per unit of energy consumption, which should be borne by the energy industry, is reduced from t_0 to t_1 with the rest (t_2) being used as government's general revenue, the level of pollution, which should have been at Q_2 because of t_1, can be regarded as being improved to the level Q_0.

As explained earlier, most of the revenue from the energy taxes has been used for expanding the foundation of fossil and nuclear energy and constructing and improving roads. This shifts the existing demand curve for energy from D_0 to D_1 over the long run. As a result, energy consumption increases from Q_0 to Q_3. For these reasons, the current energy taxes also have the characteristic that they have stimulated energy consumption and have expanded to internalize various external diseconomies including carbon dioxide emissions.

13.3. Revenues from Energy Taxes and Their Uses

13.3.1. *Revenues related to the tariff on crude oil and petroleum products, and petroleum and coal tax*

The revenue from the Tariff on Crude Oil and Petroleum Products started to be used for coal-related expenses, and in 1967 the Special Account for Coal-Related Measures was established. The special account was then modified, and the Special Account for Measures Concerning Coal and Petroleum was set up in 1972 for the purpose of resource development related to coal and petroleum.

The Petroleum Tax was created in 1979, which led to the establishment in 1980 of the Special Account for Measures Concerning Coal, Petroleum, and Petroleum-Substituting Energies. The Petroleum Tax was modified into the Petroleum and Coal Tax in 2003. The revenue from the Tariff on Crude Oil and Petroleum Products came to be used mainly for measures to support the coal industry. Also, the Petroleum and Coal Tax began to be used mainly for petroleum security and petroleum-substituting energies. In 1993, the special account was revised and became the Special Account for Improving the Supply and Demand Structure for Coal, Petroleum, and Energy, and its use for energy conservation programs and the development of new energies became emphasized. In 2002, as the support for the coal industry based on the revenue from the Tariff on Crude Oil and Petroleum Products was terminated, the special account was renamed the "Special Account for Improving the Supply and Demand Structure for Petroleum and Energy" (referred to as the "Petroleum Special Account" in the following).

In addition, based on Article 36 of the Act on Promoting Administrative Reforms to Realize Simplified and Efficient Government (aka the Administrative Reform Promotion Act), which took effect in June 2006, the Petroleum Special Account and the Special Account for Promoting the Development of Electric Power Facilities (referred to as the "Electricity Special Account" hereafter), whose main revenue source was the Tax for Promoting the Development of Electric Power Facilities, were integrated into the Special Account for Energy Measures, starting in FY2007.

The Special Account for Energy Measures is a special account created to clarify the accounting for funds used for stable fuel supply, structural improvement of energy supply and demand, power plant locations, and diversified electric power generation. It is divided into the Subaccount for Energy Supply and Demand and the Subaccount for Promoting the Development of Electric Power Facilities. The nature of this special account shows that, in effect, the former and latter subaccounts succeed the abolished Petroleum Special Account and Electricity Special Account.

Until FY2000, 20% to 30% of the Petroleum Special Account was used for providing assistance for the coal industry, and 60 to 70% was for petroleum-related measures such as oilfield development and efficiency improvements in oil production and transportation, which shows the inclination of the special account toward supporting the existing energy industry. However, in addition to such support, the share of funds used for new energies, carbon dioxide reduction projects, and energy conservation began to rise rapidly in recent years in the Special Account for Energy Measures (the Subaccount for Energy Supply and Demand), which succeeded the Petroleum Special Account.

13.3.2. *Transportation fuel tax revenues and their uses*

Prior to the termination of the Special Account for Road Improvement (FY2008 and earlier years) all revenues from the taxes related to transportation fuel had been used as special-purpose revenues for road construction and improvement along with automobile-related taxes such as the Automobile Weight Tax and the Automobile Acquisition Tax. For the taxes related to transportation fuel and automobile-related taxes, the relationship between benefits and burden is relatively clear, as mentioned earlier. Therefore, taxpayers have accepted them with little resistance.

The share of taxes related to transportation fuel in the revenue sources for road construction and improvement is 46.1% (FY2005 budget). The automobile-related taxes (for central government plus the local governments), which are the combination of this and the "automobile body taxes" that include the Automobile Acquisition Tax and the Automobile Weight Tax, amount to 61.4%. Furthermore, if the Automobile Tax, which is a general revenue source of the local governments, is added to the automobile body taxes, then the share of automobile-related taxes in road investment becomes 81.9% (FY2005 budget). That is, except for FY1995 when road-related investment reached its peak, the share of the burden associated with automobile use on the beneficiaries (users) is 60 to 80%. In contrast, in European countries, the taxes related to transportation fuel often exceed the cost of road construction and improvement. Therefore, they are well regarded as they satisfy the beneficiary pays principle. In the case of European countries, the portion that exceeds the cost of road-related projects can be interpreted as a pseudo-environmental tax, which implies that automobile users bear part of the social cost.

Environmental issues faced by roadside areas have become a social problem, and budgets for addressing such issues have been created since FY1998. These budgets are allocated to, for example, projects for improving the roadside environment by implementing measures (such as the use of low-noise pavement) for roadside areas where the level of noise or air pollution exceeds environmental standards. The budgets have been used recently for infrastructure improvement for monorail and streetcar systems, improvements for optical fiber networks along major roads, a project to place electric power cables in underground common utility ducts, comprehensive support for the revitalization of towns, maintenance of roadside trees, assistance for the purchase of equipment to remove particulates from diesel engines, assistance for creating electronic toll collection lanes, and the construction of underground shopping arcades.

In December 2005, the central government issued the Basic Guideline on the Reassessment of the Revenue Sources Designated for Road Projects, and the system governing revenue sources designated for road projects was abolished on April 30, 2009. The taxes related to transportation fuel, which had provided revenues to the Special Account for Road Improvement and the Special Account for Airport Improvement as special-purpose revenue sources, became general revenue sources. The two abolished special accounts were succeeded by the Subaccount for Road Improvement and the Subaccount for Airport Improvement of the Special Account for Infrastructure Development, which began to receive necessary funds for road and airport projects from the general account. The FY2009 revenue budgets for these subaccounts were 2.48 trillion yen and 0.53 trillion yen, respectively. Among them, funds from the general account were 1.64 trillion yen and 0.14 trillion yen, respectively. For FY2008, funds from the General Account to these subaccounts were 2.06 trillion yen and 166.2 billion yen, respectively. Therefore, funds into these subaccounts decreased after the elimination of the special-purpose revenue sources.

13.3.3. *Revenue from the tax for promoting the development of electric power facilities and its use*

The revenue from the Tax for Promoting the Development of Electric Power Facilities had entered the Special Account for Promoting the Development of Electric Power Facilities (Electricity Special Account) and has contributed to the expansion of the capacity to supply electric power by promoting the development of power plants centering on nuclear power generation. With this revenue, the structure of electric power supply, which relied heavily on petroleum, has shifted toward nuclear power. In the FY2005 budgetary standards, the share of the budgets related to the development of nuclear power generation facilities in the Electricity Special Account reached approximately 76%.

It is notable that funds from the Subaccount for Power Plant Locations of the Electricity Special Account have been used to raise people's receptivity to nuclear facilities, being designated as subsidies to promote economic and social development in regions where nuclear power plants have been located. As a result, potential risks of radioactive pollution caused by nuclear power plants or risks associated with radioactive waste disposal have increased. Also, the Subaccount for Diversified Electric Power Generation, which is intended for a shift away from petroleum-based power generation, has been used to develop a nuclear fuel cycle or construct hydropower generation facilities, increasing potential risks and social costs.

In addition to the development and promotion of nuclear power generation and hydropower generation, the development and promotion of new energies, such as solar power generation, power generation from waste, wind power generation, and geothermal power generation, have been included since the second half of the 1980s. The share of such development and promotion of new energies in the revenue from the Tax for Promoting the Development of Electric Power Facilities increased from 13.3% in FY1990 to 18.3% in FY2000, but decreased to 10.2% in 2005.

The Electricity Special Account and the Petroleum Special Account contribute to the development of petroleum-substituting energies at the level of secondary energy (electric power) and primary energy, respectively. Therefore, the two accounts play separate roles. However, intrinsically, it is difficult to separate primary energy from secondary energy. In fact, as the function of the two accounts to promote new energies was strengthened, there were problems relating to their functions becoming redundant in recent years. The two accounts were thus merged in FY2007 into the Special Account for Energy Measures.

The functions, to ensure energy security and promote the development of nuclear power generation, have been continued even after the two accounts were unified, unchanged from the pre-unification period. In short, the basic framework of accounting administration has not significantly changed for Japan's energy tax system. The reason is that, despite the recent abolishment of special-purpose revenue sources, funds enter related special accounts (e.g. the Subaccount for Road Improvement and the Subaccount for Promoting the Development of Electric Power Facilities) through the general account, and thus the energy tax system remains one that is inclined to the securing of funds for the purpose of promoting development.

13.4. Reform of Energy Taxes and Fiscal Accounts toward a Low-Carbon Economy

What are conditions and issues for reforming Japan's energy tax system toward one that promotes a sustainable low-carbon economy?

One way to realize such a tax system is to change the taxation mechanism so that it encourages low-carbon energy use. This means changing the tax base from the amount of energy to the amount of carbon contained in that energy. In fact, there is a plan for reforming the existing energy taxes with an "anti-global warming tax". Under the plan, on the one hand, the provisional tax rate of the Gasoline Tax is lowered, but, on the other hand, part of the tax comes from carbon-based taxation in upstream sectors. The current tax rate for gasoline is the highest among the energy taxes at 55.84 yen/l if the Petroleum and Coal Tax, an upstream tax, is included. Under the plan, despite the introduction of a carbon tax, the tax rate for gasoline is 50.84 yen/l, which is lower than the current tax rate because of the reduced provisional tax rate. Therefore, the plan is politically feasible. However, some criticize that under such a reform it is difficult to expect improved incentives and increased tax revenue from the carbon tax, and that the introduction of an upstream carbon tax merely increases the cost burden on industries.

Energy is a good with a considerably low price elasticity of demand, at least in the short run. A substantial tax rate is thus needed in order to induce energy conservation with only tax rate adjustments. However, a high rate of taxation can significantly increase the burden not only on firms as energy

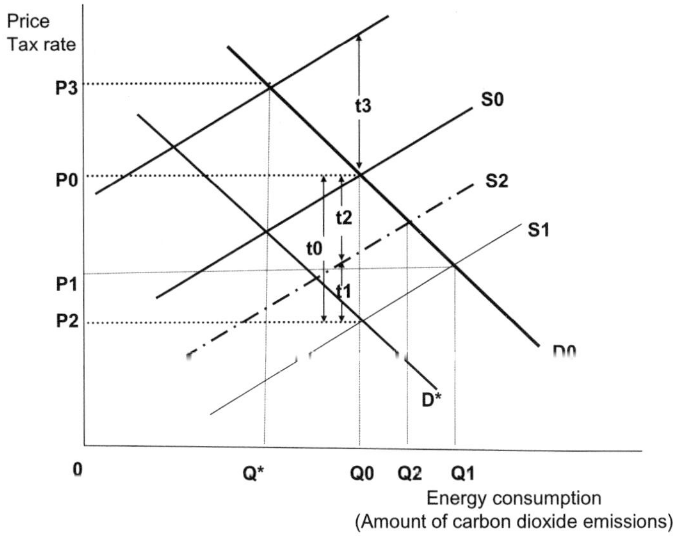

Fig. 13.3. Energy tax and energy consumption.

users or sellers, but also on ordinary consumers, which politically makes it less feasible. Therefore, taking social receptivity into account, taxing at a low rate and dedicating the use of the tax revenue to the promotion of low-carbon energies can be one option for achieving the minimization of impacts on general consumers and increasing the capacity of the tax system to induce low-carbon activities.

These observations can be summarized with Fig. 13.3 as follows. As in Fig. 13.2, the curve S_2 is the private supply curve for energy, resulting from the addition of costs (t_1) of, for example, ensuring energy security, which should be paid by energy-related industries, to the private marginal cost of supplying energy (S_1). Given that D_0 is the demand curve for energy, the market equilibrium is achieved at Q_1 and Q_2 is the corresponding quantity of energy consumption and the level of pollution associated with it.

Given that t_2 is part of the energy tax performing the internalization of externalities caused by energy use, the social marginal cost curve for energy is S_0, that is, the result of social costs (t_2) added to the private marginal cost curve S_2. One can interpret that the level of pollution was reduced from Q_2

to Q_0 due to t_2. Now, suppose that the desired level of energy consumption and carbon dioxide emissions is set at Q^*. To achieve this, as mentioned earlier, a new tax t_3 needs to be created, which is in line with Baumol and Oates' idea. The introduction of the new tax t_3 leads to an increase in the energy price from P_0 to P_3.

It is obvious from this observation that it must be clarified whether the main objective of the energy tax t_1 or t_2 is the internalization of externalities or the promotion of low-carbon energy use. If the objective is the latter, there is an option of actively using the revenue from the existing energy taxes mainly for promoting low-carbon energies. If the revenue is used for environmental measures such as the provision of subsidies to firms with low-carbon technological innovation capabilities, and if, as a result, low-carbon energy use progresses to an extent that the demand curve for energy shifts from D_0 to D_1, then it is possible to achieve Q^* without an increase in energy price.

Among Japan's seventeen Special Accounts in 2010 not one has specialized in low-carbon energies, which should be noted in discussing the energy tax system and reform of related accounts toward low-carbon activities. The subaccounts for road improvement, river improvement, and harbor improvement of the Special Account for Infrastructure Development, which is connected to public projects, and the Special Account for Energy Measures do not have a mechanism for executing projects that emphasize the protection of the environment or the promotion of low-carbon energies. The energy taxes used to be special-purpose revenue sources; that is, the use of the revenue from these taxes was pre-designated. Today, however, the revenues from the energy taxes first become general revenues, retaining no obvious sign of being special-purpose revenues on the surface, as shown in Fig. 13.4. At the same time, it is true that, in the end, little change was made to the administrative mechanism of relevant special accounts, which are the "expenditure windows" mentioned earlier.

Based on the discussion thus far, let us specifically examine several issues on the possibility of comprehensive tax reform toward secured energy sustainability.

First, funds for measures for structural improvement of energy supply and demand in the Subaccount for Energy Supply and Demand of the Special Account for Energy Measures could get separated and become a source

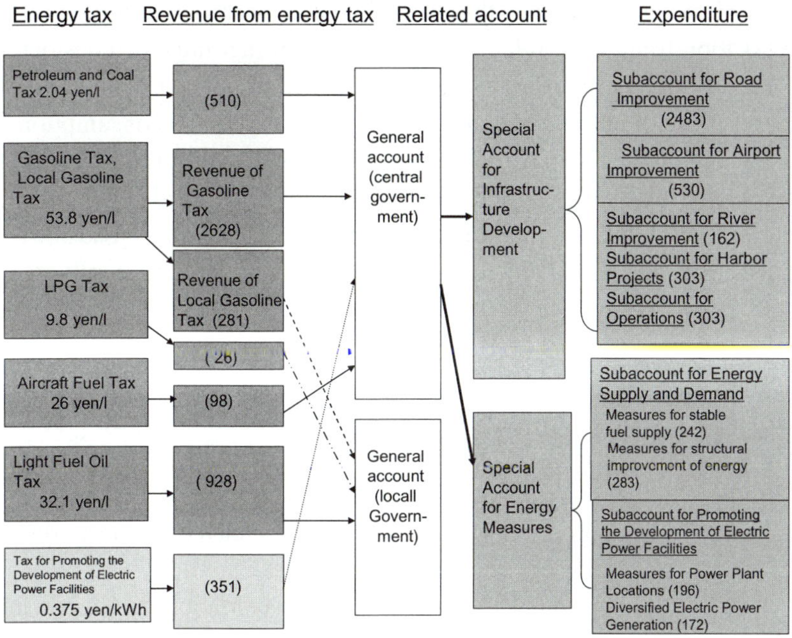

Fig. 13.4. Revenues from energy taxes and their uses from related accounts (after 2007). (unit: billion yen, budget base: FY2009).

of funds for a "Special Account for Promoting a Low-Carbon Economy" (tentative name). That is, the existing Subaccount for Energy Supply and Demand gets specialized in providing funds for ensuring the security of petroleum resources. To maintain consistency, it is appropriate to reallocate funds for structural improvement of energy supply and demand, which are currently used mainly for energy conservation projects or the development of petroleum-substituting energies, to the promotion of low-carbon energy.

Second, funds for diversified electric power generation under the Subaccount for Promoting the Development of Electric Power Facilities of the Special Account for Energy Measures could get separated and become a source of funds for the aforementioned Special Account for Promoting a Low-Carbon Economy. At the same time, the Subaccount for Promoting the Development of Electric Power Facilities gets specialized in the implementation of measures for power plant locations to promote a stable supply of electric power.

Third, issues surrounding the promotion of low-carbon activities through the taxation framework related to transportation fuel must be considered. Road construction financed by the taxes related to transportation fuel and expanded use of automobiles not only have promoted the consumption of fossil energies, but also have been major causes of damage to natural ecosystems, scenic landscapes, agriculture, forestry, fisheries, local environmental deterioration including vibration, noise, and exhaust gas which roadside residents suffer, and global environmental problems through increased greenhouse gases. For instance, according to Koyama and Kishimoto (2001), the total social cost (external cost) of automobile traffic in Japan is estimated to reach 19.7455 trillion yen (low estimate) to 60.3689 trillion yen (high estimate) annually, or 4.0 to 12.3% of GDP. Also, approximately 20% of total carbon dioxide emissions come from the transportation sector. Given the enormous external costs resulting from the consumption of transportation fuel, it would be appropriate that part of the taxes related to transportation fuel should function as environmental taxes.

A full-scale conceptual plan to reform transportation fuel taxes into carbon taxes is an issue for a longer-term examination. Instead, this paper attempts to consider a case where the promotion of low-carbon energy is conducted through the use of revenues from the Gasoline Tax. Though the basic tax rate of the Gasoline Tax is 28.7 yen/l, an additional 25.1 yen/l has been applied since 1979 to further increase the source of funds for road projects, making the total provisional tax rate 53.8 yen/l. This paper hypothesizes the use of the 25.1 yen/l added to the basic tax rate as a source of funds for the Special Account for Promoting a Low-Carbon Economy.

A reform proposal could see the Special Account for Promoting a Low-Carbon Economy being created by integrating revenues from the Petroleum and Coal Tax, the Tax for Promoting the Development of Electric Power Facilities, and the Gasoline Tax (actually, the National Gasoline Tax, with the Local Gasoline Tax being excluded) and budgets related to the promotion of a low-carbon society which are scattered within the central government. Based on the FY2009 budgetary standards, the sum of 282.5 billion yen from the funds for measures for structural improvement of energy supply and demand under the Subaccount for Energy Supply and Demand, 0.17 trillion yen from the funds for diversified electric power generation under the Subaccount for Promoting the Development of Electric Power Facilities, and

1.22 trillion yen from the National Gasoline Tax (the amount from the provisional tax rate minus the amount from the basic tax rate) totals 2.36 trillion yen. This is assumed to be the revenue source for the Special Account for Promoting a Low-Carbon Economy.

In addition, the budgets for environmental protection managed independently by central government agencies have been around 1% of the entire national budget. In these budgets, elements related to the promotion of a low-carbon economy exist in the Ministry of Agriculture, Forestry, and Fisheries, the Ministry of Economy, Trade and Industry, and the Ministry of Land, Infrastructure, Transport and Tourism as well as the Ministry of the Environment. The Petroleum and Coal Tax and the Tax for Promoting the Development of Electric Power Facilities are managed by the Ministry of Economy, Trade and Industry; the Gasoline Tax is managed by Ministry of Land, Infrastructure, Transport and Tourism. By placing under unified management the budgets related to the promotion of a low-carbon society, which are separately managed by different ministries, and making them a revenue source for the Special Account for Promoting a Low-Carbon Economy, overlapping funds or inefficient management can be prevented and can thus contribute to the integration of environmental policies.

This type of plan for restructuring revenue sources related to the promotion of a low-carbon economy will become a driving force in the future in making the entire national fiscal policy inclined toward low-carbon activities.

13.5. Conclusion

This chapter summarizes the current state of the energy tax system and related fiscal accounts and their issues from the standpoint of a shift toward a low-carbon economy and examines how the energy tax system should be reformed and how the relevant fiscal accounts should be restructured in order to realize such an economy. As a realistic solution, this chapter proposes the establishment of the "Special Account for Promoting a Low-Carbon Economy" and a conceptual plan to make the energy tax system and relevant accounts inclined toward the promotion of low-carbon activities. As mentioned in the main text, the Special Account for Promoting a Low-Carbon Economy could bring rigidity in fiscal management and thus is not desirable for permanent implementation regardless of its objectives. Also, it is pointed out that this conceptual plan shows the significance of reorganizing the energy tax system, which has long been vaguely positioned as special-purpose revenue sources or special-purpose taxes, and that the plan is feasible within the current budgetary system.

This chapter also considers funds managed separately under different accounts with different characteristics that can be used for promoting a low-carbon economy and proposes the creation of an account that integrates and manages them.

Today's measures toward a low-carbon society are diverse and complex, and any solution strongly requires coordination among different policy fields and actions overarching them. Policies on energy and the environment become effective through integration with policies on energy, transportation, industry, science and technology, trade, regional issues, and national land planning. In this context, a fiscal account aimed at realizing a low-carbon economy is expected to advance policy integration that promotes a sustainable low-carbon economy, but not the interests of individual government agencies.

References

Koyama, S. and Kishimoto, A. (2001). Estimation of External Cost by Automobile Transpotation in Japan, *Study of Transpotation Policy*, 4 (2), 19–30 (in Japanese).

OECD (2010). *Environmental Related Tax in OECD Countries: Issues and Strategies*, Paris.

Yokayama, A., K. Ueta and K. Fujikawa (2000). Green Tax Reform, *Environmental Economics and Policy Studies*, 3 (1), 1–20.

Zaisei Chosakai (Fiscal Research Group) (1990–2009). *The National Budget*, Tokyo: HaseShobo (in Japanese).

Chapter 14

LOW-CARBON ECONOMIC INDICATORS

*Hideaki Itoh**

14.1. Introduction

A society which reduces emissions of carbon dioxide (CO_2) is referred to as a "low-carbon society", and this phrase is generally used in the mass media. However, it is not adequate to use this phrase as a technical term because of its lack of accuracy. The total amount of carbon in nature is constant under the law of conservation of mass, and hence the amount of carbon in nature is maintained as a constant in various forms such as carbon as an elementary substance or inorganic or organic compounds. On the other hand, the increase in carbon amounts in society can be calculated as positive if underground resources such as petroleum or natural gases are incorporated into society. Even in such cases the incorporated carbon would be used not only as fuel, but also as important carbon sources for plastics or polymers. Such fixed solid carbon is not a direct cause of global warming.

One of today's crucial subjects all over the world is to decrease the emission of carbon dioxide, which is regarded as the main cause of global warming. Methane and fluors are other chemical substances which have higher global warming potentials, so the total amount of such "greenhouse gases" (GHG) should be reduced in order to protect against global warming. From this point of view, "low GHG society" would be a better expression instead of "low carbon society". However, the technical term "low carbon society" is used in this chapter to unify the key phrase throughout this book.

The economics needed to create a low carbon society is diverse and a variety of approaches are introduced in this book. It is important to search

*Hideaki Itoh, Professor Emeritus of EcoTopia Science Institute, Nagoya University.

for the shortest path to the low carbon society and make a technological and economic strategy which involves less international conflict. Sharing a common image for a low carbon society requires a comprehensive discussion of international issues including global warming, environmental pollution, biodiversity loss, resource depletion, the food crisis, poverty problems, regional disparities, regional conflict, etc.

The EcoTopia Science Institute of Nagoya University is proposing an assessment indicator called the "EcoTopia Index" (*ETI*) to discuss the above issues comprehensively from diverse aspects. In the present paper, the elucidation of an image for the low carbon society is described as well as the relationship between the EcoTopia index and the assessment of the low carbon economy.

14.2. Image of a Low-Carbon Society

The present status and future prospect for global warming are worth investigating to make the image of a low carbon society clear. The average temperature of our planet is anticipated to increase by 2–4°C by 2100 with increasing CO_2 concentrations in the atmosphere. The reason has been discussed by many specialists including some skeptical of global warming.[1] Excess mining and utilization of underground carbon resources in a short period of time would increase the greenhouse effects as human activities are developed. CO_2 generation by incineration or combustion of carbon compounds originating from plants is considered to be carbon neutral on account of CO_2 absorption by photosynthesis. However, carbon offset effects by this absorption is not so actual as expected because the afforestation or forest management needs the consumption of energy supplied by the underground carbon.

It is difficult to estimate accurately how the average temperature of the earth will be increased by CO_2 emissions, because many factors will influence climate change over time. In any case, reducing temperature increases due to CO_2 emissions originating from fossil resources is essential in order to prevent the environmental risk due to the global warming.

[1]Intergovernmental Panel on Climate Change; Emori *et al.* (2008); Watanabe (2010).

Regarding the environmental impacts of global warming on ecology and human activities, the following direct and indirect phenomena are predicted: melting of glaciers because of rising temperatures, sea-level rise, acidification of seawater, a northward/southward distributional shift of living things, desertification, forest destruction, the heat-island phenomenon, abnormal climate (increase in hurricanes, typhoons, tornados, localized heavy rains, floods or warm winters), biodiversity loss, etc. The causality predicted should be verified firmly by exact data. It is required for establishing a sustainable society, however, to prevent catastrophic destruction, which leads to a "dystopia" as a result of serious global warming. Therefore, a low carbon society would minimize the risk and disaster against ecology and human society, even if global warming symptoms develop.

Such efforts for a low-carbon society are required to maintain our minimum peaceful and wealthy lives in harmony with nature over time, even though it does not necessarily guarantee a utopia of mankind. Creation of a low carbon society with reduced GHG would not be attained by a limited number of individuals, local communities or nations, but by participation of all people and nations in the world. Both a common awareness of global warming risks and an international agreement for equal responsibility to CO_2 reduction are indispensable for the low carbon society, and therefore this is the most important agenda at the IPCC.

14.3. Low Carbon Society and Resource Management

Prevention of global warming is related to many kinds of resource problems. In order to reduce CO_2 emission, we need to search for alternative energy resources instead of fossil fuel, and to attain efficient conversion, transport and storage of energy as well as the reduction of total energy consumption. As for alternative energy, the utilization of natural or renewable energy (biomass, wind power, wave power, geothermal energy, etc.) and nuclear power is discussed. Intensive development of technology is required for efficient energy conversion, transportation and storage. Moreover, we have to solve another subject to keep the impartiality of convenience for energy consumption and financial responsibility, for which the paradigm shift of people's sense of value is required toward the creation of sustainable society, and the innovation of awareness for the energy consumption from every field of politics, economy, education, culture or ethics.

The security of sustainable clean energy resources is closely related to that of material resources such as industrial raw materials, biomass and food. In particular, the production of biomass energy such as bio-ethanol or bio-diesel competes against the supply of plant foods such as corn or sugar cane, so that it will drive the production of human foods or cattle feed out of the farmland. Nonrenewable petroleum is an important energy resource and also a raw material for plastic and fibrous polymers, which leads to the trade-off relationship between its resource distribution to energy or industrial raw materials. Furthermore, the usage of clean energy has an influence on the manufacture and consumption of many industrial products, i.e. the high energy efficiency of material production by technology innovation and the effort to save energy of advanced commercial goods in the society. Detoxification and re-use of wastes will facilitate thermal recycling and material circulation and reduce the environmental burden.

It is clear from the above considerations that the three kinds of resources, i.e. energy, material and biomass are closely connected in a trade-off relationship from the viewpoint of the utilization of resources. Figure 14.1 shows the correlation among resources derived from nature (energy, material

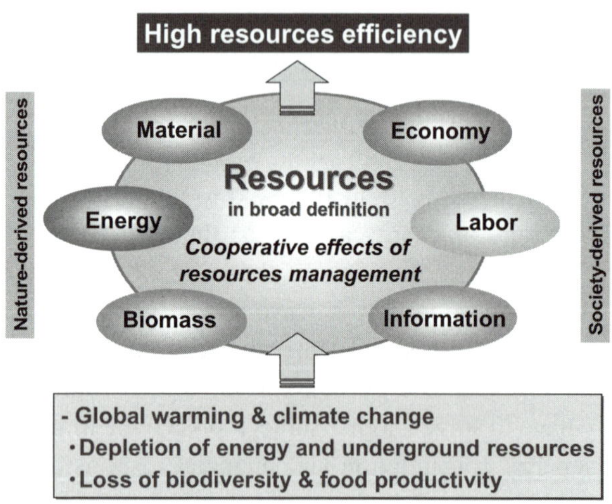

Fig. 14.1. Cooperative effects of resources management.

and biomass resources) and the society (economic, labor and information resources). The society-derived resources play the role of managing the distribution and investment of the nature-derived resources. The integration of both resource categories enables the cooperative effect of resource management. The present worldwide issues such as (1) global warming/climate change, (2) depletion of underground resources and (3) biodiversity loss and food crises should be discussed from the aspect of such resources management. Realization of a low carbon society will be attained eventually in the framework of simultaneous solutions of resource management in the broad definition described above.

Here we will introduce a new assessment index of "resource circulation efficiency", which is an indicator of sustainable or intergenerational resource supply. This index evaluates the degree of circulation of the nature-derived resources to the society-derived resources, and thus the sustainable resource circulation can be expressed by the integration of effective utilization of energy, material and biomass resources over a long period of time. Therefore, the goal of resource management is to attain the maximum resource circulation efficiency.

14.4. A Comprehensive Indicator of "Well-Being"

It is important to understand the multi-spheres structure of the global environment in considering the relationship between the environment and human beings. Figure 14.2 illustrates that the sphere of human life is a part of a bigger ecological sphere, and the latter is part of still bigger global spheres (atmosphere, hydrosphere and geo-sphere). Human beings lead a mass life as biological humankind (Homo sapiens) and also lead a social or family life as individuals. Therefore, human life encompasses both natural ecology systems and human social systems, and therefore a consistent human life sphere should be formed in harmony with both systems. This means that the coexistence of the natural environment and the social environment must be established by ceaseless efforts to preserve the natural environment and to reduce the environmental load from the human society.

For the sound preservation and development of the human life sphere, clean water, soil, air, energy and safe food are indispensable as a demand

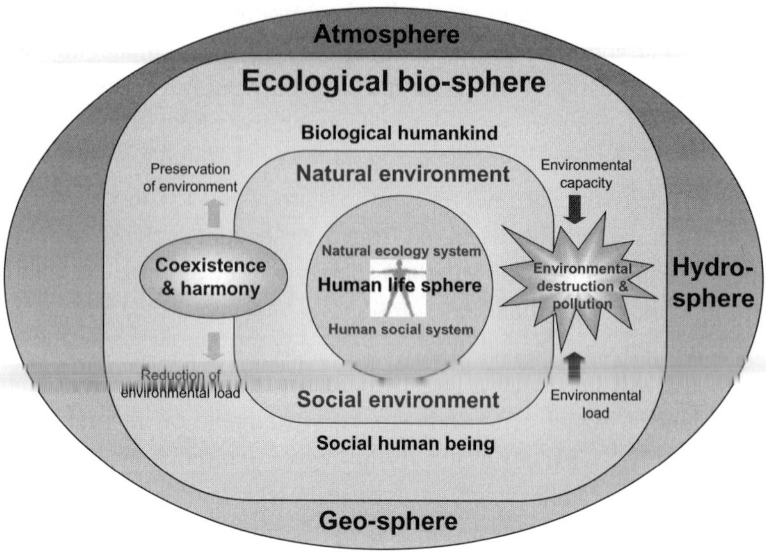

Fig. 14.2. Relationship between global environment and human life sphere.

for the natural ecological sphere, while the social infrastructure of industry, transformation, communication, construction, medical care, etc. is essential as a demand for the human life sphere by the management of both nature-derived resources and society-derived resources. The harmony and optimization of both spheres will guarantee safe, secure and comfortable human life. If environmental loads in nature exceed its capacity, however, the human life sphere will be destroyed or polluted to bring about so-called "dystopia".

As described above, a simultaneous solution of resource issues is required to establish the low carbon society, where we need to share an ideal image of human society based on the intense consciousness of the environment. Here let me put a basic question. "What is the well-being of humankind?" The word "well-being" could be replaced by "happiness" or "welfare", which is not only confined to personal health or happiness, but also represents the total happiness or welfare of the society for many generations, and indicates the coexistence or harmony with nature ecology systems. It is a matter of course that individual well-being is dependent on his/her sense of values, i.e. material/physical satisfaction and spiritual or cultural satisfaction. Now is the time to focus on the global environment, so we need to expand the concept of "well-being" or "happiness" to harmonize the natural and social environment issues and to create a sustainable well-being of the human life sphere.

14.4.1. *The EcoTopia Index*

"Ecotopia" is a compound word of "ecology" and "utopia", which means an ideal human society in harmony with the ecological environment. Originally "eco" comes from the Greek *"oikos"*, and "topia" comes from *"topos"*, both of which include the concept of the space where human beings live. Since the publication of the two books *Ecotopia* (1975) and "Ecotopia Emerging" (1981) by Callenbach, the word "ecotopia" has been well-known. These fiction books are about the independence movement of an ecotopia state from U.S.A. and the report on the state by a journalist after independence. They include interesting and suggestive stories about the new policies of the "ecotopia" society, which achieved independence from countries suffering from environmental pollution and destruction.

The main policies in the ecotopia state described by Callenbach are as follows: (1) environmental policies: conservation of wood, river and ocean; construction of smaller cities; abolition of high-rise buildings; de-motorization (use of trains or bicycles); (2) energy policies: use of solar cells, wind power, thermoelectric power and energy saving; abolition of nuclear power plants; (3) material policies: use of woody materials; no use of chemically synthesized products (plastic or textile goods); use of biodegradable plastics; prevalence of recycled goods; (4) life-style policies: natural food without chemical fertilizer; no fast-food; reducing use of the air-conditioner, etc. Apart from whether such an ecotopia could be realized or not in the present world, the image of an ideal society, which Callenbach drew in the last century, will get some sympathy from people in the 21st century. However, such a utopia movement is anticipated to miscarry and fail in fact, as verified by various similar cases in history.

The EcoTopia Science Institute at Nagoya University was founded in 2005 to pursue research and development for the establishment of ecotopia, which is interpreted in a modern style as "a sustainable society in harmony with the environment". This institute aims at basic research on nano-materials science, energy science, environmental research and information and communication sciences, and their integrated and interdisciplinary research including natural, social and human sciences. In order to make the goal clear and to promote research activities, the institute proposed the assessment indicator[2] called "EcoTopia Index" (*ETI*) in 2007. This index is defined as follows:

$$(\text{EcoTopia Index}) = (\text{Quality of Sustainable Life})/(\text{Environmental Load}) \tag{14.1}$$

According to this definition, the *ETI*, which is regarded as well-being or happiness in modern society, is determined by the ratio of the quality of sustainable life (*QoSL*) to the environmental load (*EL*) caused by human activities.

The definition of the numerator (*QoSL*) and the denominator (*EL*) can be explained as well as the reason for equation (14.1). As described in 14.3, "well-being" in deep consciousness of the environment is not assessed by the conventional quality of life (*QOL*) of a private person, but by a sustainable one over multiple generations. The assessment category of the numerator is considered as a function of many factors, but can be classified broadly into three assessment items, i.e. human (psychological/physical), social and economic items. The whole assessment for the *QoSL* can be visualized by the three axis radar-chart as shown in Fig. 14.3(a), and can be evaluated numerically by the following summation of the assessment items.

$$(QoSL) = \Sigma(\mathrm{s}_i(QoSL)_i) \tag{14.2}$$

where $(QoSL)_i$ shows the relative value of *QoSL* for item *i* corresponding to the three axes in Fig 14.3(a). s_i means the respective weighting coefficient for sustainability, a correction factor for the people's action psychologically induced for the environment.

The assessment category of the denominator (*EL*) consists of three items, i.e. global warming, ecological pollution and resource depletion as shown in Fig. 14.3(b), and can be evaluated numerically by the following summation

[2] Matsui (2007).

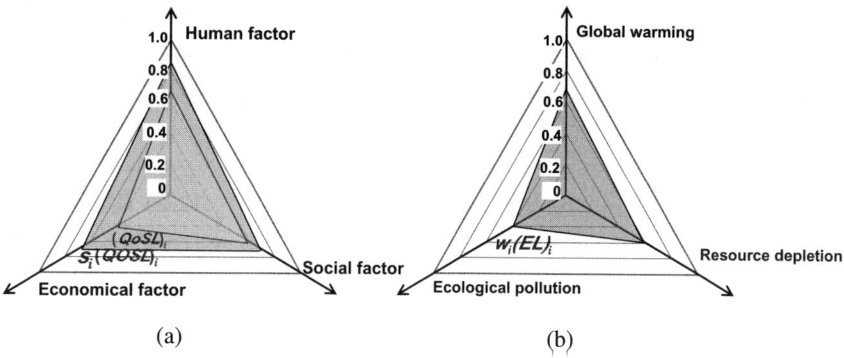

Fig. 14.3. Assessment categories and items of EcoTopia Index. (a) quality of sustainable life (*QoSL*) and (b) environmental load (*EL*). (The hatched area is virtual assessment results.)

of the assessment items.

$$(EL) = \Sigma(w_i(EL)_i) \tag{14.3}$$

where $(EL)_i$ shows the relative value of *EL* for item *i* corresponding to the three axes in Fig. 14.3(b), and w_i is the corresponding weighting coefficient. The evaluation of *EL* can be supported by existing databases, e.g. the carbon footprint for global warming, the ecological footprint for ecological pollution and the ecological rucksack for resource depletion.

Combining equations (14.1)–(14.3) yields the formulation of (*ETI*) as follows[3]:

$$(ETI) = \Sigma(s_i(QoSL)_i)/\Sigma(w_i(EL)_i) \tag{14.4}$$

Careful discussion is necessary on the methodology for formulation and visualization of (*ETI*). The mono-axis evaluation method is simple and evident, but there is a difficulty in the conversion of units and evaluation of the weighting coefficients. The multi-axes evaluation method enables a complex assessment, but makes the relative evaluation difficult among different items. A formulation of (*ETI*) as equation (14.4) gives a mathematically complete expression. However, quantitative application of equation (14.4) to actual assessment cases is likely to generate technical difficulties for numerical analysis, so minor deformations should be

[3]Itoh *et al.* (2008).

allowed. It is especially important in the general interpretation to avoid the duplication or omission of assessment items. There are some items which are unsuitable for quantification when we do the qualitative analysis according to the conceptual equation (14.1) and the quantitative analysis according to the numerical equation (14.4). They should be combined and discussed.

The EcoTopia Index has characteristics different from many other existing assessment indicators on environment, economy, society or health. First, the concept of sustainability is introduced into the conventional *QOL* by the addition of social and economic assessment axes to the personal *QOL*. Furthermore, the weighting coefficients concerning people's actions psychologically induced for the environment is multiplied for correction of $(QoSL)_i$. Second, the final assessment of (*QoSL*) is performed in reference to the global environmental load. The comparison of (*ETI*) with other existing assessment indicators, e.g. ecological foot print (*EF*), human development index (*HDI*), environmental efficiency index (*EEI*), factor X, genuine progress index (*GPI*), environmental sustainability index (*ESI*), green GDP, human satisfaction measure (*HSM*), ecological rucksack (*ER*), and life cycle assessment (*LCA*) is described, along with as the mapping of (*ETI*) and other indicators, in the references.[4]

14.5. Application of EcoTopia Index to the Assessment of the Low Carbon Society

The EcoTopia Index is an assessment indicator based on global environmental issues. The influence of a low carbon society on the healthy cultural and economic life of human beings is evaluated by each assessment axis in Fig. 14.3(a), while the environmental load in the society is evaluated by the global warming (carbon footprint) axis in Fig. 14.3(b). Comprehensive assessment of the low carbon society will be performed more adequately by using (*ETI*) because the realization of the low carbon society is closely related to resource and pollution problems, as described in equation (14.2) and equation (14.3).

Assessment case studies on some projects for low carbon society by (*ETI*) have just started, so we have to await the outcome of future research,

---

[4]Hayashi *et al.* (2009a); (2009b).

especially on quantitative case studies. Here we introduce an example case study by Hayashi *et al.* (2009a, 2009b) and Hayashi (2008) on the production project of bio-diesel fuel (BDF) from palm oil plantation in Indonesia. BDF is a biomass resource which is an alternative to diesel oil (light oil) produced from petroleum. It is necessary to take the trade-off relationship into account among the assessment items of environmental load, so that CO_2 output calculated by (*LCA*) is used for the assessment of global warming, (*EF*) is used for the assessment of the environmental capacity of ecologically affected land area, and (*ER*) is used for the assessment of resource consumption. The scope definition for data acquisition of (*EL*) is very important. The case study supposes that palm fruits are transported by truck from plantations to local facilities, where BDF is produced from crude palm oil, and then BDF is transported to Japan by oil tanker. The reference object is selected for the case that crude oil is refined on-site in Indonesia to diesel oil and then transported to Japan by oil tanker. Figure 14.4 shows the results of the environmental load of BDF on three assessment items, where that of each item for diesel oil from crude underground oil (UDF) is supposed to be unity. The detailed calculation process for each item is described in the literature (Hayashi *et al.*, 2009a). It is evident in the figure that the (*EL*) of BDF is smaller in every assessment item than that of UDF. This is because the CO_2 emission amount of BDF would be smaller than that of UDF on account of the carbon neutrality of BDF.

The assessment described above ignores the environmental impact created by the land-use change when the palm wood plantation is cultivated

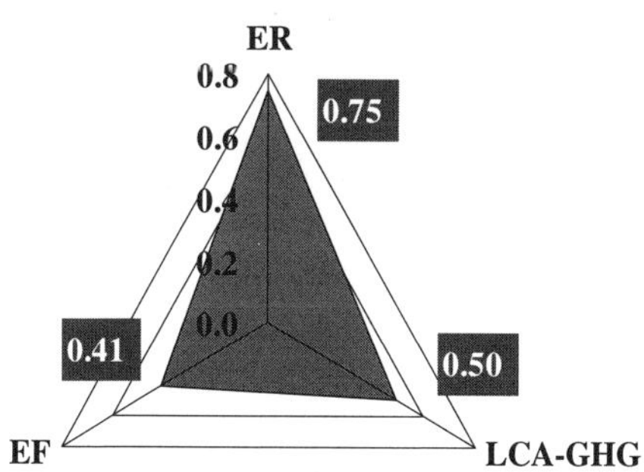

Fig. 14.4. Assessment result of (*EL*) by production of BDF (Hayashi *et al.*, 2009a). ER: Ecological rucksack, EF: ecological footprint, LCA-GHG: life cycle assessment of global warming gas emission.

from tropical forest. The difference in CO_2 absorption amounts from this cultivation should be considered from a long term perspective for sustainable clean energy development, because the development of the plantation from tropical forest has been actually implemented in Indonesia. Although the accurate estimation of this impact has not been performed, the total (*EL*) for BDF might be larger than that of UBF by this scope definition.

Quantitative evaluation of (*QoSL*) in the case study on the production of BDF requires various statistical data and questionnaires to inhabitants or citizens in Indonesia and Japan, which are now under investigation. The operation of the BDF production plant will have quite a few influences on people's economics and life styles, so that a comprehensive consideration should be done on the assessment for low carbon society.

Regarding the project for approaching a low carbon society, the present author assessed the CO_2 reduction effects by (*LCA*) for the production of cascade recycled board from used plastics and used paper/fibers (Itoh, 2010), as shown in Fig. 14.5. The recycled board is considered as an alternative for plywood board, which is imported from South-Eastern Asian countries and used in Japan in large amounts. The application of (*ETI*) assessment to this project is expected to clarify the multiple effects

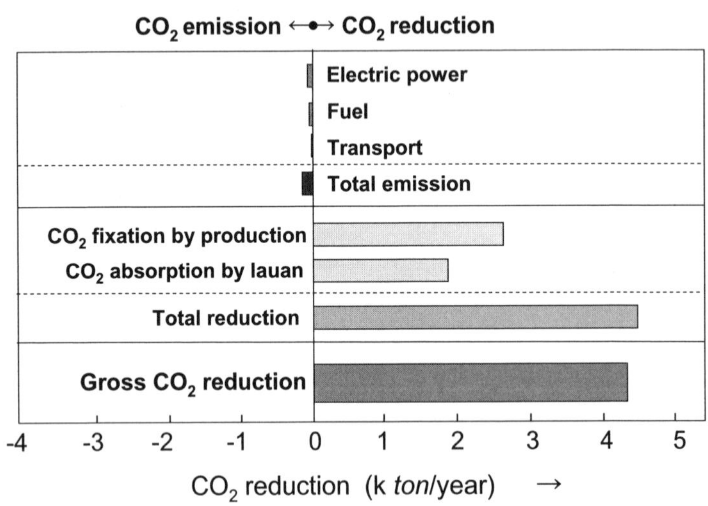

Fig. 14.5. Reduction effects of CO_2 emission by cascade recycle board. (LCA results supposing that 840 ton/year of recycle board is produced).

on global warming, resources depletion, and environmental pollution as well as economic, cultural and welfare effects, which leads to an integrated assessment of such interdisciplinary projects.

14.6. Assessment Indicator of the Low-Carbon Economy

In the low carbon society, all efforts should be undertaken to reduce global warming gas emissions, as described in equation (14.1). Actual policy is the reduction of these gases discharged from power plants and incineration facilities, and the reduction of fuel use such as gasoline, diesel oil, heavy oil and natural gas. The conversion from gasoline to electrical vehicles will result in a profound reduction of CO_2 emissions. Many kinds of policies for indirect reduction of global warming gases includes (1) the CO_2 absorption by increasing tree planting, forest conservation and the production of sea weed or algae, and (2) the production of bio-fuel (bio-methanol, bio-ethanol or bio-diesel oil) and power generation using bio-gases. More indirect reduction policies of global warming gases by power generation using natural renewable energy, nuclear energy and fuel cells will contribute to the realization of a low carbon society. In addition, great effects for the low carbon society will be attained by energy and material saving policies and by the development of efficient transportation, communication and information technology systems and social infrastructure.

Since the economy for a low carbon society is closely related to all the resource issues, the support of resource management by the society-derived resources is required, especially for economic activities like the CO_2 emissions exchange. Therefore, the increase in resource circulation efficiency reduces the total (*EL*) and increases the (*QoSL*), which will result in the increase of the (*ETI*).

A variety of action plans are designed toward a low carbon society, some of which have negative effects on economic activities, e.g. in the field of fossil fuel consumption. However, a big economic effect can be expected in the fields of developing bio-resources or natural energy and energy-saving technology. Figure 14.6 illustrates the transition scheme to ecotopia from various stages of the society. Generally speaking, large scale industries of food production, development of underground resources and clean energy development will be promoted, respectively, from an undeveloped

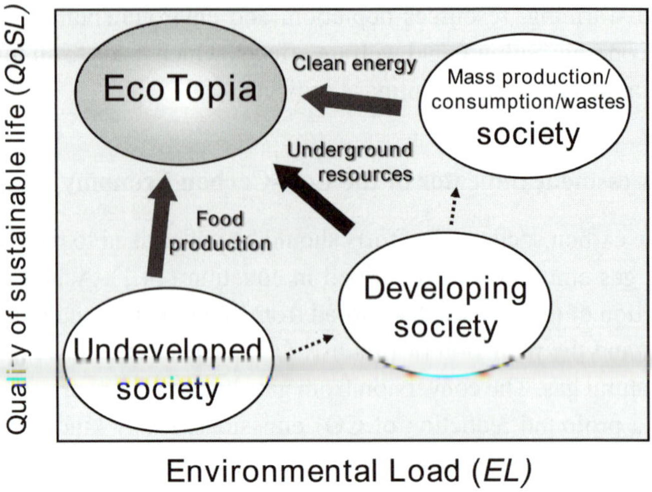

Fig. 14.6. Transition process to EcoTopia.

society, from a developing society and from a developed society (mass production/consumption/waste society) to ecotopia. International cooperation to increase the resource circulation efficiency will make a shortcut to ecotopia with the help of economic assessment by (*ETI*).

14.7. Conclusion

The assessment for creation of a low carbon society was discussed in this chapter from a standpoint of an environmental scientist or engineer. Starting from the question of what is a low carbon society, its relationship to current resource issues was described and then a new assessment indicator, the EcoTopia Index (*ETI*), was introduced with an emphasis on the importance of quality of sustainable life (*QoSL*) and environmental load (*EL*) in the definition and formulation of (*ETI*). In addition, a case study to apply the (*ETI*) for the realization of the low carbon society was presented. Finally the possibility of assessment of the low carbon economy was mentioned with reference to (*ETI*).

The significance of a comprehensive assessment by (*ETI*) will become evident as society confronts global issues on the environment and resources, though the methodology and case studies are premature. Interdisciplinary collaboration by many researchers is needed to elucidate the perspective and make a scenario for a low carbon society.

Acknowledgments

The author thanks all the colleagues in the EcoTopia Index project team in the EcoTopia Science Institute, Nagoya University for fruitful discussion, and expresses a special gratitude to Dr. Kiichiro Hayashi and Dr. Hideyuki Ito for their data provision on the case study.

References

Callenbach, E. (1975). *Ecotopia: The Notebooks and Reports of William Weston*. New York: Banyan Tree Books.

Callenbach, E. (1981). *Ecotopia Emerging*. Banyan Tree Books.

Hayashi, K. (2008). Environmental Impact of Crude Palm Oil Process Caused by Utilization of Bio-Wastes in Indonesia. (in Japanese) *Papers on Environmental Information Science*, 22, 257–262.

Hayashi, K., Ito, H., Koga, K., Tawara, Y., Katayama, A., Itoh, H., Arita, Y., Inoue, Y. and Kobayashi, N. (2009a). "Study on Assessment Indicator for Environment, Economics, Society and Human Factors — Indicator for Accomplishing Sustainable EcoTopia." (in Japanese) *Journal of Environmental Science*, 22(4), 281–289.

Hayashi, K., Ito, H., Koga, K., Tawara, Y., Itoh, H., Inoue, Y., Katayama, A., Kitagawa, K. and Matsui, T. (2009b). Study on Indicators for Sustainable EcoTopia Society from the Perspective of Environmental, Economic, Social and Human Aspects. *Proceedings CD, R'09 Twin World Congress and World Resources Forum*, Paper #228, Nagoya/Davos, Japan/Switzerland, Sept. 14–16.

Itoh, H., Tawara, Y., Hayashi, K., Koga, K. and Matsui, T. (2008). EcoTopia Index for Assessment of a Resource Circulation and Low GHG Society. In Takakuwa, S. (ed.), *Proceedings on Intern. Seminar on Manufacturing Management and Environmental Management* (S-2), (JSPS Asian CORE Program, Nagoya University, December 26), pp. 69–72.

Itoh, H. (2010). Production of Cascade Recycle Board from Used Plastics and Papers. *Proceedings of The 3rd International Conference on Engineering for Waste and Biomass Valorisation*, Paper #264C, Beijing, China.

Matsui, T. (2007). "Towards the Realization of EcoTopia 2020 (Summary)." (in Japanese). EcoTopia Science Institute, Nagoya University.

Watanabe, T. (2010). The Climategate Case — Forge Doubt of Global Warming Theory. *Chemistry*, 165(3), 34.

Chapter 15

LOW-CARBON BUSINESS MODEL

Satoshi Watanabe and Makoto Nishimura†*

15.1. Introduction

The Japanese government implemented the Eco-Point System from May 2009 to June 2010 to stimulate the demand of automobiles and household electrical appliances which have been stagnated since the Leman shock occurred in August 2008, by giving substantial incentives to consumers who bought those products during the period when the system was applied. The effectiveness of the Eco-Point System was remarkable and much greater than the expectation and those products sold extremely well despite the recession. In particular, the sales of flat-screen TV were outstanding and by far better than the sales of other products. This fact proves the hypothesis which the author has been proposing that a business model for low-carbon economy can be designed under the circumstances having four factors of (1) people's consensus of environmental improvement and energy saving, (2) regulation, (3) incentive and (4) competition. Designing the business model is very important to materialize low-carbon economy since a business is the only reliable methodology to do something consistently. This chapter focuses on how to set up this methodology. In a narrow meaning, we discuss about business realities related to our own experiences and Japan's energy savings as an application of business model, i.e. the concept of profit-generating structure. As a result, due to the difficulty in implementing a low-carbon economy under the mechanism of freely competitive market, we suggest that the internalization of the external regulations and

*Satoshi Watanabe, Assistant Professor of Graduate School of Economics, Nagoya University.
†Makoto Nishimura, Professor of Graduate School of Economics, Nagoya University.

incentives help to design and complete a business model which then derives the implementation of a low-carbon economy.

15.2. The Concept and Application of Business Model

The term "business model" is used in a very wide range of meanings including business itself, business system, business design, business style, profit model, marketing strategies and so on.[1] Firms and individuals behave rationally in order to acquire their targets of a sustainable profit, and hence, they will achieve their business targets without interruption. That is, the most suggestively realistic way to implement a low-carbon economy is to make business with the activities and targets of the aforementioned firms and individuals. In this way, our chapter employs the narrowest meaning of the business model as the so-called "a sustainable profit-generating structure" which derives the implementation of a low-carbon economy's business model. The reason is that firms and individuals trying to create businesses presuppose a scheme that raises profit over the long- or mid-term, and such scheme is the necessary and sufficient conditions for business to survive and grow.

From the viewpoints of a series of business model concepts, along with the representative revenue model given by Afuah (2004), Slywotzky and Morrison (1999) suggest the ideas of a business design, especially the phenomenon of a non-profit zone when firms and individuals could not earn profits from their activities. Then, Kagono and Inoue (2004) focus on the idea of a start from zero for a business system among many other business system strategies. Based on these references, this chapter discusses how firms and individuals can shift to the implementation of a low-carbon economy given their rationality in business targets and activities, then how to lead them to the profit zone in which they can achieve sustainable profits.

[1]According to references in the field of business model, there is a variety of meanings and concepts. However, in this chapter, we employ this concept in the narrowest meaning based on Slywotzky and Morrison (1999) and Kagono and Inoue (2004). That means, we employ the concept of business model as "a sustainable profit-generating structure."

15.3. Energy Saving Business Model of Japan after the Oil Crisis

It is often said that energy saving in Japan started from 1973, the 1st oil crisis. Figure 15.1 shows that in fact, the final energy consumption per GDP increased within 1960s–1970, but declined after the oil crisis in 1973. Average change rate of the final energy consumption per GDP within 1973–1983 was −3.1%, whereas the average change rate of GDP within same terms was 3.3%. These data show that Japan achieved both energy saving and economic growth. Energy saving after 1980s was slower than periods of just after the oil crisis, but the average change rate within 1973–2007 was −1.3%. In this way, Japan performed to the energy saving within 30 years after the oil crisis.

However, when we compare the average changing rates of final energy consumption per GDP by sectors during 1973 to 1983 (Fig. 15.2), it was −5.2% for industrial sector, −0.19% for private sector including household and service was −0.2% for traffic sector. About 10 years terms after the oil crisis, the energy saving of private and traffic sector were smaller than the industrial sector and, the rate within 1973–2007 was that the industrial sector

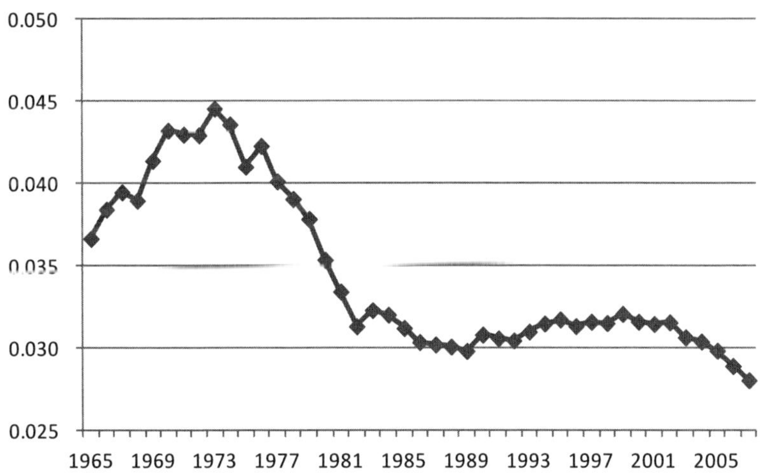

Fig. 15.1. Trend of final energy consumption per GDP, Japan. (Unit: 10^{18} J/trillion yen, 1965–2007.)

Source: Author's calculation with the data of Resources and Energy Agency of Japan (2012), Cabinet Office of Japan (2012), Institute of Energy Economics, Japan (2012).

was −2.3%, the private sector was −0.2%, and the traffic sector was −0.4%. Industrial sector compared to the private and traffic one performed to the energy saving. The energy efficiency of the private sector were improved just after the oil crisis, but were begun to worse after 1990s, especially became exacerbated than the level in 1973 after 1993. The trend of transport sector resembled one of the private sector until about 1997, but the average change rate of energy consumption per GDP after 1998 are improving at the level of −2.5%.

Furthermore, as showed by Fig. 15.3, relative coefficient between Japan and Japanese manufacturing industry shows strong positive relationship, 0.986. These facts mean that energy saving in Japan within 3 or 4 decades years contributed to the efforts by industrial, especially by manufacturing industries.

Figure 15.4 shows how the efforts by manufacturing industries contributed to Japan's energy saving. The oil dependency of manufacturing industries just after the oil crisis was about 55–58%, and it fluctuates around

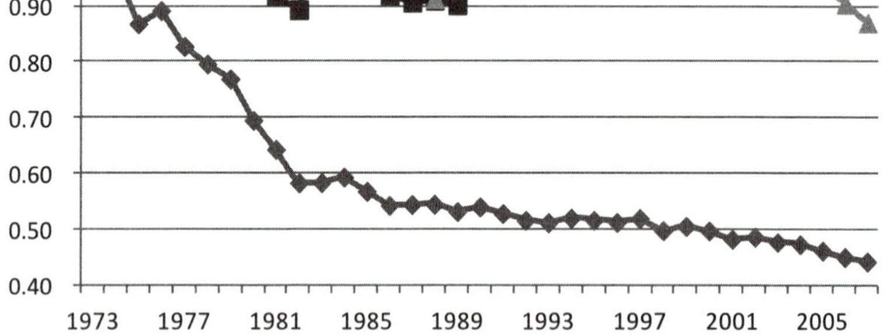

Fig. 15.2. Trend of sector's final energy consumption per GDP, Japan. (1973 = 1.00, 1973–2007).

Source: Author's calculation with the data of Resources and Energy Agency of Japan (2012), Cabinet Office of Japan (2012), Institute of Energy Economics, Japan (2012).

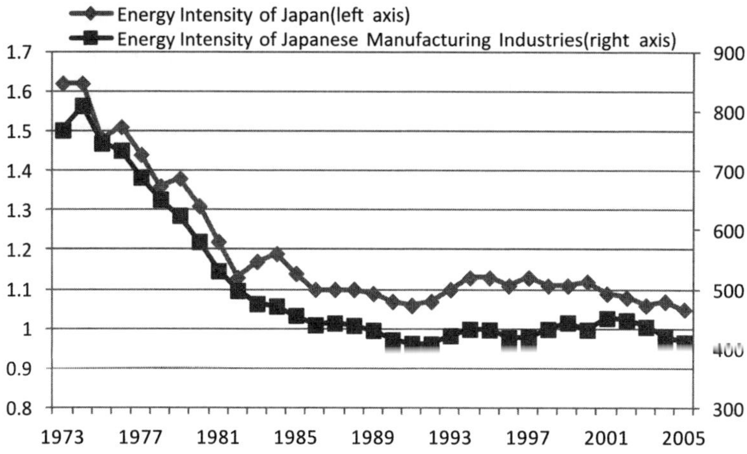

Fig. 15.3. Energy consumption intensities of Japan and Japan's manufacturing industry. (Unit: Gram of oil equivalent/yen (left axis), 10^4 kcal/million yen (right axis), 1973–2005.)

Source: Resources and Energy Agency of Japan (2012), Institute of Energy Economics, Japan (2012).

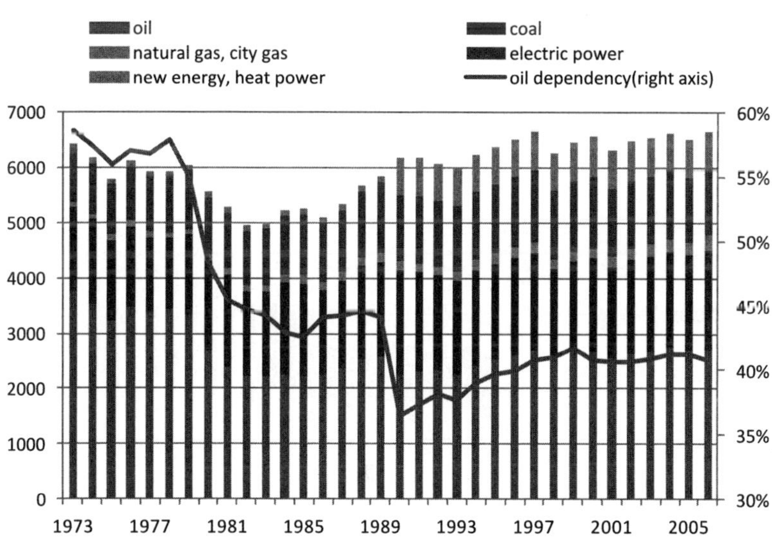

Fig. 15.4. Energy consumption compositions and oil dependency of Japanese manufacturing industries. (Unit: 10^{15}J (left axis), % (right axis), 1973–2006.)

Source: Resources and Energy Agency of Japan (2012).

40% since declining to 48.3% in 1980. On the one hands that oil dependency decreased, on the other hands that other energy dependency except oil were higher. Especially, coal or natural and city gas dependency were higher about 3–5% since 1980. In the 1990s, the share of coal were decreasing, otherwise them of natural and city gas, or new energy such as heat energy are increasing.

In these ways, what's possible to advance the energy saving are by below two reasons; (1) promotion of technological innovation for the energy saving in the manufacturing industries, (2) regulation to energy waste by government and political supports for promoting the energy saving.

Figure 15.5 shows trends of expenditures for the energy-related research and development (R&D) at each sector in Japan since 1974 to 2005. Incidentally major portion of energy R&D expenditures by government after the oil crisis was atomic power and nuclear related, accounted for 60–80% share of energy R&D, this data of the figure is not included these R&D.

Energy R&D expenditures by companies increased from 49.99 billion yen at 1978 to 242.7 billion yen at 1983, and flat trends within 1980s after peaking in 1983. The R&D within next periods since 1990 were minor

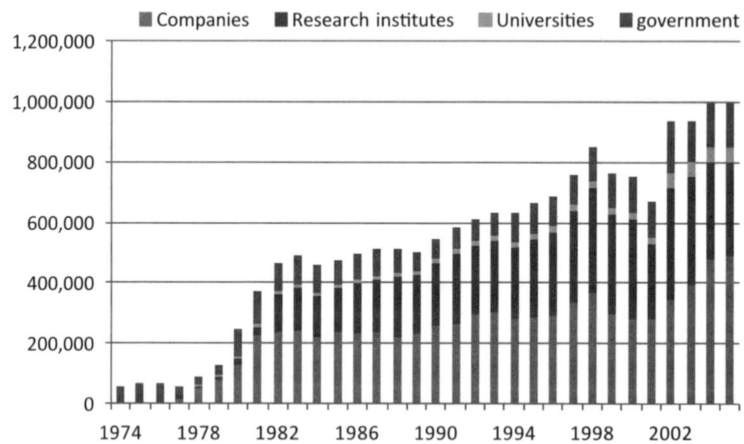

Fig. 15.5. Energy related research and development (R&D) expenditures of each sector in Japan. (Unit: million yen, 1974–2005.)

Source: International Energy Agency (IEA) (2012), The Public Management Ministry's Statistics Bureau (1974–2005).

increasing than increasing within just after the oil crisis while the periods of 1996–1998 and after 2000 were increasing.

The aspects of Japan's energy R&D expenditures since the oil crisis are that the companies played a chief role to expense. While share of the energy R&D expenditures in 1977 were 28% of companies, 19% of research institutes, 48% of government; those in 1981 were 61% of companies, 7% of research institutes, 28% of government. The central role of energy R&D in Japan shifted from government to companies within only 5 years terms.

Governments had a central role of energy R&D just after the oil crisis, while companies had a central role of it since 1980s. This means that energy R&D began to ride on the market base. These situations continued with 25 years, otherwise, after the 1990s the share of R&D expenditure of the research institutions became large.

The political actions after an oil crisis show to Table 15.1 which summarized the main policies regarding the energy saving in Japan after the oil crisis. Sunshine Project for innovation of new energies started in 1974, this year being the following year of the oil crisis; and the Moonlight Project to support the innovations for energy saving started in 1978. The low interest loan system for the promotion of energy saving investment of companies started in 1975, this system were provided for promoting energy saving investment of R&D support. Enforcement of Energy Saving Act in 1978 was after those R&D support or funds systems.

Since the end of the 1980s, Japan's government planed and enforced some policies or regulations regarding the global warming controls. For example, the government firstly made the action plan for prevention of global warming in 1990, made the global warming law in 1998, and amended partially such as obligating to report the emissions of large scale offices. Furthermore, Energy Saving Act were amended in several times such as introducing the leading runner approach in 1998, and obligations of reporting the action plan for energy saving and management to large scale factories. Otherwise, the energy R&D system were not only positioned the subsidies policies to the innovations as the energy saving, but also positioned them as the CO_2 emissions after 1990s. New Sunshine Project was created through integrating Sunshine Project for new energy innovation support and Moonlight Project for energy saving innovation in 1993. Thus,

Table 15.1. Japan's energy saving policies and related systems (1970–2005).

	Events	Law or regulations	R&D support	Investments support
1970–1974	1973 1st oil shock		1974–1993 Sunshine Project (new energy R&D)	
1975–1979	1978 2nd oil shock	1979 Energy Saving Act	1978–1993 Moonlight Project (energy saving R&D)	1975 Low Interest Loan System of Energy Saving Investments
1985–1989			1989 R&D support system related Global Environmental Technologies	
1990–1994	1992 Earth Summit (The Rio Declaration)	1990 Action Plan for prevention of Global Warming 1993 the revised Energy Saving Act (Obligation of regular report By large-scale factories)	1993–2000 New Sunshine Project	1993 Tax System for promoting the investment of implementing the energy demand-supply structures change
1995–1999	1997 COP3, Kyoto Protocol	1998 Global Warming Law 1998 the revised Energy Saving Act (leading runner approach)	1998 Energy Saving R&D Program	1998 Industrial energy saving investment support 1999 Energy Saving Action Support for Investment of Houses and Buildings

(*Continued*)

Table 15.1. (*Continued*)

	Events	Law or regulations	R&D support	Investments support
2000–2005	2002 Rio+10 (Johannesburg Summit) 2005 Enforcement of the Kyoto Protocol	2002 the revised Global Warming Law (Obligation to report emissions) 2002 the revised Energy Saving Act (Obligation of reporting the energy managements and energy saving actions) 2005 Goal Achievement Plan of Kyoto Protocol		2000 Energy Saving Promotion on areas

Source: Resources and Energy Agency of Japan (2012), Institute of Energy Economics, Japan (2012).

energy saving policies after the oil crisis implemented R&D program preferentially than regulations by policies, while the recent global warming policies introduced some regulations or policies such as Energy Saving Act or the global warming law.

Japan's energy saving policies after the oil crisis focused on the subsidies to energy saving innovations of companies than the regulation policies such as Energy Saving Act. As a result, it was increasing the new technologies of the energy saving within the late 1970s to the early 1980s in Japan. Figure 15.6 shows the trends of patent application numbers of some energy saving technologies in Japan from 1970 to 2005. The average change rate from 1973 to 1983 was increasing by 6.0% of heat exchange, 26.3% of exhaust heat recycling, and heat pump 14.2%. When it became late in the 1980s, the application number began to decrease, but when it became late in the 1990s, any technologies showed the upward tendency again.

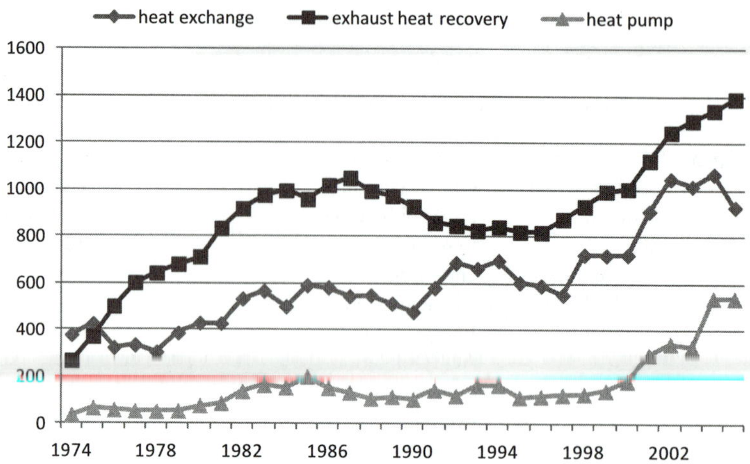

Fig. 15.6. Patent application numbers of some energy saving technologies in Japan. (Unit: Numbers of Applications, 1970–2005.)

Note: These are not all energy saving technologies but showed one of them only.

Source: Drawing by author, referred by Japan Patent Office (2012).

What factors were effective to these technologies development of energy saving? Table 15.2 shows the analysis result with econometric model to clear the effects of the oil prices or R&D expenditures to the patent applications of energy saving technologies.

As a result, oil price and R&D expenditures of government had significant and positive effects to innovate each technology. This result means that inflation of oil price by the oil crisis and R&D expenditures of energy promoted the technological development about energy saving.

This implication from the analysis result is that inflating the oil price by the oil crisis induced the technological development for saving the energy cost. This fact was based on social consensus between Japanese that using oil could be risky. And, this consensus did not only induce the development of new technologies, but also provided private firms with an incentive to innovate with providing the public R&D expenditures for supporting that consensus. Furthermore, regulations such as the energy saving act gave a competitive advantage to the innovators who produced new technologies, and this policy encouraged new entry into innovators.

Table 15.2. Analysis results (analytical method: OLS, 1974–2005).

		Coefficient α	Oil price ln(P)	Patent stock ln(K)	R&D expenditure ln(RD)	Adj. R^2	D.W. Stat.	F value (P value)
Dependent variable (i)	Heat exchange	−5.08**	0.18*	−0.16**	0.25**	0.50	1.37	7.85
	(hex)	(−12.80)	(2.45)	(−3.38)	(2.93)			(0.00)
	Exhaust heat	−6.26**	0.36**	−0.15*	0.33**	0.43	0.73	6.17
	(Eht)	(−13.77)	(3.85)	(−2.21)	(2.83)			(0.00)
	Heat pump	−9.67**	0.64**	0.08	0.42*	0.55	1.10	13.60
	(hp)	(−18.55)	(4.55)	(1.38)	(2.43)			(0.00)

Notes: Independent variable is share of patent application numbers of each the energy saving technology to total patent applications.
Dependent variables are oil import prices, patent stock numbers of each technology, and R&D expenditure of energy related by governments.
**is 1% level sufficient by t-value test, *is 5% level sufficient by t-value test.
Source: Authors' own research results.

This section reviewed the energy saving in Japan after the oil crisis. Energy saving actions and innovations did not occur autonomously, but were due largely a result of social consensus by the oil crisis, policy incentive through a public R&D, and regulations such as the energy saving act. This is a significant factor why the industrial sector could archive the energy saving, and why the private sector could not do. Because the regulations, incentives, or competition on market are not imposed on private sector, even if the private sector had a consensus of energy saving, it is difficult to be continued with the motivation to do the energy saving.

15.4. A Business Model for a Low-Carbon Economy

15.4.1. *Why does the transfer of environmental technology not get through?*

In this part, we focus on the design of the "general business model" for the implementation of a low-carbon economy in details. One of the most effective methods to implement a low-carbon economy is to practice the method of energy-savings. In Fig. 15.7, we show an international comparison in the per GDP consumption amount of primary energy. We can see that Japan is the most energy-efficient country among all. It is considered that if countries throughout the world can reach the same level of energy efficiency as Japan, a low-carbon economy will be immediately applicable. Therefore, it is suggestive that we can meet our target if we can transfer this excellent energy-saving technology of Japan abroad. However, this is only the round-table's imagination. That is, except for the exportation of fuel-saving complete Japanese cars and energy-saving Japanese home appliances, most of these technologies exportation does not succeed. In reality, the author engaged in a business of transferring the leading, most efficient energy-saving apparatuses from Japan to a Southeast Asian country from 2005 to 2007, and experienced a dreadful ending. However, this failure teaches us an important lesson when implementing a low-carbon economy; it will be mentioned in more detail in the following.

The object energy-saving apparatus is "Hydrate Slurry Air-conditioning Technology" of JFE Engineering Company which was awarded the Energy Saving Prime Minister Prize in 2006. This apparatus helps reduce electricity

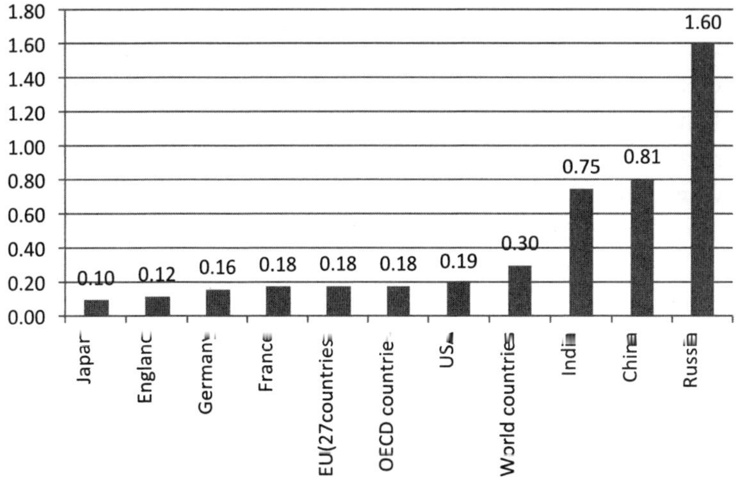

Fig. 15.7. International comparison of primary energy supply per GDP. (Unit: Oil ton/1000 U.S. Dollar, 2008.)
Source: IEA (2012a) and (2012b).

consumption by about 30–40% if we use it with the existing water-cooled type air-conditioning system. The initial investment is high, but can be paid back in 3–5 years, and after that the savings in electricity expense will become profit. It was first thought that with the high technological level, and the record of actual performance in Japan, such technology would be easily transferred to Southeast Asian countries where energy usage is inefficient. All participants in the technology introduction seminar, including governmental officials, building owners, building maintenance companies, tenants, environment-related consultants and engineering companies, were amazed at the excellent technology, so that there was no question of whether to import it or not; it was just a question of when. However, the actual condition drastically changed when we started sales operations. The problem was that it was impossible to determine who would be the actor to import the energy-saving technology. Specifically, building owners and maintenance companies get no benefit from the investment because tenants pay the electricity expense. Similarly, whether tenants benefit from that investment or not is unclear because they may move when their lease contracts expire. Government will not subsidize if the actor is not decided. Consultants and engineering companies could not provide support for sales promotions when the actor was unclear. Even though Singapore and Thailand depend on external sources of energy, at the period of the year 2005, common consumers show only one kind of energy-saving trend and it is clear that their concern to energy-savings is not as deep as in Japan.

Owing to this experience, we derive a hypothesis as follows:

To carry out energy saving, say, to purchase an energy-saving apparatus, firstly, it is necessary for firms and common consumers to clarify their high interest which is mentioned as the activity to create value in "The Profit Zone". Then, the second necessity is to regulate the one dimension of "who" purchases it in the market.

To make clear this hypothesis, let's look at Fig. 15.7 again and discuss the reason that made Japan the most energy-efficient country in comparison with not only developing countries but also the Western developed countries. Primarily, the main actor to practice the environmental improvement and energy saving is consumers, and the people. We imagine from international conferences and academic conferences that people's interest in the environment and energy saving is high, but this is just the sense among a part of experts. The relation between experts and the people can be imagined as the relation between the heart and capillaries in blood circulation. In other words, it is impossible to make blood circulate just by the heart. For example, there are about 60 trillion capillaries in a human, and these capillaries, especially in the muscles, accurately synchronize to make a human's blood flow. Similarly, related to the issue of environmental improvement and energy saving, the people react to experts' messages, share the sense of an issue, then start actions.

In Japan during the first oil shock not only the government felt panic, but the sense of emergency was felt by all the people, then energy-saving initiatives were raised by both the government and the people. In comparison with other developed countries, Japan's level of energy security, especially its low self-supply ratio relates to the magnitude and sustainability of the sense of emergency. After that, enterprises aggressively promoted the development of energy-saving apparatuses, and consumers were given priority to the purchase of energy-saving products as a response to the "top-runner method" initiated by the Ministry of International Trade and Industry. Consumers themselves also put their first priority in purchasing energy-savings appliances. That means, a low-carbon economy has been implemented in Japan during this period. As a result of this favorable circulation, Japan has become the top energy-saving nation in the world.

An example of the establishment of environmental consciousness based on shock therapy is the case of the U.S., when interest in environment quickly spread after Hurricane Katrina in 2005. After facing personal and material injuries from abnormal weather conditions and the global warming alarm, U.S. citizens as a whole have started recognizing the importance of environmental improvement. Accordingly, this conscious enhancement leads to the Green New Dealer Program initiated by President Obama's administration.

15.4.2. *Factors for the establishment of a business model*

15.4.2.1. *Social consensus*

It is highly appreciated that Japan's high energy efficiency is based on that kind of technology. However, the reason is the possibility that firms can develop, sell energy-saving appliances and can get profits from that business, i.e. the possibility for the design of the "general business model". As seen in Fig. 15.5, it actually takes 30 years for such sustainable high-energy-efficiency power to work, since enterprises must develop energy-saving apparatuses, sell them on the market, and finally get a profit. In other words, the sense of emergency that took hold all over Japan from the first oil shock enabled the "sustainable profit-generating structure of business model". This hints at a design for a business model for a low-carbon economy. That is, the development of technology for a possible low-carbon economy is important, but even more important is to construct a mechanism to sell that technology on the market. Here the point is that the oil crisis made Japanese common consumers more interested in energy savings. In other words, in the freely competitive market without the existence of such strong interest in which there is not clear clarification of "who" and "how", it is impossible to design a business model for both firm side and individual side. That is, that should be a market where there is a strong sense of emergency among common consumers instead of the freely competitive market. The difference in energy efficiency between Japan and other countries lies not only in the existence of technology, but in the difference in the sense of emergency as well. Hence, the first element for purchase of energy-saving apparatuses on the market is common consumers' strong consciousness and consensus of energy saving. We call this consciousness "Energy-saving Consensus", and we call consciousness of general environmental improvement "Environment Consensus". Once these ways of thinking are widespread enough, they become the most important elements enabling the design of a business model for a low-carbon economy. Therefore, to establish such consciousness it is necessary to teach environmental education in everyday life, not only during emergencies like the oil crisis period.

15.4.2.2. *Regulations*

It is difficult to sustain this consciousness and consensus, because the nature of consumers is to desire, but the rationale to practice environmental improvement and energy-saving restrains that kind of desire. Hence, the consciousness gradually diminishes over time. However, Japan's ability to maintain an "energy-saving mind" is an exceptional case in which the traumatized "shock" and "panic" exist both in government and among the people for a long time.

In general, such a consciousness is short-lived. Hence, it is necessary to have regulations to rationalize and keep up the consciousness. The point here is procedure: the consciousness comes first, and regulations follow. It is possible that if we set up regulations when the consciousness has not been fully established, such regulations may themselves sink the consciousness into atrophy. An example is Japan's RPS method which regulates electric power entrepreneurs' use of new energy. The RPS method contributes to the work of energy diversification and greenhouse gas reduction. However, electric power entrepreneurs, such as power companies, reluctantly accept the rule because their consciousness has not been made ready enough.

Another important role of regulations is to regulate all parties related to environmental improvement and energy saving unitarily. This helps to solve the issue of "who" in the case of transferring energy-saving technology to the Southeast Asian countries mentioned above. Regulations help determine uniformly that "who" takes responsibility of practicing energy saving, and "who" benefits from this practice. This is considered to be a crucial determinant in order to make business with energy saving.

15.4.2.3. *Incentive*

Once the consciousness is well enough established, both in government and among the people, relevant regulations are set up, and the design of a business model for a low-carbon economy will be put into effect. However, effects of various regulations not only for the above RPS method case but also other previous actions against pollution issues fall into the same trend of dead-end regulations. In other words, business models may have been designed, but their effects may be very restrictive because profits can be attained from only the one source of government subsidy. There may be a high effect on the people's consciousness regarding activities such as separation of home waste, but still, this cannot lead to business when there are more entrants. Therefore, an incentive is requested. Under the circumstances of the Eco-Point System from 2009, regardless of a sharp decline in demand just after the financial crisis, we can see the effect when real sales of eco-electric appliances and eco-cars far exceed the expectation. It can be said that because of the common consumers' well-developed consciousness they decide to purchase energy-saving products, but they can acquire a better opportunity to purchase if they receive incentives. It is imagined that such an Eco-Point System may be more effective if it is linked with some regulations controlling exhaust gas.

15.4.2.4. *Competition principle*

Owing to the Eco-point method, home appliances makers and automobile makers are induced to sell more environment-improving products, or more highly energy-saving products in the market (i.e. they have entered a competition). This leads to a favorable circulation in which consumers' consciousness becomes higher and hence increases consumption, and as a result, makers speed their mass production that leads to lower production costs. This Eco-point method is now compensated by government subsidy;

however, we can utilize it, and even give the right to only to private enterprises in designing business models. The point is to construct a mechanism to promote competition among makers, regions and enterprises. As a result, a low-carbon economy will be present in a short time.

15.4.3. *An economic model for a low-carbon economy*

This part summarizes the discussion above. It is the most difficult to design a business model for a low-carbon economy in a general market, say, a freely competitive market. Consumers are likely to purchase the cheaper products if they are commonly standardized products in a freely competitive market. In general, the merit of environmental improvement and energy saving does not match the consumption habit of purchasing cheaper products.

It is necessary that firms and individuals construct a market where the design of a business model for a low-carbon economy is feasible. We call this market structure the "economic model for a low-carbon economy". An economic model is a prior concept for a business model. It can be considered as infrastructure. This economic model is composed of four elements as follows: people's environment and energy-saving consensus, regulation, incentive and competition.

There may be a question whether technology itself is a determinant of an economic model. It is exactly that technology is a determinant which makes a business model design more feasible. Strictly speaking, however, it is completely possible to design a business model regardless of the existence of technology or not. That is, if there is a business model a business in that field succeeds, therefore, technology itself is not a determinant of a business model design.[2]

This economic model is not standardized. It reflects the diversity of countries, regions and eras; therefore, it must match the respective country, region and era. Otherwise, it will not lead to increasing consciousness in the common consumers. Unification of the central government, local governments, educational institutions, mass media and so on is needed

[2]Related to the relation between technology and business, and that between technology and business model, the author presents his paper "Business Creation and the Design of a Business Model" at the 59th Japan Information and Management Conference held at Nagoya University on November 21–22, 2009.

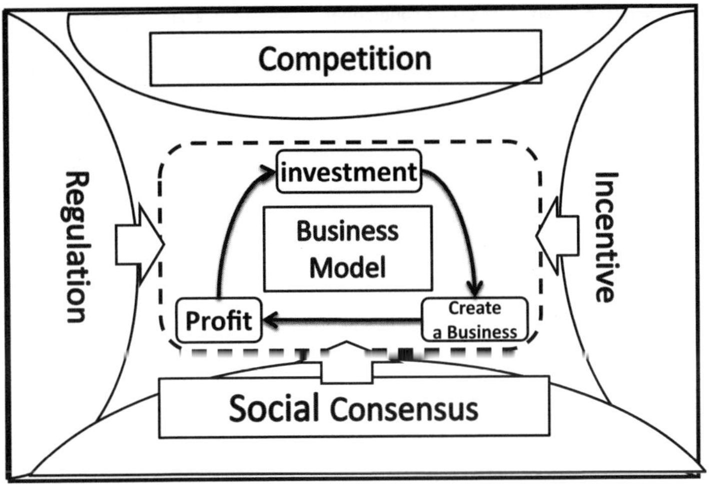

Fig. 15.8. Case for the existence of an economic/business model for a low-carbon economy.
Source: The author's creation.

for the establishment and elevation of consciousness which seems to be applied in Japan during the 1970s–1980s. Regulations and incentives are administrative tasks, but it is preferable when thinking about diversity to have cooperation from four sides: industry, academic research, government and the people. There is a likelihood of thinking of competition as an issue related to firms only. However, in the case of environmental improvement and energy saving, it is possible to create favorable circulation if there is a mechanism that induces individual prefectures and communities to compete to reach a target level of environment improvement and energy saving.

Once such an economic model is ready, firms enter the market and compete with one another in order to make the design of a business model possible. We describe in Fig. 15.8 case when there is a ready economic model.

15.5. Effectiveness of the Eco-Point System

Based on the hypothesis image above, we once again discuss about effectiveness of the Eco-Point System in Japan. Figure 15.9 show you that the

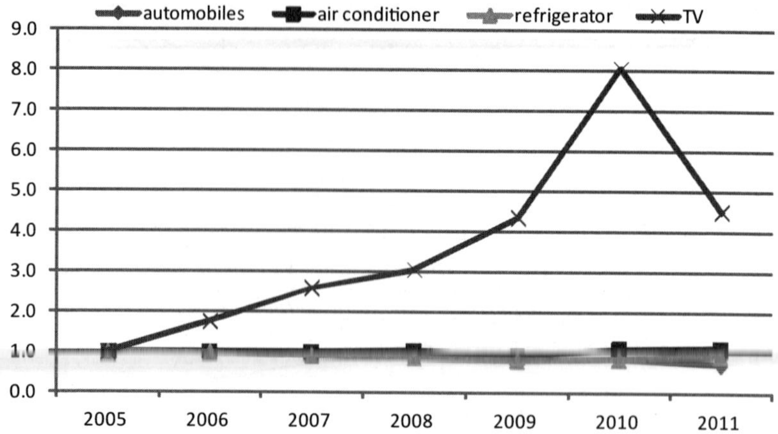

	2005	2006	2007	2008	2009	2010	2011
automobiles	5,852,067	5,739,506	5,353,648	5,082,235	4,609,256	4,956,136	4,210,220
air conditioner	7,481,272	7,520,394	7,390,038	7,749,477	6,775,383	8,242,030	8,279,040
refrigerator	292,840	288,018	265,208	251,530	233,013	243,270	264,957
TV	3,132,000	5,485,000	8,077,000	9,552,000	13,579,000	25,187,000	14,073,000

Fig. 15.9. Domestic sales of household electric goods (Unit: 2005 = 1.0.)
Source: The Automobile Information Center (2012); The Japan Refrigeration and Air Conditioning Industry Association (2012); Japan Electronics and Information Technology Industries Association (2012).

sales of every targeted products namely automobile, air conditioner, TV (flat-screen) and refrigerator increased since the Eco-Point System stimulated the consumers' appetite by its incentives. But, as shown by Fig. 15.9 the flat-screen TV sold by much more than the other products. Why did the Eco-Point System not increase to consume all electric products, and did it do the flat-screen TV only? The reason is that time frame for this Eco-Point System overlaps the ending time of analog TV broadcasting and the start time of terrestrial digital media broadcasting. Therefore, there is no other choice that consumers have to buy new TV. Such consumers' correspondence to terrestrial digital media broadcasting exactly implies effects of the so-called regulation. That is, incentive and regulation simultaneously derive effects by chance.

15.6. The Application and Reality of a Low-Carbon Economy Model

We have just discussed that it is necessary to set up an economic model which enables a business model for firms and individuals in order to implement a low-carbon economy. Will, however, a "general business model" be created based on the aforementioned argument? Japan itself has achieved its authentic general business model due to the fact that the oil crisis motivated them to give birth to the world-top-level energy-efficient products such as cars and home electrical appliances and export them abroad in the framework of a successful business model. This, then, drives the economic growth in Japan. Japanese makers who uniformly have to take responsibilities of energy saving due to regulations (Revision of the Rationalization in Energy Use Law) gain benefits from their sales of energy-saving products domestically and internationally.

However, the possibility for the creation of this business model in some representative countries in the community of Newly Industrialization Countries, say, BRICs is quite low. For, if there is possibility, the author's experienced case for transferring the aforementioned energy-saving technology abroad would have succeeded without troubles. Needless to say, the higher the cost for energy hikes the more attractive of highly energy-efficient products. However, only this attraction is not enough. During the 1970s–1980s, the exported products of Japan were famous for not only their property of energy saving but for their price competition also. That means, there is a high likelihood that the experienced "general business model" will not work even though Japan succeeds in their production of highly energy-efficient products under the implementation of a low-carbon economy. Therefore, it is yet crucial to think of a framework to export the business model as a whole.

To implement a low-carbon economy becomes as important as to diffuse this low-carbon economic model throughout the world. It is not normal to force any other country to pursue another country's economic model because such economic model is considered as the infrastructure of that country itself. However, as we discussed, sooner or later we have to implement this low-carbon economy due to the hike in energy cost when the fossil fuel tends to run out.

Therefore, if Japan succeeds in implanting this kind of low-carbon economic model, then other countries will actively emulate which may lead to the establishment of the so-called regional "Smart City", or "Smart Town". It is said to be the most ideal and effective if Japan take the initiatives in creating those kinds of establishments and spreading them abroad. Because Japan owns the most excellent environmental technologies as well as energy-saving technologies, it will be more persuasive to employ the effective model of Japan than that of any other country.

From the viewpoint of implementing the low-carbon economy globally, Fig. 15.10 suggests the ideal perspective of the cooperation of the most up-to-date "high-carbon economies" as the U.S. and China with the owner country of energy-efficient technologies Japan in order to create a global standardized low-carbon economic model.

The U.S. itself is skilled enough at making necessary regulations and incentives for a low-carbon economic model because it originally has its own achievements in economic policies as well as in economic and environmental improvements. Examples for such regulations are the creation of the Super Fund Tax for lead-free gasoline and Muskie Act. Similarly, an example for incentives is the Investment Tax Credit (ITC) employed under the regime of President Ronald Reagan which promoted the excavation of crude

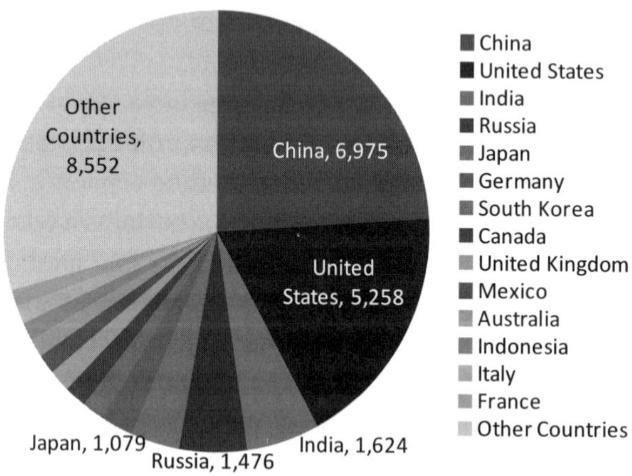

Fig. 15.10. Carbon dioxide emission by country in 2009.
Source: Institute of Energy Economics, Japan (2012).

oil and natural gas for the economic recovery. This kind of ITC incentive is comparatively easy to proceed in accordance with the production and consumption of the sustainable energy like natural energy. For instances, there is a high likelihood to set up a so-called global standardized "Eco-investment" in the framework of the low-carbon economic model when we give the wealthy elderly rights to get income-tax and inheritance-tax exemption if they agree to invest in solar-electric and harness wind power generation.

In cooperation with China, the implementation of a highly competitive model employing Japanese technology and Chinese industrial power is not enough. It is suggested to mimic and make use of the Chinese-styled special economic zones which has been accelerating under the mechanism of the large-scale Chinese modernization, then diffuse it to other developing countries.

Still, it is hard to implement this ideal low-carbon economy with the cooperation of Japan, the U.S. and China. However, the establishment of such scheme itself is a signal to declare the "responsibility of large countries". If there is any chance for it to be in practice, then human beings will be liberalized and may achieve a sustainable soft-landing society.

15.7. Conclusion

The globe can be considered as "Spacecraft Earth" flying at hyper-speed under the driving power of capitalism. However, there is a "growth limitation" in it. In other words, this spacecraft will crash once it loses speed due to a slow-down in the rate of growth, or it will crash after operating at such hyper-speed and running out of fuel. The idea of implementing a low-carbon economy was born in the 15th century and is the antithesis of capitalism, since then mankind's wealth has been increasing dramatically, and life has been made innovative and comfortable. Such a challenge is not easy for capitalism only on the premise of mankind's desire. This chapter presents an economic model in which mankind performs rationally, as well as one of the methodologies to carry out a low-carbon economy. It is naturally skeptical about the possible implementation of such scheme. However, Japan itself took its challenge in the oil crisis period as a utopia and became a wealthy country in the 1970s 1980s. It is important to bear in mind the ideas of making not only our own country but also the whole global society wealthy based on this mechanism. Therefore, to implement this low-carbon economic, we not only look forward to some environmental supremacist voluntary activities but we have to ensure the possibilities to successfully design a business model for firms and individuals as well as the possibilities to change those business activities into sustainable value-making business.

Acknowledgments

This research was supported by Grant-in-Aid for Asian CORE Program "Manufacturing and Environmental Management in East Asia" of Japan Society for the Promotion of Science (JSPS).

References

Afuah, A. (2004). *Business Models: A Strategic Management Approach*. Boston, New York: McGraw-Hill/Irwin.

Cabinet Office of Japan (2012). *National Accounting*.

Institute of Energy Economics, Japan (2012). *Handbook of Energy & Economic Statistics in Japan*.

International Energy Agency (IEA) (2012). "IEA Energy R&D Statistics." https://www.iea.org/stats/rd.asp

International Energy Agency (2012a). *Energy Balances of OECD Countries* (2010 Edition).

International Energy Agency (2012b). *Energy Balances of Non-OECD Countries* (2010 Edition).

Japan Electronics and Information Technology Industries Association (2012). http://www.jeita.or.jp/english/index.html

Japan Patent Office (2012). "Industrial Property Digital Library." http://www.ipdl.inpit.go.jp/homepg.ipdl.

Kagono, T. and Inoue, T. (2004). *Strategies for Business System Design* (in Japanese).

Resources and Energy Agency of Japan (2012). *Comprehensive Energy Statistics*.

Slywotzky, A. J. and Morrison, D. J. (1999). *The Profit Zone: How Strategic Business Design will Lead you to Tomorrow's Profits*. New York: Three Rivers Press.

The Automobile Information Center (2012). http://autoinfoc.com/index.html.
The Japan Refrigeration and Air Conditioning Industry Association (2012). http://www.jraia.or.jp/index.html
The Public Management Ministry's Statistics Bureau (1974–2005). *Report of Science Technology Research.*

Chapter 16

LOW-CARBON ECONOMY OF FOREST ECOSYSTEMS: FROM THE PERSPECTIVE OF GLOBAL CLIMATE CHANGE

*Shin-ichi Yamamoto**

16.1. Introduction

Recent estimates of the area of forest worldwide are just under four billion hectares, or 30% of total land area. Forest ecosystems are generally regarded as carbon sinks, meaning that they are net extractors of CO_2 from the atmosphere and that they store carbon. Compared with CO_2 reductions by industrial producers, the cost of forest maintenance is relatively low, but is of great benefit. Thus, forest ecosystems play a critically important role in the global carbon economy. In this chapter, the carbon economy of various forest ecosystems is discussed from the perspective of global climate change, in particular, the key role of gap dynamics for the carbon economy of old-growth forests. Current data indicate that further research is required to examine the gap dynamics of old-growth forests in relation to the global carbon economy. In order for this to be possible, old-growth forests must be strictly preserved and left intact. However, around the world such ecosystems are not protected by international treaties and have been considered of no significance in the national carbon economy as outlined in the Kyoto Protocol. Carbon-counting rules for forests should, therefore, be changed to give credit for leaving old-growth forests intact.

*Shin-ichi Yamamoto, Vice President of Okayama University, Japan. He also served as Professor of Bio-agricultural Sciences and Vice President of Nagoya University.

16.2. Forest Ecosystems from the Perspective of Global Climate Change

There are two major ways to cope with global climate change. One is the reduction of carbon emissions, and the other is the development of carbon sinks such as forest ecosystems, including the soil. Forest ecosystems are generally regarded as carbon sinks, i.e. they are net removers of CO_2 from the atmosphere and they store carbon. While the reduction of carbon emissions has a direct effect on the carbon budget, carbon sequestration is an indirect solution to increases in CO_2. The maintenance of forests is a relatively low-cost activity and has significant associated benefits, such as the preservation of biodiversity. It is also considered to be a cost-effective approach to mitigating carbon emissions. In 2005 the total area of forest worldwide was estimated to amount to just under four billion hectares, or 30% of total land area (FAO, 2005). Thus, forest ecosystems play a key role in the global carbon economy, and it is essential that they are appropriately managed and protected.

I consider plant carbon production and forest dynamics, and introduce the key role of gap dynamics in the carbon economy of old-growth forests.

16.3. Forest Ecosystems in East Asia

Forests in East Asia are present across a wide latitudinal range, distributed from the tropics, through the temperate regions, to the sub-arctic (Yu *et al.*, 2006). There is, therefore, a wide variety of forest types in this region. However, in recent years, the volume and area of natural old-growth forests has declined rapidly as a result of severe anthropogenic disturbances, particularly for tropical systems including mangrove forests.

Generally, forest ecosystems can be roughly classified into natural forests and plantation forests. Furthermore, natural forests are usually divided into old-growth (primary) forests, and secondary successional forests that are regenerating or recovering after natural or anthropogenic disturbances.

16.4. Primary Production and Carbon Economy of Forest Ecosystems

Primary production refers to the production of organic compounds from carbon dioxide (CO_2) through the process of photosynthesis. On land, higher plants are the main primary producers. Primary production can be considered in the form of gross primary production (GPP) or net primary production (NPP). NPP measures increases in organic matter after losses to respiration (Kimmins, 1996). Recently, interest has grown in the study of the carbon fluxes of an entire ecosystem, including carbon emissions from soil respiration. Net ecosystem production (NEP) is a sophisticated criterion used to evaluate carbon fixation on the basis of NPP and the rate of soil respiration (Komiyama *et al.*, 2008).

The GPP of forests peaks when they are at an intermediate age and then declines slightly, reaching a stable level that is maintained into old age. However, respiration losses continue to increase as biomass accumulates, with the result that NPP reaches a peak at an intermediate age and then declines (Kira and Shidei, 1967; Kimmins, 1996). Thus, middle-aged forests, for example secondary successional forests that are regenerating and re-growing after disturbance and plantation forests, are very important carbon sinks.

16.5. Soil Carbon Dynamics in Forest Ecosystems

Soils represent both a short and long-term carbon storage medium, and contain more carbon than all terrestrial vegetation and the atmosphere combined (Swift, 2001). Plant litter and other biomass accumulate in the form of organic matter in soils, which is degraded by chemical weathering and biological action. Forest soils are carbon stores, and act as CO_2 sinks as they get older or increase in area. Tsutsumi (1963) measured the quantity of soil organic matter at different times during forest development after bare ground, and found that soil carbon increases monotonically, eventually reaching a steady state. This process of development was modeled by Olson (1963). Thus, a vast amount of carbon is accumulated in the soil of old-growth forests.

16.6. Gap Dynamics of Old-Growth Forest Ecosystems

Traditionally, old-growth forests have been recognized as having zero NPP, because respiration equals photosynthesis and hence there is no net production and no net biomass increase (Kira and Shidei, 1967; Odum, 1969,

Kimmins, 1996). However, recent findings show that old-growth forests act as carbon sinks (NPP or the net carbon balance of the forests including soils is greater than zero) (Luyssaert *et al.*, 2008). This is an important point, demonstrating the need to maintain old-growth forests intact for the sake of the global carbon economy as well as for conserving biodiversity.

Irrespective of whether old-growth forests are carbon neutral or act as carbon sinks, they are very important for the global carbon economy. If these forests are disturbed, much of the accumulated carbon in the standing trees and soils will return to the atmosphere. Gap dynamics play an important role in maintaining the stability of old-growth forests in relation to their overall carbon economy (Bormann and Likens, 1979; Feeley *et al.*, 2007). "Gaps" are small openings created in the forest canopy by the death of one or a few canopy trees (Yamamoto, 1996, 2000).

When gaps are formed, the dead canopy trees begin to decay and gradually release CO_2 but young regenerating trees grow rapidly in the gaps and quickly begin to assimilate more CO_2 than they produce.

The canopy of old-growth forests is not completely closed; there tends to be a variable number of small or large gaps. Yamamoto (1996) reported that, in Japan, the percentage gap area was 17.0% in warm-temperate forests, 13.8% in cool-temperate forests and 7.3% in subalpine forests. Using aerial photographs taken at different times, Fujita *et al.* (2003a, b) estimated rates of gap formation and closure in an old-growth evergreen broadleaved forest. While they found large temporal variation as a result of typhoon disturbances, the overall mean rates of gap formation and closure (0.58 VS. 0.50%/year) were nearly balanced.

Thus, old-growth forests are temporally and spatially dynamic, and are maintained by the process of gap formation and closure. The carbon economy scenario associated with gap dynamics in old-growth forests can, therefore, be estimated. Such gap dynamics involve the continuous sequence of tree death (biomass loss and gradual release of CO_2) and tree growth (biomass gain and rapid fixation of CO_2). The carbon economy associated with such gap dynamics is assumed to function as follows. While the NPP is approximately zero in the closed canopy part of the forest (i.e. the old and mature part), it is greater than zero in the gaps (i.e. the young and regenerating areas). In the gaps, dead trees decay slowly and release CO_2 gradually, whilst regenerating or regrowing young trees rapidly assimilate CO_2.

Moreover, Manabe *et al.* (2009) found recently that gap formation events produce vertically stratified layers within an old-growth forest, indicating an efficient primary production structure (thus absorbing more CO_2).

Mangrove forests are found in tropical and sub-tropical tidal areas. Some of their productivity may flow into adjacent ecosystems, or conversely, they may receive organic matter from estuarine or oceanic ecosystems (Kristensen *et al.*, 2008). While it is not clear whether old-growth mangrove forests are maintained by gap dynamics (Imai *et al.*, 2006), known patterns of biomass allocation, NPP, and soil respiration indicate high net ecosystem productivity, making mangrove forests highly efficient carbon sinks in the tropics (Komiyama *et al.*, 2008). Mangrove forests are also important for the protection of coastal areas against the damage caused by tsunamis.

16.7. Secondary Successional Forests and Plantation Forests

The secondary successional forests and plantation forests are very important carbon sinks for the reasons outlined earlier. Secondary Successional forests are also important for conserving or restoring native biodiversity.

Active reforestation programs are required in denuded and degraded areas. Although most countries in the East Asia and Pacific region have lost forests since 1990, those losses have slowed since 2000. China and Vietnam are the only countries in the region where forest cover is increasing (World Bank, 2010). The World Bank (2010) reported that China's forest cover has grown from nearly 12% to more than 18% in the past 20 years as a result of a concentrated program and a countrywide ban on logging (in order to protect the remaining natural forests).

16.8. Conclusions

Gap dynamics play a key role in the carbon economy of old-growth forests, irrespective of whether those forests are carbon neutral or act as carbon sinks. Further research is required to examine the gap dynamics of such forests in relation to the global carbon economy. In order for this to be possible, old-growth forests must be strictly protected and left intact. However, around the world such forests are not protected by international treaties and were considered to be of no significance in the national carbon budgets outlined in the Kyoto Protocol. Forest conservation activities aimed at avoiding deforestation or degradation, which would result in CO_2 emission reduction thorough the conservation of existing carbon stocks, are not eligible for inclusion in countries' carbon budgets at this time (Manguiat *et al.*, 2005). Much of the carbon within old-growth forests, even the soil carbon quantified by Tsutsumi (1963), will return to the atmosphere if old-growth forests are disturbed. Therefore, carbon-counting rules for forests should give credit for leaving old-growth forests intact.

Secondary Successional forests and plantation forests are very important as carbon sinks, because NPP reaches a peak when forests are middle-aged. Such forests should be treated as carbon sinks and appropriate management should be promoted. For both types of forests, suitable and sustainable forest management is essential.

Finally, forest conservation and actions to prevent deforestation, which would prevent additional CO_2 emission by conserving existing carbon stocks, should be eligible to be treated as certified emission reductions. Therefore, the plan for developed countries to pay developing ones to halt destruction of tropical forests, proposed at the COP15 in Copenhagen, 2009, is good news (Theil, 2010). While the participating countries could not reach a general agreement, it was the first time that they focused on halting deforestation as a way to reduce global climate change, as pointed out by Theil (2010).

References

Bormann, F. H. and Likens, G. E. (1979). *Pattern and Process in a Forested Ecosystem.* New York: Springer-Verlag.

FAO (2005). Global Forest Resources Assessment 2005.

Feeley, K. J., Davies, S. J., Ashton, P. S., Bunyavejchewin, S., Supardi, M. N. N., Kassim, A. R., Tan, S. and Chave, J. (2007). The Role of Gap Phase Processes in the Biomass Dynamics of Tropical Forests, *Proceedings of the Royal Society B*, 274, pp. 2857–2864.

Fujita, T., Itaya, A., Miura, M., Manabe, T. and Yamamoto, S. (2003a). Canopy Structure in a Temperate Old-growth Evergreen Forest Analyzed by Using Aerial Photographs. *Plant Ecology*, 168, 23–29.

Fujita, T., Itaya, A., Miura, M., Manabe, T. and Yamamoto, S. (2003b). Long-Term Canopy Dynamics Analysed by Aerial Photographs in a

Temperate Old-growth Evergreen Broad-leaved Forest. *Journal of Ecology*, 91, 686–693.

Imai, N., Takyu, M., Nakamura, Y. and Nakamura, T. (2006). Gap Formation and Regeneration of Tropical Mangrove Forests in Ranong, Thailand. *Plant Ecology*, 186, 37–64.

Kimmins, J. P. (1996). *Forest Ecology*, 2nd ed. USA, Berkeley: Prentice-Hall.

Kira, T. and Shidei, T. (1967). Primary Production and Turnover of Organic Matter in Different Forest Ecosystems of the Western Pacific. *Japanese Journal of Ecology*, 17, 70–87.

Komiyama, A., Ong, J. E. and Poungparn, S. (2008). Allometry, Biomass, and Productivity of Mangrove Forests: A Review. *Aquatic Botany*, 89, 128–137.

Kristensen, E., Bouillon, S., Dittmar, T. and Marchand, C. (2008). Organic Carbon Dynamics in Mangrove Ecosystems: A Review. *Aquatic Botany*, 89, 201–219.

Luyssaert, S., Schulze, E.-D., Börner, A., Knohl, A., Hessenmöller, D., Law, B. E., Ciais, J. and Grace, J. (2008). Old-growth Forests as Global Carbon Sinks. *Nature*, 455, 213–215.

Manabe, T., Shimatani, K., Kawarasaki, S., Aikawa, S. and Yamamoto, S. (2009). The Patch Mosaic of an Old-growth Warm-temperate Forest: Patch-level Descriptions of 40-Year Gap-forming Processes and Community Structures. *Ecological Research*, 24, 575–586.

Manguiat *et al.* (2005). Legal aspects in the implementation of CDM forestry projects, Series: IUCN environmental policy and law paper, No. 59.

Odum, E. P. (1969). The Strategy of Ecosystem Development. *Science*, 164, 262–270.

Olson, J. S. (1963). Energy Storage and the Balance of Producers and Decomposers in Ecological Systems. *Ecology*, 44, 322–331.

Swift, R.S. (2001). Sequestration of Carbon by Soil. *Soil Science*, 166, 858–871.

Theil, S. (2010). Forests Save the Day. *News Week*, January 11, p. 6.

Tsutsumi, T. (1963). Forest structure and tree species distribution in relation to topography-mediated heterogeneity of soil nitrogen and light at the forest floor. *Bulletin of Kyoto University Forest*, 34, 37–64.

World Bank (2010). http://data.worldbank.org/news/increa.

Yamamoto, S. (1996). Gap Regeneration of Major Tree Species in Different Forest Types of Japan. *Vegetation*, 127, 203–213.

Yamamoto, S. (2000). Forest Gap Dynamics and Tree Regeneration. *Journal of Forest Research*, 5, 223–229.

Yu, G.-R., Wen, X.-F., Sun, X.-M., Tanner, B. D., Lee, X. H. and Chen, J.-Y. (2006). Overview of China FLUX and Evaluation of its Eddy Covariance Measurement. *Agricultural Forest. Meteorol.*, 137, 125–137.

PART 2

STRATEGIES AND POLICIES FOR A LOW-CARBON ECONOMY

Chapter 17

LOW-CARBON ECONOMIC POLICIES

*Kazuhiko Takemoto**

17.1. Introduction

Japan decided on the "Action Plan for Achieving a Low-Carbon Society" in July 2008, based on the successful outcome of the G8 Hokkaido-Toyako Summit. The overall framework of policies and measures toward a low-carbon economy and society has been set out. To reduce GHG emissions, two key laws were enacted, namely the Climate Change Law (1998) and the Energy Saving Law (1979), which enabled the implementation of many policies and measures. Taking into account the important role of local governments in addressing climate change, the government has extended its support to their continuing efforts for low carbon cities.

17.2. Low-Carbon Society as a National Policy

17.2.1. *Action plan for achieving a low-carbon society*

In July 2008, the G8 leaders agreed that global GHG emissions should be halved by 2050 from its current level, and that this long-term goal should be shared with the international community. Based on this agreed outcome, the Government of Japan formulated the "Action Plan for Achieving a Low-Carbon Society" as a Cabinet Decision.

---

*Kazuhiko Takemoto, Senior Fellow at the Institute of Advanced Studies (UNU-IAS) of United Nations University. He has served as Vice-Minister for Global Environmental Affairs at the Japanese Ministry of the Environment, Japan, and involved in the designing of the Kyoto Protocol, developing policies on international environmental cooperation and global environment and climate change.

This Action Plan set the framework to implement the policies as follows:

(a) Setting mid- and long-term (economy wide) targets,
(b) Strengthening cooperation with developing countries in addressing climate change,
(c) Enhancing R&D on innovative technologies and disseminating existing advanced technologies,
(d) Considering possible introduction to economic instruments such as carbon taxation and emission trading systems, and
(e) Further developing local initiatives.

As for a long-term goal, Japan has been committed to reducing 60–80% of its GHG emissions by 2050 from its current level (2005), taking into account that the G8 Hokkaido-Toyako Summit agreed on at least 50% reduction of global GHG emissions by 2050 from the current level.

According to the timetable proposed by the Action Plan, Japan officially announced a mid-term target in June 2009. Major elements of the mid-term target are as follows:

- 15% reduction in 2020 from 2005, which is equivalent to an 8% reduction from 1990 levels,
- This target is based only on domestic efforts,
- Increasing energy efficiency by 33%,
- Introducing solar power generation systems 20 times more than current level,
- Doubling the number of environmentally friendly vehicles such as fuel cell cars, electric cars and hybrid cars up to 50% or more of new cars in market.

The Plan pledged included the experimental introduction of domestic emission trading systems and reconfirmed further consideration of carbon taxation systems and economic analysis of environmental policies. The local initiative was highlighted in the Action Plan on the grounds that local governments and communities should play a very important role in addressing climate change. The government is committed to its financial and technical support to participating local governments.

17.2.2. *Legislations on the climate change issue*

- **The Climate Change Law**

The Climate Change Law was enacted in 1998, based on the outcome of the COP3 of UNFCCC, which was adopted in the Kyoto Protocol. The Law aims to develop a national plan to achieve the Kyoto targets and to implement policies and measures in a systematic manner.

The Law specifies the roles and responsibilities of the government, local governments, business sectors and citizens in addressing climate change. The government is responsible for planning and implementing policies and measures on GHG emissions reduction and carbon sinks. In particular, the government is required to develop a plan to achieve the Kyoto targets, which covers a wide range of policies and measures and encourages the participation of all stakeholders.

The Law was amended in 2005 and 2008. The 2005 Amendment introduced a regulation system for business sectors to measure their GHG emissions and to report their performance to the government annually. The 2008 Amendment requested local governments to develop a plan to reduce GHG emissions in their areas and to strengthen their linkage with urban planning policies.

Based on the Climate Change Law, the Action Plan to Achieve Kyoto Targets (Action Plan) is formulated to decide policies and measures to achieve targets. The Plan includes policies and measures by sectors (manufacturing, transport, service and households). It is important to note that the Plan shows a clear policy direction to maximize the effect of GHG reduction through reforming Japan's economic and social structure, for example, by designing low-carbon cities and transport systems.

Along with this policy direction, the Plan promotes longer-life houses, buildings and infrastructures, and develops the idea of a Compact City, which is designed to enhance the concentration of cities' functions. Forest management and urban greenings should be further developed to secure carbon sinks under the Action Plan. In addition, the introduction of carbon taxation and emission trading systems should be further discussed as useful policies to address climate change. The Plan also stresses the importance of public awareness and public relations. At the same time, it is expected that local governments play a more important role. In order to monitor the implementation of the Plan, the government is to review the performance of policies and measures every 6 months.

- **The Energy Saving Law**

The Energy Saving Law was enacted in 1979. The Law aims to promote energy saving activities to overcome oil crises and plays a very important role in promoting industries' actions to reduce GHG emissions by their enhanced efforts to save energy. The Law covers the following four sectors: factories, energy-consuming appliances, houses and buildings, and transport. The owners of factories and office buildings with large energy consumption are required to promote energy saving activities and report their performance to the government on a regular basis. The Law was amended to expand its coverage to shops and restaurants in 2008.

The Law has adopted a unique approach, the so-called "top runner approach", which applies stringent energy consumption standards to electricity appliances and automobiles. The owners of houses and buildings with large energy consumption are asked to introduce measures on insulation and to install energy-efficient equipment. Their development plans and performance in saving energy will be reported to the government regularly. The owners of freight and logistic companies are also requested to make their energy saving plans and to report their performance to the government.

17.2.3. *Emission trading system*

Emission trading systems (cap and trade) have either been introduced or considered to ensure cost-effective GHG emission reduction in many countries and regions such as the USA and the EU. The International Carbon Action Partnership (ICAP) was established in 2007 to explore global linkage between systems of countries and regions.

Japan started the experimental introduction of domestic emission trading systems in October 2008. The MOE has implemented the Japan Voluntary Emission Trading System (JVETS) since 2005, which is a voluntary scheme with the participation of the private sector. This system is designed to extend financial support to enterprises who wish to reduce GHG emissions by setting their own targets.

17.2.4. *Low-carbon city planning*

Local governments play a very important role in addressing climate change. The Climate Change Law (2008 Amendment) has strengthened the role of local governments by requesting that they expand the scope of their action plans to actively introduce renewable energies, public transport systems and urban greens. At the same time, the 2008 Amendment required local governments to incorporate the concept of low-carbon cities into urban and agriculture development plans.

The Government of Japan has extended its support to local governments in order to ensure the smooth implementation of a wide range of policies and measures as follows:

(a) Environmental Model Cities and Municipalities: The government selected cities and municipalities which have taken actions to advance environmentally sound policies and measures by setting ambitious GHG emission reduction targets. There are 13 cities and municipalities who have now been selected. They are required to formulate their own action plans and implement policies and measures based on those plans. The government is to support these policies and measures by making full use of programs of the relevant governmental agencies.

(b) Compact Cities: The MOE supports local governments who would like to advance policies and measures such as the introduction of public transport and renewable energies, toward becoming low-carbon cities. The MOE has extended its financial support to local governments and, at the same time, the Ministry of Infrastructure, Land and Transport (MILT) is providing technical guidance.

(c) Cool City Pilot Project: The MOE has supported enterprises who wish to install facilities to mitigate heat island effects and to reduce GHG emissions. The MOE has extended its financial support up to half the costs for projects such as painting with high reflection rates, roof greenings and watering systems.

17.3. Innovation for Green Economy and Society

The MOE formulated the "Innovation for Green Economy and Society" in April 2009, which aims to overcome the current economic situation and to strengthen Japan's future economy and society through its environmental investments. The Innovation was developed to restore the active and energetic economy of Japan by innovating economic and social structures and, at the same time implementing environmental policies and measures to address mid- and long-term global environmental problems such as climate change and biodiversity. The MOE intends to shift Japan's current economy and society to a more sustainable one by introducing these dynamic policies and measures.

The Innovation includes the following policies and measures:

(a) Innovation for green consumption: energy saving houses and buildings, eco cars and energy-saving appliances by introducing an "Eco-Point" system,
(b) Innovation for green capitals: eco schools with solar panels, green procurement by the public sector and compact cities,
(c) Innovation for local governments and communities: introduction of Local Green New Deal Funds, active participation of multi-stakeholders and capacity building,
(d) Innovation for green investment: greening taxation systems, environmental financing, green IT, further installation of solar panels and the introduction of renewable energy,
(e) Innovation for green technology development: technology for renewable energy such as solar power generation, solar heat, bio-fuel, wind power, small hydraulic power, geo-thermal and Carbon Capture and Storage (CCS). Energy-saving refrigerators, freezers and air conditioning units using low Global Warming Potential (GWP) refrigerant and non-fluorocarbon and high-performance insulation, and
(f) Contribution for green Asia: Clean Asia Initiative, co-benefit approach, Asia 3R Forum, Satoyama Initiative and environmental model cities.

It is estimated that the market scale of environmental business will be expanded to \$1.2 trillion in 2020 from \$0.7 trillion in 2006 and that job opportunity in environment fields will be increased to 2.8 million in 2020 from 1.4 million in 2006.

17.4. International Cooperation

To address global environmental issues such as climate change, it is crucial to further develop international cooperation. Both developed and developing countries have to work together for their common goals. Climate change is one of the top agendas at the G8 Summit and the Major Economies Forum on Energy and Climate (MEF). The outcome of these world leaders meetings is summarized here, focusing on mitigation and different technologies. Japan has extended its support in a variety of ways to developing countries for their efforts in addressing climate change. The Cool Earth Partnership and the co-benefit approach are among them.

17.4.1. *G8 Summit and MEF Leaders' Meeting*

The MEF leaders also held a back to back meeting at the G8 summit. The member countries of the MEF included G8 member countries, China, India, Brazil, South Africa and Denmark as the COP15 Presidency. The G8 Summit and MEF discussed a wide range of issues on climate change and agreed on their respective declarations. The agreed outcome on mitigation is as follows:

(a) Both leaders' meetings recognized "the scientific view that the increase in global average temperature above pre-industrial levels ought not to exceed 2°C." It is meaningful, in the context of future global actions, that the world leaders of the G8 and MEF have shared this kind of recognition,
(b) G8 leaders reiterated "willingness to share with all countries the goal of achieving at least a 50% reduction of global emissions by 2050," as agreed at the G8 Hokkaido-Toyako Summit of 2008. The MEF leaders agreed to "work between now and Copenhagen to identify a global goal for substantially reducing global emissions by 2050,"
(c) G8 agreed supporting "a goal of developed countries reducing emissions of GHG in aggregate by 80% or more by 2050 (compared to 1990 or more recent years)." This outcome is advanced from the results of the

G8 Summit of 2008. By this strong message, the world leaders made it clear that developed countries should take the lead in reducing GHG emissions, and

(d) The mitigation action of developing countries is one of the top agendas of COP15. The MEF agreed "major developing countries will promptly undertake actions whose projected effects on emissions represent a meaningful deviation from business as usual in the midterm by support." The actions of developing countries are crucial in advancing international discussions on global mitigation actions.

The MEF leaders also agreed to establish "a Global Partnership to drive transformational low-carbon, climate friendly technologies." They shared their strong will to "dramatically increase and coordinate public sector investments in research, development and demonstration of these technologies, with a view of doubling such investments by 2015." They welcomed the efforts among interested countries to advance actions on technologies such as energy efficiency, solar energy, smart grid, carbon capture, use and storage, advanced vehicles, high efficiency and low-emissions coal technologies, bio-energy and other clean technologies. It was agreed that lead countries would make their report on action plans and roadmaps by November 15, 2009.

17.4.2. *Cool earth partnership*

Japan has promoted the "Cool Earth Partnership," which supports developing countries for their actions on mitigation measures, adaptation measures and access to clean energy. This program makes on the scale of $10 billion available over the 5 years since 2008 to developing countries aiming to reduce GHG emissions and achieve economic growth, working to contribute to climate stability. Japan extends its aid in the form of ODA loan, grants and technical cooperation. Aid is allocated to forest conservation and natural disaster prevention in response to climate change, co-benefit measures and measures toward energy conservation and new energy development.

The scope of this assistance program is as follows:

(a) Adaptation: to assist developing countries, which are vulnerable to the adverse effects of climate change such as African and Pacific island countries, to take adaptive measures against disasters related to climate change such as droughts and floods, planning of adaptation activities,

(b) Improved access to clean energy: to promote economic development through utilizing clean energy in developing countries which have insufficient access to modern energy supply, such as the electrification of rural communities by introduction of solar power generation and small-scale hydro power generation,

(c) Mitigation: to mitigate the effects of global warming by reducing GHG emissions, such as the improvement of energy efficiency of electric power plants.

17.4.3. *Co-benefit approach*

The concept of co-benefits is to implement simultaneously both the reduction of GHG emissions and resolution of environmental problems, such as air water pollution as well as waste management. At the G8 Hokkaido-Toyako Summit of 2008, the importance of a co-benefit approach was shared.

Japan is committed to taking the lead in advancing the following areas:

(a) reducing local environmental pollution in a mid-term perspective and the mitigation of climate change effects,

(b) helping developing countries take part in the international negotiations on climate change in a more effective manner, and

(c) contributing to the international community by extending its technologies, management systems and systematic know-how developed by experience in overcoming serious pollution since the 1950s through the early 1970s.

Addressing mitigation of climate change has often been regarded as a negative factor to economic growth. However, the co-benefit approach has resolved this concern. Japan has started co-benefit projects in the transport sector and the industrial sector by using ODA and CDM.

17.5. Conclusion

It is a common goal for the world to realize a low-carbon economy and society. The world leaders have committed themselves to making every effort to reach an agreed outcome at the COP15 in Copenhagen this December. Japan is committed to developing a framework to achieve a low-carbon society at the highest political level and has implemented a wide range of policies and measures under this framework. At the same time, Japan has taken deliberate actions on the innovation of a green economy and society in order to recover its economy through intensive investments in the environmental areas. Japan has to take the lead in further developing a green economy together with other major economies.

It is very important for the government to deliver a clear message on the future direction of the nation so that the private sector can take steps forward with confidence. To realize a low-carbon economy and society, all the stakeholders have to commit themselves to active participation, based on respective responsibilities and roles, which are specified by the laws. In this context, it is useful to introduce reasonable regulations by the relevant laws and standards. Taking into account the important role of local governments, it is very important to develop systems to encourage them to further their existing commitments.

Global actions are crucial to realize a low-carbon economy and society. In this context, it is important to enhance international cooperation between developed and developing countries. Japan has extended a wide range of cooperation programs and has to make full use of these programs to address global environmental issues in a more effective manner.

References

Takemoto, K. (2009a). "Climate Change, National-Local Policy Frameworks and Actions." International Symposium: Towards Low-Carbon Cities, Nagoya.

Takemoto, K. (2009b). Outcomes of G8 Summit and MEF and Future Perspectives on Climate Change Negotiations. *Environment and Society*, 8.

Takemoto, K. (2009c). "Innovation for Green Economy and Society." Global Environmental Forum 2009. Incheon.

Takemoto, K. (2009d). "Policies and Measures toward Global Environmental Era." Japan Architecture Association, Sendai.

Chapter 18

LOW-CARBON SOCIETY

*Shuzo Nishioka**

18.1. Introduction

With the deterioration from the world greenhouse effect, international discussion over the rising temperature has become more and more intense. The Copenhagen meeting, held at the end of 2009, is considered to have raised common concern to a new high. Almost all the countries in the world have already realized the emergency and necessity of abating greenhouse gases emissions in order to stabilize the climate and maintain life. Japan, famous for its advanced environmental consciousness, has been engaged in controlling this issue for years. In 2004, Japan started research on "Japan's Low-Carbon Society Scenario", which gathered more than 60 experts from related areas to work on the establishment of a Low-Carbon Society. Later, in 2007, the result was published: "By 2050, supposing all the regular national demands are met, it's technically feasible to reduce CO_2 by 70% from the base year of 1990". This view served as a foundation for the 80% long-term abatement goal put forward by the Hatoyama regime in 2009.

This chapter tries to expound the necessity of establishing a low-carbon society and putting forward the long-term target of cutting greenhouse gas emissions by 70% by 2050 compared with 1990, a goal that has been

*Shuzo Nishioka, one of designers of the low-carbon society in Japan. He served as Climate Policy Project Leader, Institute for Global Environmental Strategies, and Executive Director of the National Institute for Environmental Studies, Japan (NIES) as well as project leader (Climate Policy) of the Institute for Global Environmental Strategy (IGES). He also served as a member of the Environment project of Science and Technology Council of Japan, and of Central Council for Environment of Japan. In the international arena, he has served as coordinating lead author of IPCC Reports since 1988 and Review Editor of Millennium Ecosystem Assessment since 2004. Currently he is Senior Advisor for IGES, Japan.

demonstrated to be technically feasible. Furthermore, schemes and means to achieve this goal are also discussed here. The chapter comprises the following sections. In Section 18.2 the research achievement of "Japan's Low-Carbon Society Scenario", especially the 70% abatement target, is introduced. Section 18.3 clarifies the technical feasibility of the long-term target mentioned above as well as explains the way to construct this scenario. Schemes besides technical means for low-carbon promotion, such as improvements of social capital and infrastructure investment, reform in the transportation area, low-carbon city planning as well as popularizing insulating buildings, are discussed in detail in Section 18.4. Section 18.5 appeals to us to get ready for the coming change of the social system, while Section 18.6 concludes the whole chapter.

18.2. Low-Carbon Society and the Evaluation of the Scenario in Japan

18.2.1. *The concept of a low-carbon society*

The concept of a low-carbon society was first mentioned in the above research. Compared with the "low-carbon economy" suggested by the British, this expression means: under the guidance of sustainable development, every member of the society should make efforts to extend the use of clean technologies, reduce energy use and minimize CO_2 emissions as much as possible, in order to realize the mutual development of society and the environment. The 70% abatement goal, mentioned above, is thought to be the premise as well as a crucial part of the "Low-Carbon Society Scenario" and can be demonstrated as feasible from a technical view. Then the question here is how this seemingly tough task can be accomplished. There is no doubt that it is the use of energy that generates most of the CO_2. Thus, the task of abatement should mainly rely on the reduction of energy demand (40%–45%). Under the assumption of 2 types of society — energetic society and relaxed society — energy-efficient technology must be applied to minimize energy exhaustion as well as fulfill the national demands for energy services. The technology transfer cost accordingly is expected to be 500 billion–1 trillion yen (about 0.1%–0.2% of expected GDP). In addition to the technology aspect, social capital accumulation and infrastructure investment are also supposed to move in the low-carbon direction. Meanwhile, it's reasonable to make use of cooperation and competition in environment technologies across areas to adjust the industrial structure to become environmentally friendly. Furthermore, reform in transportation focusing on restrain of vehicles, city design which enables citizens to live on foot, and promotion of insulating buildings can all be important factors in achieving the 70% abatement goal.

Heading for the low-carbon society, we need a giant change to the current situation. Development in energy-efficient technology, upgrade of technical systems, reduction in energy demand, application of renewable energy, together with industry transition, introduction of the "value of climate stability" in the economic system, and conversion in lifestyle will make the new society look quite different. The establishment of a low-carbon society is of profound meaning.

18.2.2. *Japan's low-carbon society plan and legislation*

Can the low-carbon society in which greenhouse gases are abated drastically really come true in Japan? What and how we should change to achieve this goal? The "Study on Japan's Low-Carbon Society Scenario", started in 2004, gathered 60 experts from various related areas to work on this issue together.[1]

Nowadays, in the international political world, the UNFCCC is engaging in conducting negotiation over the framework for greenhouse gas abatement post-Kyoto; meanwhile, the G8 summit has already declared climate change as a great issue to human society and put it on the order of the day every year. The time is almost ripe for worldwide carbon reduction aimed at climate stability. The IPCC Forth Assessment Report, published in 2007, enlightened the following: the development of the greenhouse effect has undoubtedly begun to cast bad effects on ecosystems and global systems around the world; the greenhouse effect can be mainly attributed

[1]Results of the study are showed in (Nishioka, 2008; Fujino *et al.*, 2008; 2009; National Institute for Environmental Studies, 2008; 2009; Hanaki, 2009; Center for Commercial Research, 2007; Skea and Nishioka, 2008). Financed by Global Environment Research Fund, this study was conducted during 2004 to 2008 and the writer served as the project leader of this study.

to man-made emissions of greenhouse gases; for climate stability, it is necessary to implement greenhouse gases abatement; abatement is technically available, etc.

Compared with the EU countries' quick actions in the low-carbon area, in Japan, there wasn't a long-term abatement plan, but only the "Plan for Kyoto Protocol Target Achievement" (2005) based on "Laws for Global Warming Measures" (1998). At the G8 summit meeting held in England in 2005, the long-term measure of climate change was first discussed; the conclusion was submitted in 2008's Toya Lake G8 summit. Since then, Japan became more and more concerned about the long-term abatement problem.

In February, 2007, the study published as a report to the government from the team this achievement: "By 2050, there exists the technical potential to enable CO_2, which accounts for the main body of greenhouse gases, to be abated by 70% compared with 1990." This first long-term prospect of CO_2 abatement was presented to the minister of environment and serves as a foundation for Japan's environmental policies that were formulated later.

In the eve of the G8 Heiligendamm summit in 2007, Prime Minister Abe put forwarded the proposals of "Cool Earth 50", declaring the willingness to construct a "low-carbon society" though the implementation of the world's 50% abatement target in 2050 and domestic development of energy-saving technologies as well as reconsideration of lifestyle. Later in May, 2008, Prime Minister Fukuda pointed out in his "Fukuda Vision" that "in Japan, abatement of 60–80% based on the current situation should be proposed as a long-term target; we're aiming to be the world's largest low-carbon society." In this way, Japan's long-term abatement target was first set forth.

Here came the expression of "low-carbon society". This concept was first mentioned in the joint study between "Project Team on Japan's Low-Carbon Society Scenario" and the British. It originated from the concept of "low-carbon economy" introduced earlier by the British; however, considering the fact that measures taken to cope with climate change will turn out to influence not only the economy, but also the whole of society from individual action to social foundations, the phrase "low-carbon society" came into being. The "low-carbon society" was defined as follows in this joint study.

The Low-Carbon Society is a society in which people

- Take actions that are compatible with the principles of sustainable development, ensuring that the development needs of all groups within society are met
- Makes an equitable contribution towards the global effort to stabilise atmospheric concentration of carbon dioxide and other greenhouse gases at a level that will avoid dangerous climate change through deep cuts in global emissions
- Demonstrate high level of energy efficiency and uses low-carbon energy sources and production technologies, and
- Adopt patterns of consumption and behaviour that are consistent with low level of greenhouse gas emissions

The results of our study were published as "Japan's Low-Carbon Society Scenario: The way to Abate CO_2 by 70% in 2050" in Nishioka (2008). Nevertheless, domestic attention to this long-term outlook was fairly weak and some sectors of industry even considered it as talking in one's sleep. Nonetheless, consensus on this future course is gradually being formulated among those low-carbon technology-concerned industries. Our study summarized various schemes and measures that enable the realization of the scenario in "12 schemes heading to low-carbon society" published in Fujino *et al.* (2009) and National Institute for Environmental Studies (2008).

Seeing that the 2020 abatement targets of each nation or region were planned to be discussed in 2009's COP15 meeting held in Copenhagen, the Aso cabinet arranged a discussion over how to set the mid-term abatement target for 2020; the team of "Study on Japan's Low-Carbon Society Scenario" also took part, with the methods and means developed in the study. Accordingly, Japan announced its mid-term target of a 15% reduction compared with emissions in 2005 and the determination to carry out a "low-carbon revolution" to reach this goal at the L'Aquila G8 summit in July, 2009.

Furthermore, at the UN Summit on Climate Change convened in September, 2009, Prime Minister Hatoyama announced Japan's mid-term target of 2020 as "to abate CO_2 emission by 25% with the base year of 1990". Based on this, later in June, 2010, the cabinet submitted to congress "Basic Law of Measures to Cope with the Greenhouse Effect" in which

Japan's long- and mid-term abatement targets were set up to be 80% and 25%, respectively, when compare with 1990. This 80% long-term target can also be found in the bill of the Liberal Democratic Party as well as the New Komeito.

Although Japan has stated its long-term target in clear figures, still there remain problems like by what means will the target be reached, which has not been mentioned formally so far. At this moment, "Japan's Low-Carbon Society Scenario" and "12 schemes heading to low-carbon society" is being developed further to formulate the abatement schemes of advanced city, local governing body and corporations at the Central Environment Council. The following scenario has become the textbook forming the basis of Japan's low-carbon society construction.

18.3. Japan's Low-Carbon Society Scenario

18.3.1. *Possibilities to reduce by 70%*

People are always inclined to imagine a society where life is tough and uneasy when they hear the hard words "low-carbon society". It's true that we ought to reconsider the comfortable life we have lived when energy and resources are used wildly; however, there's no need to worry too much since a low-carbon society can enable us to live in an even better way. Due to the development of technology based on human knowledge, we don't need to worry about climate change anymore and can live a better life.

According to "The Study of Low-Carbon Society Scenario", when the low-carbon society is realized, the services requested in the so-called energetic society or relaxed society where the economic growth rate is kept at 1–2% should be met. Based on this condition, the study clarified: "There is the technical potential to meet all the expected demands in 2050 while reducing CO_2 emissions by 70% compare with 1990", and concluded that "the low-carbon society is realizable".

While Japan currently has its long term target at 80%, in the very beginning of this study, 2004, that target was considered so unrealizable that we had to settle for 70% in the end, which was still doubted as dream words at that time. However, this 10% difference itself makes no essential problem; what matters greatly is the big transition of society towards the substantial reduction of greenhouse gases.

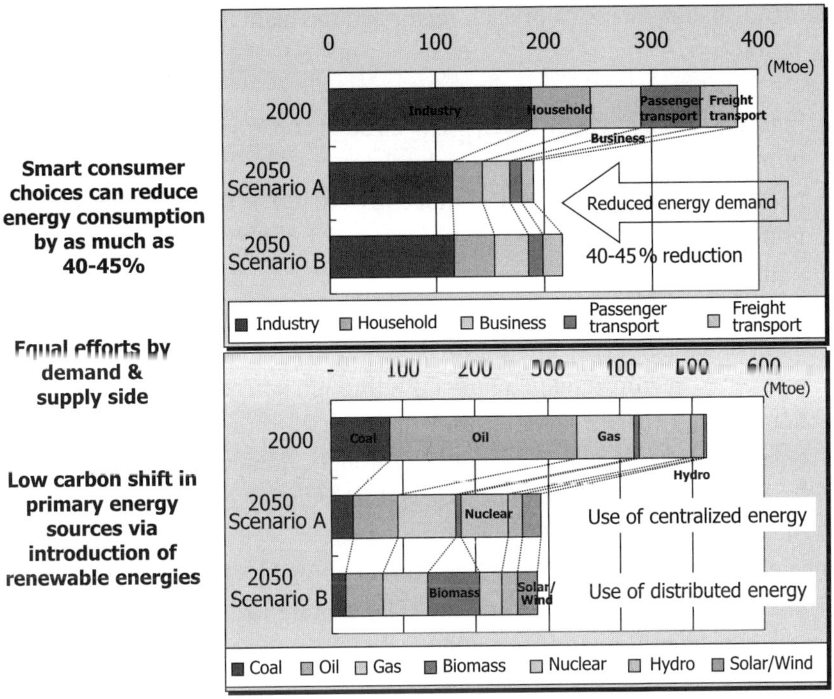

Fig. 18.1. Scenario of abating CO_2 by 70% in 2050.

Figure 18.1 depicts the expected construction of energy in 2050. The conclusions are: (1) it is technically feasible to meet all the expected demands in 2050 while reducing CO_2 emission by 70% compare with 1990; (2) the abatement task should mainly rely on the reduction of energy demand (40%–45%); (3) the combinations of various primary energy resources are available to keep CO_2 emissions below 70%, relative to year 2000, when abated energy demand is met; (4) the technology deployment cost (additional technology cost) accordingly is expected to be 500 billion–1 trillion yen (about 0.1%–0.2% of the expected GDP); (5) reforms on social infrastructure and social systems ought to be carried out as early as possible in order to extend the use of new technology.

Seen from the demand side, while increasing energy demand is observed in several sectors, decreasing population and shrinking energy service demand resulting from reasonable energy use as well as improvements in energy efficiency may, at the same time, reduce energy demand. Energy demand abatement ratios of each sector and the main measures applied are as follows:

- industry sectors: reduce by 25–40% through construction conversion and introduction of energy-saving technology
- passenger transport sector: reduce by 70–80% through appropriate utilization of national land, enhancement of energy efficiency together with modal shift
- freight transport sector: reduce by 40% through strengthening management of supply chain and improving energy efficiency of vehicles
- household sector: reduce by 50–60% through spreading high insulating housing and energy-saved instruments
- business sector: reduce by 40% through rebuilding current office buildings to high insulating types and introducing energy-saved instruments
- energy supply side: achieve low-carbon goals by choosing low-carbon energy resources properly (taking CO_2 capture and storage technology into consideration) and enhancing energy transform efficiency

18.3.2. *Way of constructing the scenario*

The scenario of 70% abatement in 2050 was constructed in the following way.

First, we consider 2 kinds of society: an energetic society where GDP growth is maintained at 2% comparing with a relaxed society with a 1% GDP growth rate. Regarding the outlook of Japan, different people have different views. Some of them believe that with the development of technologies we will construct a low-carbon renewable society which is also an energetic society where everyone lives in a positive and vigorous way, while others think that, with the background of an aging society, people will be more likely to enjoy their leisure time and thus create a relaxed society. Considering these 2 visions allows us to show that no matter which type we choose, the 70% target can be reached all the same.

The change of industrial construction under the economic growth rates supposed above are estimated with economic models, as well as energy service demand (utility) generated from economic activities like production and consumption. Different from energy, energy service is the activity which consumes energy. In other words, it's like how much light we need (brightness × hours), how much warmth we want (temperature × hours) and how much distance we want to move (person × kilometers), rather than electricity or gasoline. Technologies such as electric lights, air-conditioning and cars join the utility and energy together, and the more advanced the technology the less the amount of energy needed to maintain the same utility level. Consequently, introducing advanced technologies into society to cut down energy use which results in reductions of greenhouse gases when all the requested utility is fulfilled is quite reasonable.

To be concrete, we pick up about 600 technologies covering almost all current energy uses. Taking technical progress into consideration, we compare the technology-introducing cost based on the expected energy price between various technologies and choose to start with the cheapest one.

If we add together the energy consumed by energy services and technologies, Japan's entire energy demand and the way it is demanded can then be estimated. Correspondingly, the energy supply side must choose to apply the most suitable technologies. CO_2 emissions are calculated from the consumed primary energy resources (Fig. 18.2).

The way to a low-carbon society is found using a back casting method with the 70% abatement target fixed beforehand in the scenario. When facing huge reform like the establishment of a low-carbon society, the resolution

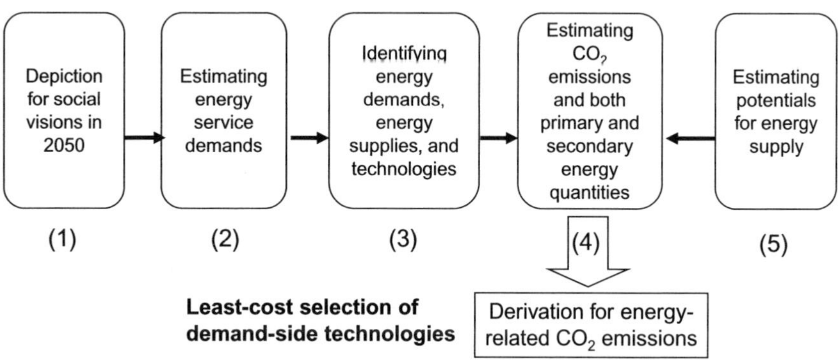

Fig. 18.2. Procedure to estimate CO_2 emissions.

of devoting all possible resources and labors to accomplish this task is extremely necessary. What's more, we can determine the best means to reach our goal after considering questions like whether the target can be achieved within the expected limits of technology; how much cost will be generated; to what extent will energy demand and supply change; what social barriers hinder reform; how much technical progress is required; etc. This kind of backcasting method, which follows the order of future-to-now, is indispensible in finding the way to the final goal. However, if we turn to the view of now-to-future, that is to say, use the common forecasting method to settle on an action plan, very likely the abatement action will be drawn with existing systems and will get trapped in a chorus of "impossible", "infeasible" and "harmful". Then precious opportunities will be missed easily and essential reform will hardly appear.

18.4. Schemes besides Technical Means for the Low-Carbon Plan

18.4.1. *Low-carbon society based on cooperation and competition among various areas*

Although decreasing population is anticipated technical innovation only is far from sufficient to achieve the low-carbon society; conversion of social means like infrastructure and social systems is also required to enable those technologies to be used properly. The expected industrial reconstruction and national infrastructure should be adapted to the low-carbon course simultaneously. Furthermore, we must accelerate the development and application of energy-saving technology as well as low-carbon technology. Amid the construction of the low-carbon society, government must show strong leadership to popularize the target early, establish policies with respect to social and technical innovation, and implement powerful measures to realize CO_2 abatement potential, promote the implementation of sound government investment, and encourage private investment.

Above all, it's necessary to break away from the high energy technology; the key is the application of energy-saving technology on the demand side. Heading for the low-carbon society, domestic and international competition on technology has become more and more cutthroat; the leapfrog phenomenon is also observed, resulting from the entry of different areas into the new technology system. With the premise of a smart grid in the future, cooperation among battery, automobile, and renewable energy together with the information area has already started. We must adapt the social capital that utilizes new technology to the land and city construction of a dwindling birthrate and aging society.

18.4.2. *Change of transportation and city*

For example, what will happen in the transportation area by 2050? (Fig. 18.3)

In the area of automobiles, which are responsible for most of the CO_2 emitted in passenger transport, electric cars able to cut down emissions per moving distance to only a fourth of the usual amount have already been developed and put into market. However, the problem here is people have to pay 2 times more than the price of a common gasoline car. This problem will be solved in the future with a drop in price due to battery efficiency being enhanced by technical progress and mass production of electric cars. As is showed in Fig. 18.3, this technical progress will enable energy efficiency improvements to take up a great part of the energy reduction needed.

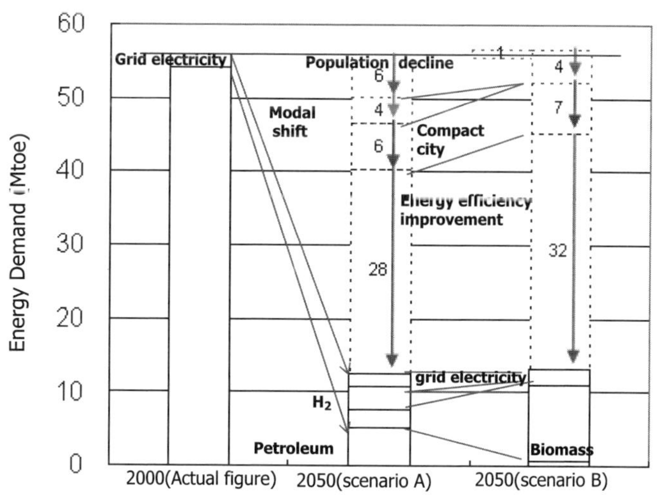

Fig. 18.3. Reduction in passenger transport sector.

In addition, if we construct our cities into a compact type as well as make the streets convenient and comfortable enough that people can live by walking, travelling by car can be reduced drastically. In Japan, it is impossible to go shopping without a car, let alone to the suburban shopping mall. To formulate a city policy when confronting the fastest aging society in the world, we should focus on the hollowing central part of the city where markets and new types of plazas must be rebuilt to attract young people. Blocks where people can live on foot are necessary. On the other hand, introducing Light Rail Transit and securing the bus-only lane, together with the European policy that enables passengers to take whichever bus they like within 1 hour using only one ticket, the utility ratio of common travel instruments taking the place of private cars will be enhanced significantly. The cruel working conditions of freight transport such as long distance driving at midnight will be solved by the conversion to train transport and the CO_2 emissions can also be abated accordingly. However, the barrier now is the capacity constraint of railways. The solution to this problem is to give priority to public investment in railways over road construction to achieve a modal shift of transportation.

Japan's population is expected to dwindle by 20–30% by 2050; at the same time, the population will become more and more concentrated in cities with the intensified aging problem, especially the mega cities and provincial central cities. However, country areas, which serve as the supplier of biomass and a source of CO_2 supply, should be administrated properly; in addition, stable and strong production ought to be maintained to prepare for unusual events in food supply resulting from climate change. Hence, we should make production and dwelling stable in country areas.

In urban areas, high insulating and energy-saving designs in which energy-saving controls are available should be applied widely to rebuilt houses and office buildings. Air conditioners equipped with energy-saving effort display function together with BEMS and HEMS, used in offices and houses, respectively, which can adjust energy use to the optimal state, are necessary. These efforts will cause industrial construction to shift gradually from the energy supply industry to the information measurement industry. On the whole, land reform that prevents buildings from pencil-type (be jumbled up close together) is in emergency, and we should prepare for a huge systematic revolution. We can make use of renewable energy simply by itself or connect it with the existing electricity grid as well as take in those unused energy resources like rivers and rubbish. To construct the district-wide energy-saving town as has already been carried forward in the Tokyo area, local corporations are required to coordinate beyond business types as a matter of course.

Since the insulating technology for home use has been fully developed, the necessary thing now is to standardize this technology and oblige people to adopt it when rebuilding. In addition, loans with low interest can be offered to those low-carbon energy system projects involving solar heater, photo-voltaic and fuel cell, etc. However, there still exist bottlenecks in insulating buildings, which is technicians for design and construction work are in great short supply irrespective of the rebuilding cycle. Therefore, to convert to new industry, personnel training must be arranged simultaneously with corresponding educational institutions.

Adapting social infrastructure including city, road, building etc. to a low-carbon pattern is an urgent need. Otherwise, once the high energy-consuming infrastructures are fixed, the property will last more than 50 years.

18.5. What Must be Changed?

18.5.1. *Readiness for change is necessary*

Aiming at a low-carbon society, we must also get ready for the coming zero emissions 100–200 years later. We all know that there is a positive relationship between the intensity of climate change and the CO_2 density in the air. And it has been noticed that CO_2 emissions surpass absorption by 2 times. Hence, our initial task is to reduce CO_2 emission by half by 2050, and then limit the emissions within the ability of absorption. As has already been shown scientifically, a low-carbon society, engaging in carbon reductions, is reasonable and inevitable. Now that there is no escape for human beings, why not make a determination as early as possible, take quick actions, and push forward with the construction of a new society? This is the new challenge we are facing now.

18.5.2. *Transition towards a low-carbon society*

A huge change will be happened to technology. We now living in the center of a technical society: it is impossible to maintain our life without technology which is considered as the accumulation of human intelligence. However, most of the technologies and products take energy, nearly all of which is provided by fossil fuels; hence, greenhouse gases are emitted. For climate stability, there must be schemes to promote the development of technology, taking it into social systems as well substituting energy with low-carbon resources.

The burdens on the demand side have been strengthened. Even the world (IEA, 2008) or Britain's 80% abatement target (UKERC, 2010) emphasized the limitation of energy supply side's abatement efforts; reducing energy demand by half by 2050 is required at the same time. As symbolized by stand-by electricity, so far the function of technology is to provide utility and convenience irrespective of energy cost. From now on, the essential work of technicians is to examine how much utility is really necessary and spare no efforts to minimize energy consumption in fulfilling this utility. The present way of enlarging energy use by monopolized energy supply and the technology of high energy-consumption on the demand side should be rectified and the demand side ought to make clear the necessity independently. In addition, introducing energy-saving technologies as well as taking responsibility to save energy is also required on the demand side. Only with the coordination of both sides will the abated energy use be fulfilled with low-carbon emissions.

Technical competition must be intensified. Taking energy/GDP as an indicator, the energy-technology progress rate, computed as around 1.25% per year so far, should be doubled to reach the 70% abatement target. After referring to several models, Japan is expected to keep this rate at 1.7%–2.4%, while with rates of 2.8% for Britain and 2.4% for Germany; European countries are striving to convert to an energy-saving mode.

The leapfrog phenomenon can be observed in technical system. Taking the electric car as an example, it's not an extension of the gasoline car, but a kind of electric good started by the motors installed in the wheels. Thus, technology will have its core shifted from engines to batteries, car bodies and steering gear. What's more, the energy supply system is also supposed to transition from gasoline stations to household charging and the battery will be incorporated into a smart grid, which enables systematic electricity adjustment. We'll see the technical system develop beyond industrial classification, forming an unprecedented look at a low-carbon technical system.

Renewable energy is unavoidable since only renewable natural resources are available in the ultimate zero-emission world. Nuclear power is applied widely at this moment; however, a system of geographically dispersed energy like solar energy, wind power and biomass, needs to be established sooner or later. For now, connecting these dispersed resources with an electricity system is one of our tasks.

For industries and corporations, great transformation is inevitable. Corporations that provide service in the form of goods are facing the question of how the energy consumed in production and utilization can be reduced. Taking this as a start, we should also extend our consideration to questions like "what is the service a firm ought to provide originally?" as well as "should the energy and goods in the production process of a low-carbon society be the same as that of the present society?" It is convenience and good travel service rather than a car itself that consumers need. If so, why don't the car company and the railroad company coordinate to provide service that enables passengers to enjoy reading time while traveling? As discussed above, cars can still move even without engines; analogously, travel service can still exist without cars.

The concept of "value of climate stability" must be taken into the economic system. To tackle climate change, carbon emissions need to be abated while carbon absorption should be enhanced simultaneously. Things will not change if everyone just sits on their hands; there should be some who are willing to devote themselves to abatement. Besides, a social award mechanism must be established to award those who are making efforts while laying burdens on those who aren't doing anything. We need to take the value of stable climate as well as the reverse side — the hazard of carbon emissions-into the economic system. For instance, it is reasonable to add a credit price and carbon tax decided in carbon trade markets onto energy prices as a kind of climate fee. On the other hand, the carbon tax revenue can be used on energy-saving technology R&D as well as mitigating employment friction caused by social conversion and other social welfare, to say, return to the people again. Thanks to the development of energy-saving industry and energy-saving actions on the demand side, money for energy imports can be saved to promote the creation of new value in Japan.

Change may also happen in people's lifestyle. Ways of showing off through squandering energy and possessing huge materials are likely to be replaced by the immaterial concept. In terms of materials, considerations like "satisfactory" will undoubtedly become more and more common.

18.6. Conclusion

Of course, a low-carbon society itself is not the ultimate form we are longing for. However, considering the great share and effect of the environment in human society, low-carbon efforts, as the first step to secure our comfortable and easy life, must surely be accomplished. This is, at the same time, consistent with the policy of sustainable development.

Conversion towards a low-carbon society cannot be achieved without fundamental social reforms. Whereas there often exist those who insist on the present system, sooner or later we will persevere in overcoming the difficulties and take a step forward under the guidance of science. The time we encounter challenges is the time our intelligence is tested.

References

Center for Consumer Research IT Team Dentsu Tokyo RCAST (2007). "The Social Lifestyle of Against Global Warming: An Eco-design for IT Society 2050." Dentsu.

Fujino, J. and Nishioka, S. (2008). *A Featuring and Implementation Scenario toward a Low-Carbon Society*. Vol. 12, No. 2/2007, Global Environmental Research. The International Association for Environmental Research.

Fujino, J., Iwabuchi, Y. and Ehara, T. (2009). *12 Measures Aim at a Low-Carbon Society* (in Japanese). Nikkan Kogyo Shimbun, Ltd.

Hanaki, K. (2009). *Analysis of the Realization of a Low-Carbon City* (in Japanese). Department of Urban Planning, University of Tokyo.

IEA (2008). "Energy Technology Perspective 2008." OECD.

National Institute for Environmental Studies (ed.) (2008). "12 Measures Aim at a Low-Carbon Society." https://2050.nies.go.jp.

National Institute for Environmental Studies (ed.) (2009). "Japan's Low-Carbon Society Scenario 2050: Feasibility Study 70% Reduction of Greenhouse Gas." https://2050.nies.go.jp

Nishioka, S. (2008). *The Pathway to the 70% Carbon Dioxide Reduction Scenario for Japan's Low-Carbon Society* (in Japanese). Nikkan Kogyo Shimbun, Ltd.

Skea, J. and Nishioka, S. (2008). Low Carbon Society (LCS) Modeling. *Climate Policy: Modelling Long Term Scenarios for Low Carbon Societies*, 8, s5–s15.

UKERC (2010). UKERC Energy 2050 Scenario.

2050 Japan Low-Carbon Society Scenario Team (Feb. 2007). Japan Scenario Towards Low-Carbon Society (LCS).

Chapter 19

TOWARD A LOW-CARBON CITY

*Keisuke Hanaki**

19.1. Introduction

Population in cities is expected to increase substantially over the world in the future. Therefore in order to suppress emission of greenhouse gasses causing climate changes, forming low-carbon city is required. Low-carbon city doesn't merely aim at suppression of emission of CO_2. It also aims at circulation of resources, symbiosis with nature, high quality of life and vitality of society. To realize low-carbon city, a systematic redesign of energy policy for both supplier and consumer and the structure of city is needed. In case we secure energy demand by limited renewable energy, suppression in demand is necessary in increase of renewable energy ratio. Different strategies should be applied among metropolitan, medium cities and small cities. Proactive city-planning focusing on the future is necessary in the realization of a low-carbon city.

19.2. Necessity of Carbon Emission Reduction

Climate changes has been already causing temperature rise and it is inevitable that this trend will continue because of the huge scale of the inertia of the ocean and the land that form climate systems of the earth. Scientists have argued to what extent we can allow the temperature rise in the world average from several viewpoints such as food production, influence

*Keisuke Hanaki, Proferssor at Graduate School of Engineering, The University of Tokyo, Japan.

Table 19.1. Increase of urban population.

	1970	2000	2030 (projection)	2050 (projection)
World average	36.1	46.4	59.0	68.7
Developing region	25.3	40.0	55.0	65.9
Japan	53.2	65.2	73.0	80.1
China	17.4	35.8	61.9	73.2
Thailand	20.9	31.1	45.8	60.0
India	19.8	27.7	39.7	54.2

Source: United Nations (2009), China Statistical Yearbook (2009). Definition of urban population varies among countries.

on health, the ecosystem and sea level rise. In the COP15 of UNFCCC held in 2009, there was an agreement that limits maximum temperature rise in the world average to 2°C. To achieve this goal it is required to reduce emission of greenhouse gasses by half at the year of 2050. Considering growth of population, industry and changes in lifestyles in developing nations to some extent, developed nations need to reduce 70–80% of CO_2 emission. Population in city is expected to increase both in developing and developed countries in the future. Especially in developing countries the growth rate is high.

Table 19.1 shows the transition and projection of the city population ratio in several nations. The population of the city in China, Thailand and India which was about 20% in 1970 has been increasing remarkably and the ratio is estimated to amount to 73%, 60% and 54% respectively in 2050. The ratio of Japan will rise up from approximately 70% to 80% in 2050.

In developing countries, since both total population and the ratio of urban population ratio are growing simultaneously, the actual urban population is increasing drastically. In Japan, since the urban population ratio increases despite the total population decline, urban population will not decrease much while non-urban population will decline remarkably. In any cases we can see from this comparison that how to restrain emission of CO_2 in cities will be the biggest issue.

On the other hand, the city area where human activities are intensive has the biggest emission of greenhouse gasses and therefore there is still much room for improvement.

19.3. Necessity of Realization of Low-Carbon City

In Fig. 19.1 we compared the CO_2 emission per year per capita in 2006 of the national average (Japan Energy Conservation Center, 2009) and that of the city (Greater London Authority, 2007; Environmental Bureau of the Tokyo Metropolitan Government, 2006, 2009; Kawasaki City, 2010; Dhakal, 2009) as for the UK, Japan and China (China Statistical Yearbook, 2009) according to several existing statistics and researches. It is well known that about the national average, the figure of Japan and the UK are at the same level and that of China is approximately 45% of Japan. However, we can find out several aspects from comparisons by cities. The figure of Tokyo is less than half of that of the Japanese average. While that of Kawasaki, which adjoins Tokyo and has many commuters to Tokyo, is the double of the national average, and the quadruple of Tokyo. There is little difference in lifestyles in both cities despite degree of dependence on automobiles in the transportation sector in Kawasaki is higher compared to Tokyo. The manufacturing industry causes

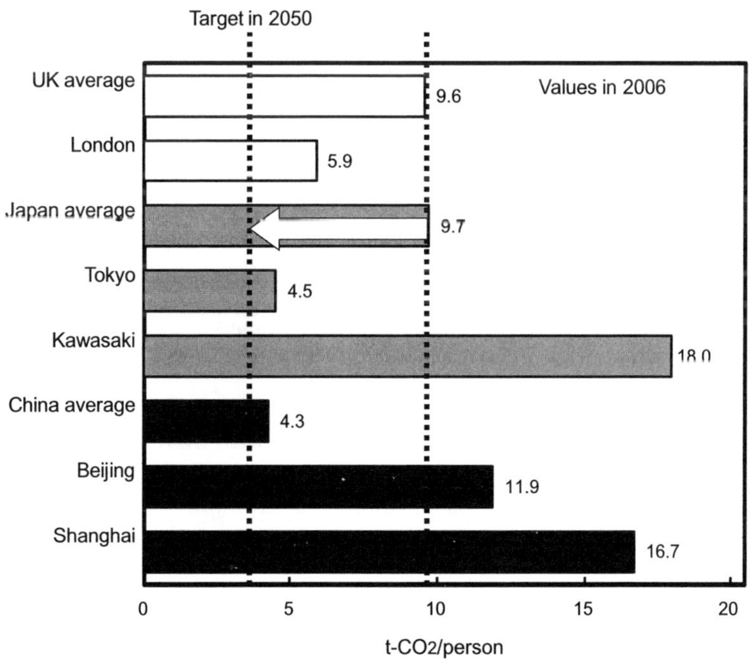

Fig. 19.1. Carbon emissions per capita by country and city.

this difference. As for Kawasaki, industries with high energy consumption such as the steel industry locate mainly in coastal areas, while Tokyo has few such industries but amounts in the head offices of enterprises and the major industry is the service industry. Actually the ratio in the GDP of the secondary industry, mainly the manufacturing industry, is 16.3% (Tokyo, 2011) in Tokyo, while in Kawasaki it amounts to 29.8% (Kawasaki City, 2011) with the attribution of the material industry. The emission in London is lower than the national average like Tokyo due to similar reason. On the other hand, the amount of CO_2 emissions per capita in Beijing and Shanghai where the economy is growing rapidly are 2.8 times and 3.9 times as the national average in China. The present ratios of the secondary industry in GDP in 2006 amount to 26.8% in Beijing (the figure in 2007 quoted from JETRO, 2010) and 48.5% in Shanghai (JETRO, 2011), showing that especially in Shanghai the energy consumption by manufacturing is enormous. However, the ratio of the national average is the same level as Shanghai, and we can't explain by the location of the manufacture industry that these 2 cities shows much larger emission, so we can infer that there are the differences in lifestyles between rural and urban areas.

It should be noted that the emission in Beijing and Shanghai is much larger than that of Japanese national average, not to mention that of Tokyo. That means the advanced areas with high economic growth in China already has larger emission of CO_2 per capita than Japan. If the urbanization continues in rural area in China at the present rate, there will be enormous amount of CO_2 emission. In order to reduce the emission by half in 2050 over the world, Japan reduces the total amount of emission in 1990 by 70%. The CO_2 emission per capita must be 3.5 tons per year.

Figure 19.1 shows this goal. The present figures in Beijing and Shanghai are much higher than this target, and that of the national average in China. This means not only cities in Japan but cities with high growth in China need to start for reduction of emission per capita.

We noted on only Japan and China here, but the situation of other developed nation is similar to Japan, and that of other developing nations is similar to China despite there is divergence in scales, which means that we need to proceed toward low-carbon society to solve the pressing issue. As mentioned above, the emission of CO_2 per capita in Kawasaki is 4 times higher than that in Tokyo. And the figure in Tokyo is similar level as the desired

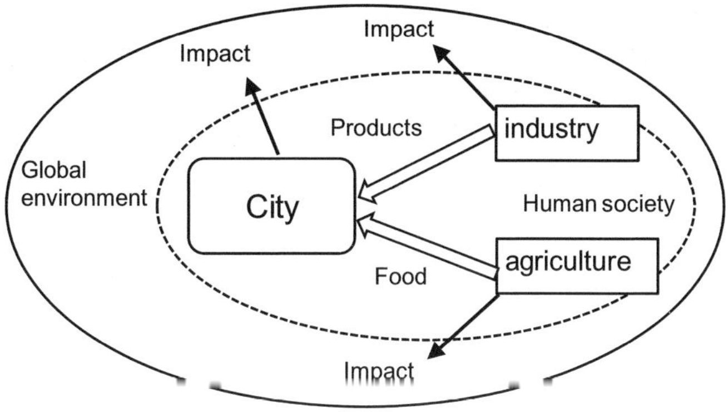

Fig. 19.2. Induced impact of urban activities.

value of the national average in 2050. However this does not necessarily mean that we can achieve this goal by merely imitating Tokyo in all other cities in Japan. This is because Tokyo depends on other cities in commodities. Figure 19.2 shows the material flow and environmental impact of city like Tokyo. Products and foodstuffs are supplied from other industrial city and rural area to sustain urban activities. Environmental loadings from such industrial city and agriculture are also attributed to the city like Tokyo.

Suppressing consumption of commodities which produce greenhouse gasses in manufacturing process will be important since greenhouse gasses have the same effect regardless of the location of emission. In this way we are required to consider the reduction of resource consumption in urban activities as well on constructing low-carbon society.

There are still more other issues that the future cities should solve. As far as environmental issues are concerned, symbiosis of nature and human activity is also a big issue. Moreover, in order to secure sustainability of activities in cities, which is the ultimate purpose, we need to consider the issue from economic and social aspects as well as environmental aspect. It is required that construction of sustainable society as a whole should be achieved through formation of low-carbon, resource circulating and nature harmonious societies, at the same time considering on sustainability of economic and social aspects. On considering sustainability from environmental, social and economic aspects, it is necessary for us to consider them in both regional and global scale (see Fig. 19.4).

Environment in region and global have mutual relationship, and local community and the global, or international society have close relationships. Needless to say, local economy and global economy have close relationships as well. Therefore, sustainability of the whole world has mutual relationships with that of the local community.

19.4. The Blueprint of Low-Carbon Society

The term "low-carbon city" derives from the city in which emission of CO_2 is low. However, as mentioned above, the achievement of construction of sustainable cities in the future calls for not only dealing with the CO_2 emission issue singly but building the image of idealistic society in aspects including circulation of resources, harmonization with nature, society and economy (Fig. 19.3).

Various technologies have been developed to solve the climate change issue so far, and currently they are under further development and lowering of costs. However merely introducing such technologies does not lead to remarkable reduction of CO_2 emission required in low-carbon cities.

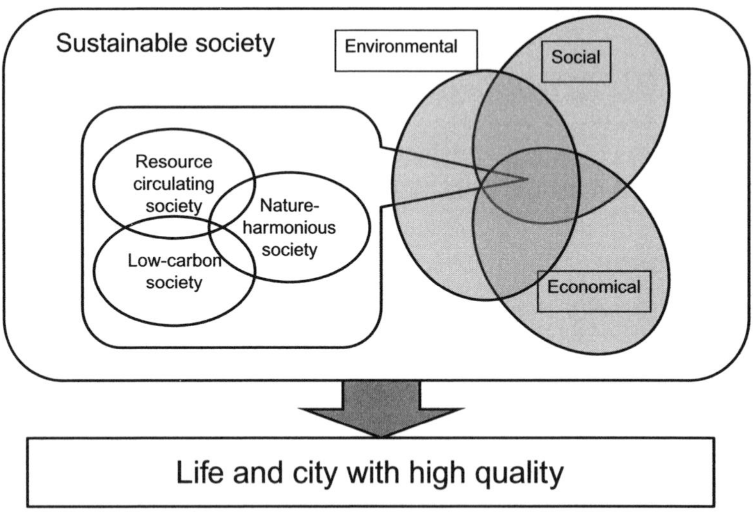

Fig. 19.3. Formation of sustainable society.

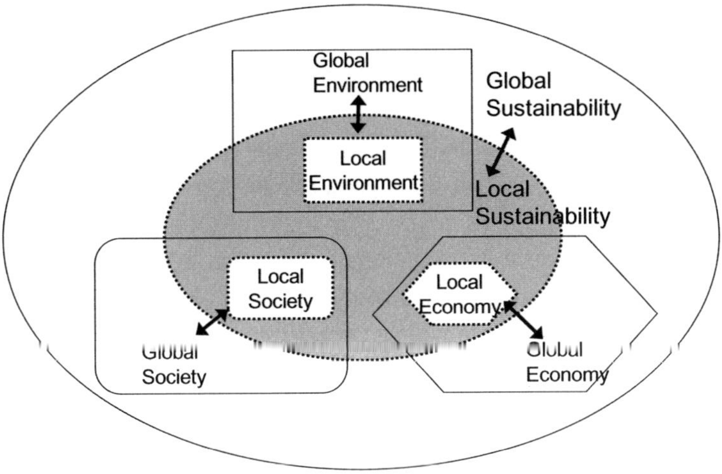

Fig. 19.4. Local and global sustainability.

Formation of ideal cities in the future with the help of these technologies is needed. Therefore such goals include other issues than the climate change issue. Formation of cities of recycling and harmony with nature from environmental aspects will be the goal. The targets are to achieve high quality of life and to solve the ongoing issue of decaying vitality, which is a large problem in current Japan, and to form a lively society. If we only aim at reduction of greenhouse gasses such as CO_2, in an extreme case, the policy will be to restrict the consumption of fossil fuel strictly and to raise their prices notably.

However it will be hard to form an ideal society by introduction of such restriction and price policies without changing the structure of current society. We should rather achieve reduction of CO_2 as a trigger to aim at forming desirable society. Problems the Japanese currently face vary and range from aging, general welfare, low growth of the economy and deterioration of infrastructures.

The demand for solution of these issues is large and it is easier to obtain social approval and participation. Changing society toward low-carbon one should include solutions for other social issues, otherwise the feasibility of the policy will be low.

19.5. Strategy for Realization of Low-Carbon Cities

Then, what sort of strategies should we take to realize the low-carbon cities in urban areas?

There is no bright idea that can singly deal with all these social problems concerning low-carbon. It is required to combine several types of measures in the strategy. There are 3 strategies: (1) measures from energy supplier side; (2) measures from energy consumer side; and (3) the forming of cities. The first strategy will be the decrease CO_2 emission of energy being supplied. It means to lower the carbon intensity, namely, the emission of CO_2 per energy. Renewable energies such as solar power, wind power and biomass produce no CO_2 on its consumption. In case of solar power, the higher introduction costs are a big obstacle although it is already possible to be introduced in small scales. Feed-in tariff system, which allows the purchase of generated power at high price, is about to be introduced to settle this issue. However there are still technical problems such as harmonization of electrical powers generated by general power companies and solar powers when it comes to large-scale introduction. In addition, the technology of battery is necessary in solar power generation. In case of biomass, the problem is the limits of the resource.

In cities wastes and sewage will be the source of energy. Although powers are generated through incineration of solid wastes over 300 facilities in current Japan, the power generation efficiency is only a little over 11% (Ministry of the Environment, 2010), which is the quarter of general power generation. And what is more, the total amount of wastes is limited, and in forming recycling society the limits will decline further. Unlike these urban biomass, forest biomass or wind power resources do not exist so much in urban areas.

On the other hand, there is not enough demand in rural areas like mountainous areas though the area abounds in these resources. This means these renewable energies are not utilized though there are potential opportunities. In this case, the rural mountainous areas like and urban areas can co-work on an undertaking.

In diffusion of new technologies, it is effective to combine technology push through engineering development and demand pull that spreads technologies along with demand.

Urban areas create new demand and can contribute in utilizing these energies in the mountainous areas. For instance, Tokyo has planned the "Renewable Energy Strategy Tokyo" that will cover 20% of energy demand by renewable resources by 2020, and it is assumed in the scheme that Tokyo may purchase renewable energies from outside the area. These cooperation are promoted in small scales in public enterprise plants. We can see social and economic factors are the keys in diffusion and practical application. Another big social issue is about policy on utilization of nuclear power. Nuclear power emits no CO_2 and therefore areas of high utility of nuclear power have been successful in maintaining low-carbon intensity.

However, the viewpoint from security on utilization of nuclear power generation has been a critical obstacle since the accident in March 2011.

The second strategy is on saving of energy consumption. It has been promoted as one of the energy policies so far. Various technologies have been developed such as household electrical and business appliances and automobiles. However in this field the activities of human play a large role. No matter how higher the energy efficiency of the appliances develops, it will not lead to effective energy conservation unless people waste energies and introduce various unnecessary machineries. In household sectors, social factors are much important in fundamental aspects. The consistent trend in Japan is the decline in number of residents per household. This means that shift from big family to nuclear, and single families has been consistently proceeding.

Figure 19.5 shows the change in trends of situation of residences from 1968 to 2008 (Ministry of Internal Affairs and Communications, Bureau of Statistics, 2009). The number of people per household has dropped from 4 to 2.5, and floor space per capita has almost doubled. It is obvious that these trends are the factors of increase in energy consumption per capita.

On the other hand, the growth of floor space of these buildings attributes to the increase of CO_2 emission in offices and commercial business sectors. There is the background of shift to the tertiary industry in this issue. In transportation of humans, automobiles are the major CO_2 consuming factor and in freight transportation tracks are the major factor. Here social factors play a large role as well.

Though it is important to enhance energy transmission efficiency and achieve energy conservation in individual scales through introduction of

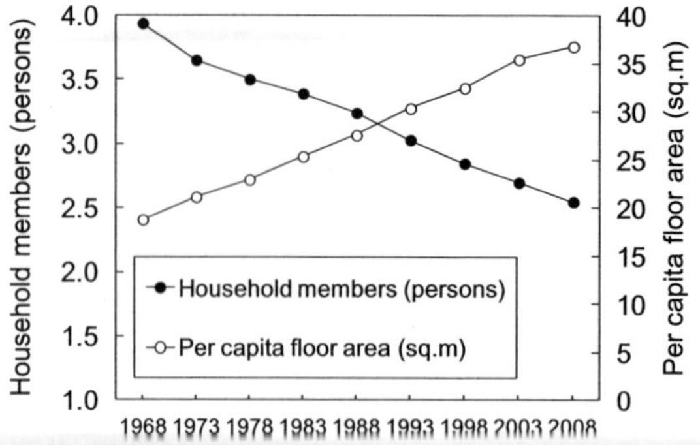

Fig. 19.5. Residential condition in Japan.

hybrid and electric vehicles, it is essentially necessary to basically restrict mobility of people and commodities themselves. Restricting this mobility without lowering quality of lives is a big issue indeed. The third strategy is the improvement on city structure. The aimed ideal city is the "compact city". We consider 2 different cases in scales in Fig. 19.6.

In first case, the strategy is to scheme a compact district especially in urban areas, and this enables introduction of district heating/cooling system.

These methods can be implemented in large redevelopment of urban cities.

It will be more effective if we can utilize sewer heats and incineration heats from wastes which are unused at present. The second strategy is about density distribution of the whole city, and it is about enhancing density of the city center, so to speak, converting into a compact city structure.

If we can model and create cities with high density and relevant scale, it will be able to move without depending too much on automobiles inside cities, and create a convenient and high-quality life environment. Especially about transportation, we can expect the effect of the compact city. Enhancing the city density enables railway service economically and environmentally feasible and automobiles are replaced. The transportation distance becomes shorter in using automobiles when necessary.

On the other hand, CO_2 emission in household and business sector is not necessarily reduced through compact cities. Energy consumption does not decrease unless the conversion of residence style such as conversion from detached houses into apartment starts along with building compact cities.

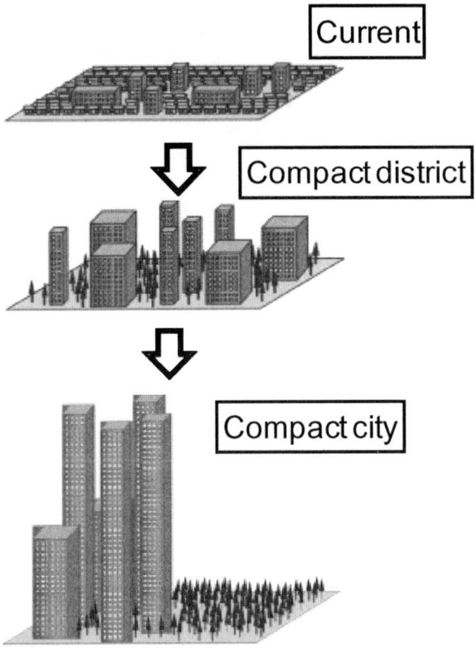

Fig. 19.6. Compact city.

The concept of compact cities has been focused on and aimed at not only in Japan but in European cities.

However, since metropolitans like Tokyo and Osaka already have high population density, despite the promotion of creating compact city blocks, further enhancement of population density of the whole city may not necessarily be an effective solution. We should rather take this measure in cities of middle scales, where cities spreads with low population density and railway transportation has not been introduced yet. Compact cities bring about effects other than low-carbon. We especially expect the contribution on improvement and maintenance of quality of lives.

As Fig. 19.7 shows, with maintaining expanded area during the past economic would lower the level of medical service, education and cultural activities whereas environmental loading from the transportation would be unchanged in the middle to small cities. This is currently happening as a matter of fact.

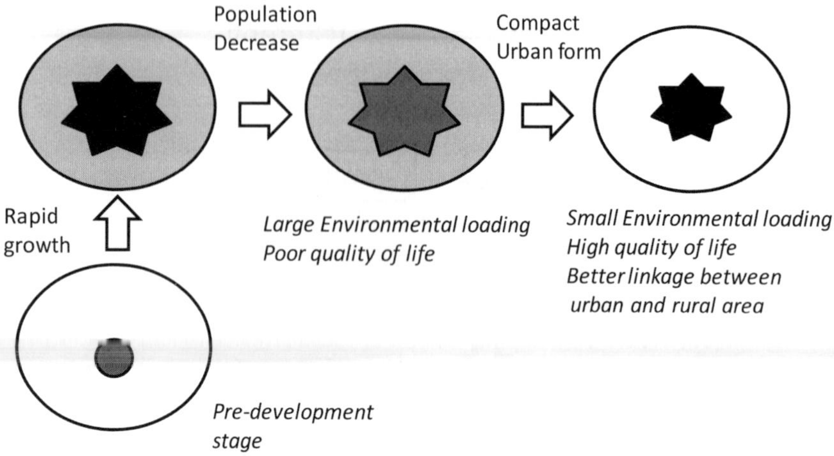

Fig. 19.7. Compact city under population decrease.

During the rapid economic growth, the agricultural and forest were converted into residential area and land uses became artificial including the vicinity. Restoration of the natural environment cannot be expected as long as population decreases in non-organized manner.

On the other hand, if the urban areas shrink along with the decreasing population with a proper plan, we can expect to maintain service qualities to the residents since a proper density will be maintained in the area, and also expect the restoration of natural environment that leads to reconnection of cities with Satochi and Satoyama (hinterland forest).

The 3 strategies, wise energy supplier, wise energy consumer and proper urban form have interactions. In case we meet energy demand by renewable energy whose resources are limited, we can enhance the share of renewable energy use when we successfully restrict demand, on the other hand the share remains low if the demand is not controlled. Conversion of single houses into apartments, restriction on size of business and commercial spaces and change of means of people's transportation, which are measures of restriction on energy demand, have strong relationship with compact cities.

With proper planning, the compact cities could provide land for wind and solar power generation with proper planning. We considered the aptitude on difference in scales, and Fig. 19.8 shows a proposal of future strategy

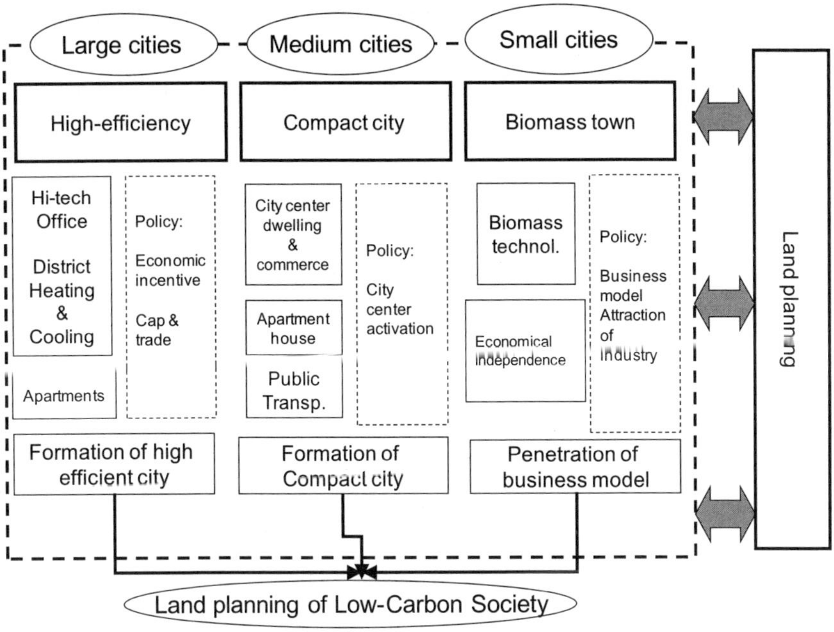

Fig. 19.8. Strategy for low-carbon city with various scales.

toward large, middle scale and small scale cities whose hinterlands have agricultural and forestry spaces.

In large cities, the basic strategy is to increase the efficiency of functions of the houses and offices. As a means to achieve this, it will be effective to construct offices of high quality, introduce district heating/cooling system and promotion of apartment living. Adding economic incentives is effective. As a pioneering instance, Tokyo Metropolitan Government has introduced a reducing obligation of total CO_2 emission and a emission trading system for large enterprises from 2010. In many of middle-scaled cities town areas spread in relatively low density, which forms a traffic system depending on private cars. In such areas the strategy will be to designing the city compactly.

On the other hand, in small cities with farmland and forests in their hinterland abound in biomass though the demand is small. The strategy is to promote utilization of biomass, and in order to achieve this, it is necessary to enhance local economies by commercialization along with adoption of biomass technologies. Firstly, we must show successful business models since there are few precedent about utilization of biomass energies.

It is required to promote the national land planning toward low-carbon society along with various strategies of different scales each city takes.

19.6. Roadmap Toward Low-Carbon City

As we have noted, it is hard to realize low-carbon cities only by improving various technologies that currently available and under development. We are required to change way of lives as humans, and change the cities that we habit and buildings and infrastructures that construct it. However the problem here is that the life of the buildings and infrastructures are approximately 50 years at least. Since we have only 40 years until 2050, which is generally target year, in order to form desirable cities by 2050, we adopt new ideas to current design of buildings and infrastructures.

As we have shown in Fig. 19.9, in maintenance of society and usual environmental policies, reactive policies that cope with problems after the occurrence of the issues are taken.

However, forming low-carbon cities by such process may take more than 100 years to form desirable cities. So it is necessary to determine desirable form of cities in advance and proceeds good city-planning in residents and cities with proactive policies. To achieve this, it is necessary to draw future blueprint of the cities, or desirable form of the society though there are uncertainties. It is important for the supplier of energies and technologies,

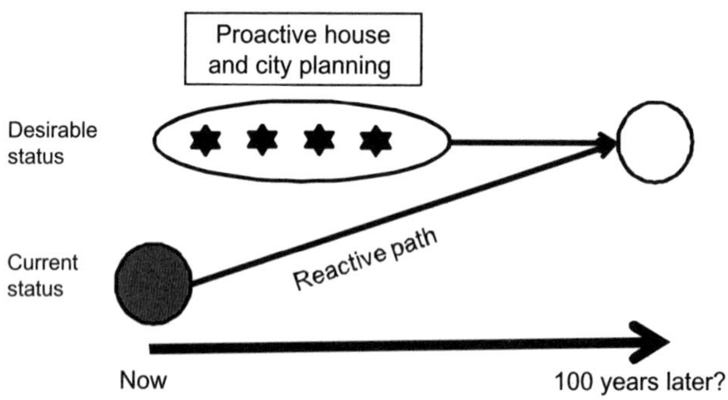

Fig. 19.9. Proactive formation of low-carbon city.

residents and policy decision makers to participate in to determination the city plan, and start new city-planning from present.

References

China Statistical Yearbook (2009). Cited from Japan Science and Technology Agency, Science Portal China.

Dhakal, S. (2009). Urban Energy Use and Carbon Emissions from Cities in China and Policy Implications. *Energy Policy*, 37, 4208–4219.

Environmental Bureau of the Tokyo Metropolitan Government (2009). *Comprehensive Survey of Greenhouse Gas Emissions in the Metropolis (FY 2006)* (in Japanese).

Greater London Authority (2007). *The Mayor's Climate Change Action Plan*, Chapter 3.

Japan Energy Conservation Center (2009). *EDMC Handbook of Energy & Economic Statistics in Japan* (2009 version) (in Japanese).

JETRO (2010). "Beijing Outlook and Investment Environment 2010." http://www.jetro.go.jp/world/asia/cn/north_west/pdf/beijing1006.pdf (visited on Dec. 14, 2011).

JETRO (2011). "Shanghai Outlook." http://www.jetro.go.jp/world/asia/cn/central_east/pdf/shanghai_1111.pdf.

Kawasaki City (2010). *Planning Measures to Prevent Global Warming* (in Japanese).

Kawasaki City (2011). *Economic Data of Kawasaki City (FY 2010)* (in Japanese).

Ministry of Internal Affairs and Communications, Statistics Bureau (2009). *Preliminary Version of Housing and Land Survey 2008.* (in Japanese).

Ministry of the Environment of Japan (2010). *Waste Disposal in Japan (FY 2008 version)* (in Japanese).

Tokyo Metropolitan Government (2006). *The Use of Renewable Energy Strategy* (in Japanese).

Tokyo Metropolitan Government (Economics Department) (2011). *Annual Report of Tokyo Metropolitan FY 2009.* (in Japanese).

United Nations (2009). "World Urbanization Prospects: The 2009 Revision Population Database." http://esa.un.org/unpd/wup/index.htm (visited on Dec. 14, 2011).

Chapter 20

LOW-CARBON GREEN GROWTH STRATEGY

*Yong-Ho Choi**

20.1. Introduction

The Korean economy, which depends highly on overseas markets was impacted by the global financial crisis that first occurred in 2008 U.S. financial crisis. The credit crunch broke out into international financial markets. The international economy fell into a temporary crisis. The growth rate of real economy in the fourth quarter of 2008 dropped to −5.1%. What is fortunate is that, to stimulate the economy, the framework among the world's major economies had been enhanced, and rapid financial market stabilization policy and expansionary monetary and fiscal policy gradually began to appear in the first quarter of 2009. Thanks to this, in the first half of 2009, credit contraction in the financial market eased and speedy recovery of economic activities in export, industrial production and the services industry was also seen. Economic growth rate forecast for 2009 were revised upward to −1–0.5% recently and some countries reached the stage to discuss a partial export strategy.

The global, "financial tsunami" (financial crisis) has now somewhat subsided and concerns in the developed countries seem to be all geared toward a low carbon society. To anticipate a green lead market, the so-called Green Race is becoming popular among the developed countries and is represented by the United States. The regime after the Kyoto Protocol has already been discussed and the 15th Conference of Parties (COP15) of United Nations Framework Convention on Climate Change (UNFCCC) that was held in

*Yong-Ho Choi, Korea KNU Emeritus Professor, Director-General of Industry Research Institute Republic of Korea. He served as chairman of the presidential committee of Low-carbon Green Growth for Korea.

Copenhagen in December 2009 is also beginning to work. In Korea, after declaring "low-carbon green growth" as a new paradigm for the nation, various plans to achieve a green country are being made and implemented. This chapter discusses the green growth problems and reveals the best features of this Korean industry as a result of the global financial crisis.

20.2. Background of Green Growth Policies and Promotion Process

The core of green growth is to turn the pattern of economic growth to a eco-friendly one while sustaining economic growth. It means an economic transition from yellow economy to green economy. In Korea, the green growth policy began being promoted almost at the same time as the global financial crisis. So it also has a strong influence on economic recovery policy.

Internal and external factors in promoting the policy are complex. Externally, we have to emphasize our position to deal with the global environmental changes of global warming and energy crisis actively. Internally, when the conversion to the new paradigm, which lights the stage of the development of Korea's economy, is coming at an inevitable time, new growth engine, job creation and economic stimulus policy are required urgently.

20.2.1. *Responding to global warming and the energy crisis*

Causing severe weather disasters and ecological destruction, global warming has been identified as the crisis that threatens the survival of humankind. The increase of global demand for energy, the rapid depletion of energy resources and the soaring energy prices are factors which increase the sense of crisis.

To solve the problem of global warming and the energy crisis, developed countries and the UN have conducted conventions on climate change between developing countries and developed countries and promoted alternative energy and low carbon technologies since the 1990s. Recently, some developed countries, like the EU, broke away from that carbon-dependent economy paradigm and concentrated their national powers to securing the global market in green technologies and green industries.

However, Korea remains in the situation that 97% of energy consumption is dependent on import (they are the world's sixth largest oil importer and second largest LNG importer) and more than 80% of that is fossil-fuel.

It is said that Korea's countermeasures for this international crisis are too simple, specifically, they are especially lacking in policies for greenhouse gas emissions. In this case, however, Korea will probably remain as a country with large obligation of carbon emissions reduction when the first commitment period of greenhouse gas reduction obligations (2008–2012) ends in applying the Kyoto Protocol. Therefore, it is time for Korea to establish extreme policies for developing a green industry and improve the level of green technology.

20.2.2. *Conversion of the paradigm and necessity of creating a new growth engine*

Since the industrialization of the Korean economy in the 1960s, the so-called "high-growth-type" compression has been achieved. However, this type of rapid growth is compressed against a serious problem. Korea's economy went into low growth accompanied by "jobless growth" in the 1990s (San Francisco, 2009; Oh, 2009). Potential growth rate also had a continuously downward trend. Price-competitiveness oriented manufacturing, export oriented growth strategy and the energy-intensive development paradigms had reached a limit. Therefore, Korea's economy has come to a new development direction that can solve energy and environmental problems while simultaneously spreading a new growth engine (The Ministry of Environment Korea, 2009).

20.3. Promotion Process of Green Growth Policy

Korea's green growth policy began in August 2008 around the time that President Lee Myung-bak presented "low-carbon green growth" as a national issue to be addressed for the next 60 years. During this time, Korea established the "National Energy Plan" to increase independence, "Master Plan for Climate Change" to present climate changing and economic revitalization as immediate tasks in September, followed by the "Green New Deal Project Development Plan" to create jobs. In November, in order to make a systematic promotion system for green growth, Korea inaugurated the presidential "Green Growth Committee" by integrating the "Climate Change Committee", "National Energy Committee" and "Sustainable Development Commission" (Fig. 20.1).

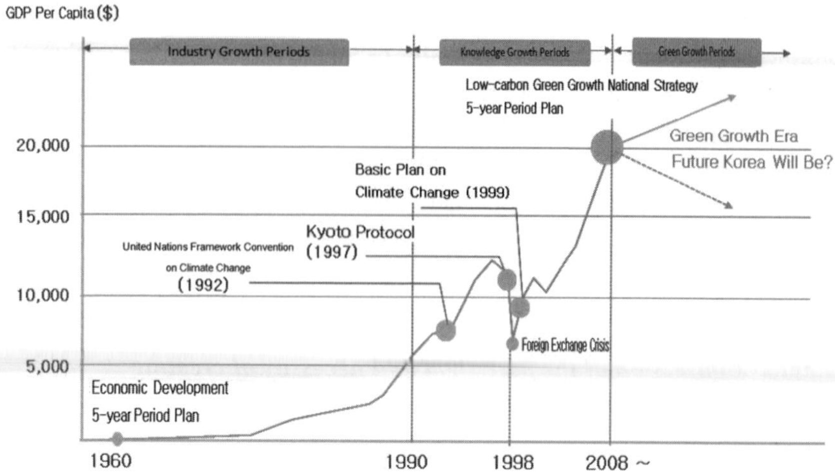

Fig. 20.1. Korea's economy and social development process and new growth era.
Source: National Strategy Committee for Green Growth (2009).

Korea also drew up "Rules Concerning the Establishment and Operation of Green Development Committee (President Instruction)" and set up the "Green Growth Planning Team" for business support. The "Comprehensive Plan for Green Technology Research and Development" was also established to develop green technology and the "New Growth Engine Vision and Development Strategy" to develop green industry. In addition to this, high value-added service industries were also being prepared simultaneously. Korea also confirmed the "Low Carbon Green Growth Basic Law (government bill)" and submitted it to Diet (yet to be approved). The "Green Growth Committee" which the prime minister is a co-chairman of, in addition to a chairman from the private sector, was also officially launched (UNEP, 2009).

With the inauguration of the committee, the drafting of "Green Growth National Strategy (2009–2050)" as a long-term strategy and "Green Growth Five-Year Plan (2009–2013)" as a Medium-term strategy, additional research was needed. For this work, national research institutes, like KDI and KIET, were mobilized. Universities and private professional organizations also participated.

Meanwhile, the "Green R&D Plan" and "Green IT National Strategy" were announced in May of the same year and the "Green Growth Strategy and the National Five-Year Plan" was summarized as a report to be utilized at the meeting for allocation of financial resources of cabinet members in order to strengthen the cooperation within the national financial management plan. The draft was created and presented (through the process of a consensus of stakeholders) in June to the committee of green growth and at the cabinet meeting in July (Ministry of Knowledge Economy, 2009; National Technical Committee, 2009).

20.4. Green Growth National Strategy and Five-Year Plan

20.4.1. *Vision system of green growth and promotion strategy*

Korea's green growth policy has three directions. The first one is a virtuous cycle of environment and economy. In order to achieve this, increasing energy efficiency in the existing competitive industries and striving to develop environment-friendly materials, so-called "greening of leader industries", has been made the top priority. In order to develop low-carbon and green industry, while remaining committed to redesigning the existing industrial structure and to promoting knowledge service industries, the greening of industry-wide value chains should be promoted (Jo, 2008).

The second direction is the improvement of the quality of people's lives and the green-revolution way of life. That is, ensuring that a green lifestyle can proceed smoothly in all living spaces, like countries, cities, buildings, parks and houses. For this purpose, Korea is doing its best to establish a good land development plan and green transportation system for environmental protection, as well as the activation of more public transportation, induced carbon taxes and "energy efficiency compensation systems" to encourage green consumption.

The third direction is the establishment of an appropriate national pride toward international expectations. Korea should participate actively in international co-challenges such as climate change initiatives, accelerate development in the green growth sector and speak firmly for less developed countries (Lee, 2009).

The Korean government has set three major strategies (as shown in Table 20.1), 10 major policy directions (the core element of success) and

Table 20.1. 3 major strategies of green growth and 10 policy directions.

3 strategies	10 policy directions
Adaptation to climate change and energy independence	1. Efficient greenhouse gas reduction 2. Strengthening petroleum energy independence 3. Strengthening adaptive capacity to climate change
Creation of a new growth engine	4. Development of green technology and growth power 5. Greening industries and development of green industries 6. Advancement of industrial structure 7. Infrastructure development of green economy
Improvement of life quality and enhancement of the national status	8. Formation of green land and transportation 9. Green revolution of living 10. Achievement of a worldwide green growth model

Source: Green Growth Commission, "Green Growth National Strategies." (2009).

selected 50 major-practice issues under that. With three directions, the vision that Korea could be in the top 7 green countries of the world by 2020 and top 5 green countries of the world by 2050 was presented in the formulated "Green Growth National Strategy".

20.4.2. *Outline of Green Growth Five-Year Plan (2009–2013)*

According to the "National Strategy of Green Growth", the plan reflects concrete forward momentum subjects and budgets in the Five-Year Plan of green growth. There are 107.4 trillion won (83.6 billion US dollars) of finance investment planned for investment during the medium-term plan. This investment level is twice that of the "Green Economy Initiative of the United Nations" recommended green investment level (1% of GDP). Table 20.2 shows organized national strategies, from the Five-Year Plan, by 10 major policy directions, the target levels of major indexes in 2013 (the year the five-year plan finishes) and in 2020 (the year Korea plans to be in the top 7 green countries of the world).

20.5. Green Industry

20.5.1. *New growth engine for green technology industry*

Korea's green technology has been evaluated as 50–70% of developed countries. To raise this level, the Korean government defined "technology of

Table 20.2. Main indexes of emissions reduction.

(1) Efficient greenhouse gas reduction

Major index	Unit	2009	2013	2020
Greenhouse gas reduction target	—	Goal setting	Promoting reduction	Promoting reduction
Carbon information publicity (CVI)	—	Indicator development	Setting plan	Setting plan
Carbon content storage of mountain and forest	Million tCO_2	1,452	1,613	1,854
Reforested area of North Korea	10,000 ha	—	5	40

(2) Strengthening petroleum energy independence

Major index	Unit	2009	2013	2020
Energy intensity improvement	toe/1,000$*	0.317	0.290	0.233
Renewable energy penetration	%	2.7	3.78	6.08
Nuclear power plant share	%	24	27	32
Energy independence	%	32	42	50

*toe/1,000$: Amount of energy consumed to produce 1,000 U.S. dollars GDP.

(3) Strengthening adaptive capacity to climate change

Major index	Unit	2009	2013	2020
Monitoring climate change	%	40	80	90
Advancement of predictive capability				
Eco-friendly agricultural production	%	4.5	10	18
Strengthening disaster response capacity	%	40	60	70
Total national forest resources	Million m^3	862	953	1,087

(*Continued*)

Table 20.2. (*Continued*)

(4) Development of green technology and growth power

Major index	Unit	2009	2013	2020
Share of green technology investment	%	16	20	25
Share of green technology products in the world	%	2	8	10
Use of green technology information system	10,000 items	Less than 0.2	5	10
Green technology overseas talent	Person	25	100	250

(5) Greening industries and development of green industries

Major index	Unit	2009	2013	2020
Recycling rate	%	15.0	17.0	17.6
Main industries green products export	%	10	15	22
Green partnership between large and small companies	—	685	1,500	2,900
Building for green industry zone	—	5	10	20

(6) Advancement of industrial structure

Major index	Unit	2009	2013	2020
Broadcasting and communications industry exports	Billion U.S. dollars	520	1,237	2,400
Nano fused innovative products creation	—	—	20	30
Number of foreign patients to be treated	10,000 persons	2.7	20	35
Number of MICE* tourist	10,000 persons	58	158	320

*MICE: Meeting, Incentive tour, Convention, Exhibition.

(*Continued*)

Table 20.2. (*Continued*)

(7) Infrastructure development of green economy

Major index	Unit	2009	2013	2020
Domestic carbon market creation	Trillion won	—	0.5	2
Green technology and industry guarantee	Trillion won	2.5	7	8
Energy poor family proportion	%	7.3	5.0	3.5
Number of social green enterprises	—	40	300	500

(8) Formation of green land and transportation

Major index	Unit	2009	2013	2020
Protected area	10,000 ha	10	12	15
UN excellent green cities	—	—	2	5
Railway passenger share	%	18	22	26
Bicycle traffic share	%	1.5	5	10

(9) Green revolution of living

Major index	Unit	2009	2013	2020
Green home	10,000 families	16	30	150
Carbon labeling products certification	—	50	500	1,000
Green purchasing price	Trillion won	2.5	4.0	10
National Green Village Center Opened	—	—	100	500

(*Continued*)

Table 20.2. (*Continued*)

(10) Achievement of a worldwide green growth model

Major index	Unit	2009	2013	2020
Green ODA* share	%	14	20	30
Overseas volunteers	Person	3,730	4,150	(setting plan)
International ranking of environmental performance index	Place	51	30	20
GEF† Providing Level	%	0.23	0.25	0.46

*ODA: Official Development Assistance, †GEF: Global Environment Facility.

high efficiency and low cost silicon solar cells" in 27 technologies as key developing technologies, then selected and trained six green technology industries as new growth engine industries (Makino, 2001).

(1) Renewable energy industry

In the renewable energy industry, strengthening the industrial base of solar, wind power, as well as fuel cell industry was focused on and the investments in these fields are currently active. In the sunlight field, large companies specializing in semiconductor and display production have begun to enter the world market. In wind power, large shipbuilders are endeavoring to enter the market by creating, merging with or acquiring wind power companies overseas. In the fuel cell industry, commercial production for domestic users and technology development for electricity generation and transportation uses are being invigorated. At the same time, commercialization of ocean energy, waste energy and fusion technology of biomass are being promoted as well (Hosada, 2008; Nu & Chang, 2009).

(2) Carbon reduction energy industry

Carbon reduction energy industry is an energy technology fundamental industry that has no or minimal emission of CO_2, which is representative of greenhouse gases. In Korea, CO_2 recovery technology is government-led. Recycling and the heightening add-value of CO_2 applying technology industry are private sector-led. However, these roles are shared, like before. The amount of domestic nuclear power share in 2008 was 35.7%. Korea also has a plan to add to total construction by 2030 and to localize core technologies which are not self-sufficient (for example MMIS) early to promote the exports of nuclear power plants.

(3) Advanced water treatment industry

The advanced water treatment industry is the "Blue Gold" industry which has been predicted to go beyond the oil industry. Application of membrane filtration technology systems that optimize water management, expansion of wastewater treatment and reuse of water and advancement of seawater utilization technology are being promoted.

(4) LED application industry

LED is the "shiny semiconductor", a device which can switch electricity to light energy. Because of the high growth prospect, large and medium-small enterprises enter this market actively. The prices of stocks are also rising. In order to create a demand in the LED application area, the Korean government is recommending the use of high efficiency LED lighting in public institutions and large construction projects. Government is also choosing to support the technology and is concentrating on R&D investment and promoting LED specialized clusters in the three big subfields of LED (EPI/chip/packaging, material/modules, application).

(5) Green transport system

Compared with conventional vehicles and ships, the green transport system is a futuristic environment-friendly transport system that is efficient and has low fuel costs and CO_2 gas emissions. Green cars, next generation ship systems (WIES-Ship), advanced railway vehicles and city-type maglev trains all belong to this category. Korea is planning to be in the top 4 in green car, top 7 in railway and first in the world for marine system (Gyeongsangbuk-do, 2009).

(6) Advanced green city

Advanced green city is the traditional city inosculated and compounded with IT technology and ecological technique; including ubiquitous city, U-city construction, spatial information industry, Intelligent Transportation Systems, etc. A Korean-style U-city comprehensive plan has been developed. Research and development of the core technology for Intelligent Transportation Systems (ITS) for the next generation and spatial technology are being promoted (Future Planning Committee, 2009).

20.5.2. *Greening all industries and value chain*

Besides the green technology based new growth policy, greening all traditional industries and value chains is also one of the national issues in becoming a green country. The training of advanced interdisciplinary industries and the knowledge service industry is also important. Because of the energy-intensive industrial structure, the industry sector, as a whole, consumes 57%, 94% of which is in manufacture.

First, Korea plans to build an environment-friendly, low cost production infrastructure, and to introduce a green business certification. Korea is focusing on the change to eco-friendly projects and material supplies in steel, petrochemical and fiber. In the steel field, the investment in low energy-consuming equipment is increasing, process technology innovation like PainEX is being developed and energy efficient materials are also being developed. The integrating-resource-management network project to collectivize eco petrochemical industry in petrochemical fields, production stage reducing process development and ultra-light green new fiber development field are being promoted.

Korea established their fusion IT industry centre with an epoch-making CO_2 emission reduction target in the automobile, shipbuilding and machine fields. Securing the core technology of green car and technical capacity building in the components industry for automobile field, and obtaining the de facto standard for green ship technology for the future of the shipbuilding field are urgent priorities for Korea. Securing the core technology, developing the core material and commercialization to produce green production facilities in the machine field are also being promoted.

Korea is aiming to be the leader in the creation of new industries such as renewable energy, international environmental model development and entering developed country markets in the semiconductor and display fields. Korea is especially focused on promoting private-public joint next-generation memory R&D, obtaining the patents and the de facto standard for next generation displays and mobile communication in IT fields like semiconductors and displays.

Then, the entire process of the value chain, "production — distribution — marketing — services — recycling", must be friendly to the environment. So "Green by IT," which is the orienting development and new industrialization of green IDC (Internet Data Centre) model, is very important. Simultaneously, the improvement of productivity through the standardization of logistics, distribution of information, widespread distribution of CO_2 reduction circulation and the development of the green consulting industry are also required.

What's more, urban mining, which re-uses scrap metals in the recycling industry, is also recommended. In order to spread circulatory aggregate made of construction waste, promotion of construction work with the duty to use it is increasing. To green the construction industry, durability and stability of construction standards, energy savings and greenhouse gas reduction targets in each stage of construction have been set quantitatively.

Further, to reduce the use of chemical fertilizers to a maximum, Korea's policy is to pursue incubation of biological agricultural chemicals (e.g. predator), organic fertilizers, eco-friendly materials and new material industries. At the same time, ensuring the diversity of agricultural genetic resources and exporting in the industrialized seed industry through breeding new varieties are being promoted.

20.5.3. *Development of the advanced fusion industry and knowledge service industry*

The advanced fusion industry creates demand for broadcasting and communications industries like IPTV, which helps build power-saving broadcasting and telecommunication infrastructure. Securing technology sources and intellectual property rights for core technologies for IPTV, DMB and DTV in these growing fields is important.

R&D and human resource development in next-generation RFID and USN (Ubiquitous Sensor Network) technology elements along with fusion technology in IT fusion systems are being promoted preferentially. In the new materials nano-fusion field, "nano-fusion 2.0 project," is being promoted. Professional training and standardization are being emphasized. Drugs like antibody treatment, stem cell treatment and gene therapy agents in biopharmaceutical (resources) fields are being developed. Korea has focused on increasing the technical level of these fields.

With the purpose of developing high added-value service industries, global health care, global education services and green finance, contents software (SW) are especially remarkable. Institution building is being developed in the green healthcare field to support the hospital service industry and attract foreign patients. In this process, the maintenance of hospital central medical cluster which is in cooperation with basic research to practical application is being conceived. Clinical validation and commercialization of U-Health sector (e.g. disease separation monitoring combined with IT, early diagnosis, health care, etc), are also emphasized.

The creation of an environment in which quality foreign educational institutions can be opened in free economic zones, and the task of attracting foreign students to the global education service field is currently being carried out. Further development of core resourcing commercial technologies and contents is being pursued internationally in order to reinforce U-strengthening and industrial base.

The basic plan for an emissions trading system and the establishment of an emission trading exchange is being scheduled in order to activate green finance. The preferential loan scheme for green companies (green prime rate, fee reductions, etc.) has already been implemented. Institutionalization of the green growth fund and joint participation in public and private sectors is also in progress.

To develop a content SW industry, core CT (culture technology) in six avenues such as game, video and new media, are committed to development. A global game hub centre is also being planned. Core technology of SW fusion in 5 major strategic industries (automobile, shipbuilding, national defense, medical care and construction), source technologies in the knowledge service sector and the development of digital design are also important businesses.

20.6. Achievements of Green Growth Policy and Issues for Embedding

20.6.1. *Results of green growth policy after having been implemented for 1 year*

It has only been a year since Korea declared a national vision for green growth policy. Though this period is too short for evaluation, industrial circles, science and technology academic circles and other circles have a widespread perception of green growth. Infrastructures which promote green growth, such as organizations and legal systems are being built comparatively. The Korean government is improving the green growth promoting organizations. "Basic Law on Low-Carbon Green Growth" is also being established.

In response to this, industrial circles are increasing their investment in green industries. According to the Korea Development Bank in June 2009, the estimation of green business investment from 400 major companies is 3.12 trillion won (averaging 7.8 trillion won per year) during the period 2009–2013 and the annual average is expected to increase by 14.7%.

Samsung is going to invest 5.4 trillion won in green industries such as LED and secondary battery. Hyundai and Kia Motors are also planning to invest 4.1 trillion won in eco-car development and greenhouse gas reduction by 2013. Many large investment plans have been announced recently in Korea. LG is constructing a glass plant for LCDs and an Automotive Battery Factory. SK is focusing on, "U City Business," which is a future city project. POSCO's plant for fuel cell equipment production has already been completed. They have also entered the clean coal energy business which produces synthetic natural gas as well as coal. Additionally, some small companies, who have core technologies of lithium batteries, solar and wind, are also entering the world market.

Based on the above, Korea's green growth policy received a positive assessment. However, the results of this first year are unsatisfactory. Except for a few small business groups and large companies, investment and recognition in green industries is still very low. After further change of administration green growth policy will be promoted subsequently. To further develop Korea's economy, there are many mountains to be climbed and rivers to be crossed.

20.6.2. *Issues of government, companies and citizens to be resolved*

As noted above, future directions have been decided by strong government policies. A roadmap and institutional framework have also been provided. But if we look closely, there are still potential threats present. In order to solve these problems, we discuss what government, companies and citizens should do in this section.

Above all, it is the role of government to lead green growth.

First, it is necessary to recheck the propulsion system. So far, green growth policies have been promoted from central government and passed downwards. But should we change the direction that started at the local and private sector, when central government only emphasizes its coordinating function? It is hard to achieve any real effect without local and private sector's creativity or voluntary cooperation. Developing a Korean-style green growth model, based on understanding the correct reality and current situation of the Korean economy should be focused on.

Second, a system which can coordinate smoothly between government departments and make decisions rapidly is very important. This year, many policies about green growth have been created. Similar policies have been made by many departments. There is selfishness in some cases and confusion in many other cases. Though the Presidential Committee on Green Growth is coordinating overall, it does not yet seem to function correctly.

Third, we should be cautious of the green bubble. Korea is currently experiencing a green wave. The tendency to capitalize on this wave is high. There are also some cases where policies are wrapped in green disguise, but which are actually polluting the nature of green growth. There is some fear that a situation, like the redundant investments in the heavy chemical industry in the 1970s or venture bubble in the early 2000s, could be reproduced.

Next, for opening a green industry era, the importance of the role of companies should be emphasized. The largest part of green growth are private sector companies. Green growth without private sector companies' creative and passionate work would be difficult. These are all very promising signs that investment and R&D activities in green industries are expanding vigorously, and that the greening of major industries are being promoted actively. But the majority of these activities are still mainly large companies with only a few small companies. Many small companies are lacking awareness and do not care about the energy crisis or greenhouse gas issues. That is to say, even though green growth has been agreed with, practical preparations are still insufficient.

Finally, we want to emphasize the importance of people's awareness about green growth and energy saving in daily life. The way to green growth cannot be avoided. It is inevitable. If people are able to empathize with the necessity of green growth and have great faith in the policy direction, the dream that Korea becomes a great green nation one day will come true.

According to the data gathered by the Public Energy Management Association, the value of energy wasted by households and offices together with automobiles in Korea is 708 million won (5.6 billion US dollars) per year, which accounts for 4% of the 141.5 billion US dollars (2008) of energy import. It is embarrassing that about 11% of the energy consumed by households is stand-by electricity.

Consequently, a campaign for basic national consciousness transformation and investment toward new energy, as well as education and technical R&D for energy saving, is emergent. We must make it clear that the easiest green movement, to say energy saving and rubbish recycle, is at the same time the first step toward green growth.

References

Bae, J. H. (2008). Raw material for a new energy policy direction of Low-Carbon Green Growth. *Energy Focus*, 5(4), 1–11.

Bernanke, B. (2009). The Fed's Exit Strategy. *Wall Street Journal*, 7. 20.

Future Planning Committee (2009). 5 IT Large Strategies for the Future Korea. Seoul, September 2009.

Gyeongsangbuk-do (2009). *Gyeongsangbuk-do Promoting Plan Toward a Low-carbon Green Growth*. Korea: Gyeongsangbuk-do.

Hosoda, E. (2008). *A Recycling-oriented Society* (in Japanese). Tokyo: Keio University Publishing.

Jo, M. (2008). Implementation of Green Growth and Green Society. *The Land*, January.

Lee, J. (2009). A Reconsideration of Korea's Energy Policy Paradigm: A Focus on Overseas Resource Development and Green Growth. *International Relations Research*, 1(14), 1–8 and 23–31.

Makino, N. (2001). A New Age for Business Environment. *Economic Community*, 7–16.

Ministry of Knowledge Economy (2009). "A Plan for Strengthening the Renewable Energy Industry." August 2009. "A Vision for the Future Industrial Strategic and Green Industry." July 2009.

National Technical Committee (2009). "Vision and Development Strategy for New Growth Power." January 2009.

National Strategy Committee for Green Growth (2009). *National Strategy for Green Growth: Five-Year Plan of Green Growth (2009–2013).*

Nu, G., Sang, N. & Chang, Y. B. (2009). *Policies and Institutions of a Recycling-oriented Society*. Seoul, Korea: Shisuma Press.

Oh, S. H. (2009). Possible to Reproduce the American Jobless Recovery. *Overseas Economic Information*, 59, 34–49.

San Francisco (2009). "Jobless Recovery Redux?" *FRBSF Economic Letter*, 18, June 5, 2009.

The Ministry of Environment Korea (2009). *An Implementation Plan for Green Growth in the Environmental Aspect.*

UNEP (2009). *Overview of the Republic of Korea's Green Growth.* Switzerland, Geneva: International Environment House.

Chapter 21

CARBON EMISSIONS TRADING SCHEME*

Peter Heindl† *and Andreas Löschel*‡

21.1. Introduction

This chapter deals with designing emissions trading in practice. After a short introduction to the general idea of emissions trading, practical requirements for the introduction of an emissions trading scheme are considered, including the temporal and spatial dimension as well as administrative requirements and the role of markets. Historical developments regarding emissions trading are discussed. Currently, the largest trading scheme is the EU Emissions Trading Scheme (EU ETS) that aims to reduce greenhouse gas emissions in the European industry by 21% until 2020 compared to 2005 levels. Because of its prominent role, the basic design and the process of introducing the EU scheme is reviewed in more detail. Finally, the impact of the EU ETS on the regulated entities is analyzed based on an annual survey among German companies regulated by the EU ETS which is conducted by the Centre for European Economic Research (ZEW) in a common project with KfW Bankengruppe.

Emissions trading is a key component of climate and environmental policy in Europe. First experiences with emissions trading schemes were made

---

*A shortened version of this chapter previously was published (in Chinese language) as Heindl, Peter and Löschel, Andreas (2011), Designing Emissions Trading in Practice — General Considerations and Experiences from the EU Emissions Trading Scheme (EU ETS), in: Xue, Jinjun and Zhao, Zhongxiu (eds.), *Annual Report on China's Low-Carbon Economic Development (2012)*, Beijing: Social Science Academic Press, pp. 360–378.

†Research Fellow, Centre for European Economic Research (ZEW), Mannheim, Germany. E-mail: heindl@zew.de

‡Head of Research, Department of Environmental and Resource Economics, Environmental Management at the Centre for European Economic Research (ZEW), Mannheim, Germany and Professor of Economics, University of Heidelberg, Germany. E-mail: loeschel@zew.de.

in the late 1970s in the U.S. Various applications for regulation by tradable permits occurred afterwards, mostly in the U.S. While emissions trading is under certain assumptions equivalent to a tax on emissions, it offers flexibility for policy-makers when designing policies for environmental protection. The European Emissions Trading Scheme (EU ETS) is the largest GHG regulation scheme worldwide and started in 2005. As the EU Emissions Trading Scheme shows, emissions trading can be successfully employed to incentivize firms to reduce CO_2 emissions from fossil fuels. Emissions trading is also in the center of the Low-Carbon Development Pilot Program (LCDPP) in five selected provinces and eight cities of China.

Emissions trading offers the option to achieve a given emissions reduction target at least costs. Pigou (1920) introduced the idea of a tax to reduce negative side-effects of economic activities on humans, such as environmental pollution. The tax can be designed, so that the negative impact of pollution can be reduced in a way that harms consumers and industry as little as possible while maximizing social welfare. In the case of a Pigouvian tax, the actual amount of achieved emissions reduction is a priori unknown due to unknown marginal abatement costs of polluters. Emissions trading can be regarded as the counterpart of a Pigouvian tax on emissions. In an emissions trading scheme, policy makers exogenously set an overall limit for a pollutant and assign permits to each unit of the pollutant. Emissions permits can be traded freely between regulated firms. As a consequence, a price for each unit of emissions occurs on the market. Crocker (1966) and Dales (1968) discussed tradable permits for the cases of air and water pollution. Baumol and Oates (1971) and Montgomery (1972) proved theoretically that regulation under tradable permits can yield an efficient allocation of resources, i.e. is able to cut back effluents or emissions from various sources under an exogenously given overall quantity-restriction at least costs. The basic difference between a tax and an emissions trading scheme is, that in the case of a tax a price per unit emissions is announced by policy makers, while in the case of an emissions trading scheme, an overall amount of allowed emissions is set by policy makers. As shown by Downing and White (1986), both schemes can offer adequate incentives for emitters for green innovations and to reduce emissions efficiently.

In this chapter, we look at the designing of emissions trading in practice. We briefly describe the elements for the use of emissions trading and the European experiences with its emissions trading scheme. Section 21.2 describes the general aspects of emissions trading and summarizes the requirements for a successful introduction in practice. In Section 21.3, we briefly discuss the evolution of different emissions trading schemes, e.g. in the USA or Chile. In Section 21.4, we describe the evolution of the EU emissions trading scheme and its most important aspects. Section 21.5 presents an overview of firms responses to the EU emissions trading scheme, based on the experiences made in Germany in the past years. Finally, in Section 21.6, a brief conclusion is drawn.

21.2. Designing Emissions Trading in Practice

A central feature of quantity regulation is that the regulator defines a fixed amount of allowed emissions for the whole economy (or the regulated area) for a certain period of time. The quantity restriction is often referred to as "cap". The period in which emissions are restricted is often called "compliance period" or "commitment period". To assure that the overall quantity restriction is not exceeded, the regulator assigns permits to each unit of emissions. These permits are distributed to regulated entities and must be surrendered (handed in to the regulator) for each emitted unit.

21.2.1. *Allocation of permits*

Permits can be sold or auctioned by the regulator or can be partly or fully distributed for free (known as "grandfathering"). In the case of selling or auctioning, the regulator generates revenues which can be "recycled" or "redistributed", for example to reduce other taxes. It was shown that revenue recycling can lead to an increase in social welfare compared to non-recycling (Goulder *et al.*, 1997). Free allocation of permits (grandfathering) in contrast is an implicit subsidy for regulated entities. The allocation of permits distributes the cost burden of the required emission reduction among firms, sectors and households, which are affected through prices on emission in the emissions trading system. With auctioned permits the scarcity rents go to the government and the distributional consequences are largely the same as with an emission tax. With grandfathered permits, the scarcity rents are given to the firms. This offers an incentive for companies to accept an enforceable constraint on emissions. The allocation of permits, however, has no direct impact on the permit price on the market under general assumptions, i.e. if firms that receive free allocation are aware of the opportunity costs that would arise from not valuing the free allocation at the market price.

21.2.2. *Emissions trading in time*

While compliance is organized within a certain period (e.g. one year), meaning that regulated entities must surrender permits for verified emissions at the end of the period, banking or borrowing of permits between the compliance periods can increase flexibility in the trading scheme. Banking of permits occurs when regulated entities are allowed to hold back permits for compliance in the future. Intermediaries in the permit market can generate special products like futures or options when banking is allowed. Banking can not only occur at the level of regulated entities or intermediaries but also at the level of the regulating authority. The regulating authority can bank permits over time and build a permit reserve. If permits are freely allocated, a reserve might be necessary to provide free allocation to entities that newly enter into regulation. If permits are auctioned or sold to regulated entities, the regulator can build a permit reserve and sell additional permits to the market in times of high demand to avoid large upward price volatility. Borrowing is the counterpart to banking. In that case permits from future compliance periods are borrowed for current compliance. Borrowing incurs the risk of excessive borrowing which could lead to non-compliance in the future. Because of the perverse incentives that can stem from borrowing, it is often starkly restricted in practice. While, for example, banking is allowed within the EU ETS over the whole period 2008 to 2020, borrowing is restricted from year to year. Regulated entities receive free allocation in February each year, while permits must be surrendered in April. Hence, implicit borrowing from next year's free allocation for current year's compliance is possible. The modest borrowing provision in the EU ETS increases flexibility while preventing excessive borrowing.

21.2.3. *Emissions trading in space*

Spatial aspects of environmental regulation dependent on the nature of the regulated emissions. For example, the concentration of effluents in a river or the concentration of SO_2 in the air is perceived locally (downstream of the river or in the periphery of a coal fired power plant). When local pollutants (like SO_2) are addresses, emissions stay in a certain area, and hence, must not exceed a specified amount in a certain period. In such a case, a trading scheme has to be designed in a way that assures that the concentration of a pollutant or emissions from regulated sources in a regulated area (e.g. county) stay below a targeted level, which can imply restrictions on purchasing additional permits from outside the regulated area as well as banking or borrowing within the region. Emissions trading systems may otherwise violate local air quality goals and may lead to the concentration of emissions in certain areas (hot spots). Examples are the U.S. Offset-Policy, starting in 1976, and the U.S. Acid Rain Program, starting in 1980. In the former, achieved reduction of pollutants were tradable. In the latter, emission permits were allocated to sources and banking was allowed (see Chapter 3). In contrast, the emissions of greenhouse gases, like CO_2, are expected to have an impact on global climatic conditions. When global pollutants, like greenhouse gases are addressed, the spatial and timely occurrence of emissions is of less importance. Greenhouse gases, for example, stay in the atmosphere for a long time and are widely diffused (fully mixing). They are harmless on the local level and at the time they are emitted, at least if the concentration is not extremely high. However, the global stock of greenhouse gases in the atmosphere is expected to have an impact on overall climatic conditions in the long run.

21.2.4. *Administrative requirements*

For an emissions trading scheme to be maintained, a regulating authority is required that allocate permits or verifies emission offsets, assures that emissions are properly monitored and reported, manage a permit registry to assure that each permit is surrendered only once so that the overall "cap" is not exceeded, trace transactions in the registry to document property rights, and enforce compliance, e.g. imposes a fine on regulated entities in the case of non-compliance. Compared to a regulation by prices (unit tax), a quantity approach (tradable permits) is more complex in terms of administration. The most important difference is the permit registry which is needed for regulation by quantities. Within the registry each market transaction of permits must be recorded. In the case of emissions offsets (reducing emissions compared to a baseline) the authority must verify the offset and allocate the offset permits.

21.2.5. *Role of markets*

Quantity regulation by emissions trading is often referred to as a market based approach. The price that incentivizes emitters to reduce economy-wide emissions is generated by supply and demand on markets. Reduction efforts by companies are not defined by the regulating authority as under a command and control regime. Hence, liquid and transparent markets are of great importance. Transparency in markets (about prices and traded volumes) is beneficial as it provides information to all market participants. In the U.S. SO_2 trading program, prices for private transactions were unknown to others and hence assessing a "fair" price was accomplished with relatively high informational costs. The market was not very liquid meaning that finding a potential seller/buyer was accomplished with search costs. These factors lead to relatively high transaction costs in general and hampered the efficient exchange of permits and the efficiency of the emissions trading system as such. Since the price is generated at primary auctions or sells (initial permit allocation by authority) and at secondary markets (exchanges, over the counter trade), liquidity and transparency within the market is crucial for minimizing transaction costs and facilitating efficient exchange of permits. Allowing intermediaries to be active in permit trading can play a crucial role here. Also markets for machinery and equipment to achieve emissions reductions can be of importance. While in the case of SO_2 trading, retrofitting of existing plants to reduce emissions was relatively easy. Technical solutions for the reduction of CO_2 are more complex because of the non-existence of end-of-pipe technologies for CO_2 emissions. If markets for energy efficient machinery and equipment are sticky, transaction costs can hamper the effective transformation of the economy (Heindl, 2011).

21.2.6. *Regulation under uncertainty*

If there is full information, i.e. no uncertainty about the benefits and costs that are related to the regulation of a harmful substance, there is no difference between price (tax) and quantity (emissions trading) regulation. In practice this would imply that it is perfectly known what price (unit tax) yields which quantity (emissions reduction) and vice versa. Since there is usually uncertainty about the exact costs and benefits of regulation, the decision is made under uncertainty. Weitzman (1974) presented a model that accounts for uncertainty in the choice of the regulation scheme. Basically he finds that prices (unit tax) should be preferred over quantities (tradable permits) if the slope of marginal costs for providing environmental quality is steeper than the slope of marginal benefits and vice versa. The steepness of marginal costs and benefits defines the relative importance of one variable over the other. Newell and Pizer (2003) showed that regulation by prices (unit tax) would be preferable in the case of greenhouse gas regulation. Roberts and Spence (1975) were the first who considered mixed regulation by prices and quantities for the case of effluents. They conclude that a mixed system of permits and prices is more efficient than a scheme with either permits or prices alone. They also point out that the additional fee motivates polluters to approximate the expected damage function more closely. Pizer (2002) argues that a mixed system of price and quantity controls is preferable to a system with pure quantity controls in the case of greenhouse gas mitigation. In a first-best setup, a price approach should be chosen. However, recognising the importance of quantity targets for the political feasibility of international agreements (Nordhaus, 2006), a quantity targets with a certain "trigger price" for the issuance of additional permits (to restrict prices) would improve welfare compared to a pure quantity approach.

21.3. Past Experiences with Emissions Trading

The first practical experience with quantity based regulation schemes for pollutants were made in the USA. In the early 1970s the U.S. Environmental Protection Agency (EPA) evaluated the feasibility of market based approaches for environmental regulation. In 1976 the EPA started the "Offset Policy", an early approach with tradable offset permits amongst regulated sources in regions that did not meet ambient air quality standards, so called nonattainment areas. Firms were allowed to create emissions reduction credits (ERCs) by reducing its emissions below a baseline level. ERCs were tradable between firms and within the firm, meaning that emission permits are transferable from one installation to another. Because targeted emissions were local pollutants, the USA was divided in 247 air quality control regions. States had to develop State Implementation Plans (SIP), which must be approved by EPA. The SIPs defined how the state intended to meet EPAs ambient air quality standard, which sources were regulated and defined the amount of a given pollutant that a regulated source was allowed to emit (Hahn and Hester, 1989).

In addition to the Offset Policy, the Bubble Policy allowed firms to include various sources of emissions under one regulatory scheme. The Bubble Policy hence helped to reduce overall regulatory costs. Firms were also allowed to bank permits, which increased flexibility. Netting allowed firms to increase emissions at one source within a plant while avoiding the classification of the source as a "major emitting source" when emissions were decreased at another source and the net increase in emissions did not equal to a major source. The EPA also applied a scheme of tradable permits to phase out lead of gasoline (Hahn and Hester, 1989; Nussbaum, 1992). As a consequence of the Montreal Protocol that targeted the phase-out of Chlorofluorocarbons (CFCs) and Hydrochlorofluorocarbons (HCFCs), the USA implemented a trading scheme in 1988.

Beginning in 1990 the sulfur trading scheme under Title IV of the "Clean Air Act", an U.S. law specifying environmental policies, entered into force. The sulfur trading program extended existing command and control policies and delivered sizeable increases in regulatory efficiency (Hanemann, 2009). The Nox Nudget Programm aimed to reduce seasonal Nox emissions in the Northeastern USA and started in 2003. Also regional trading schemes were implemented like the "Regional Clean Air Incentives Market" (RECLAIM) in California or the "Chicago Emissions Reduction Market System" (Tietenberg, 2006). Howe (1994) reviewed environmental regulation in the USA and Europe. The USA tended to use market based approaches to environmental regulation and mainly quantity based approaches rather than taxes since the mid 1970s. The European Countries, on the other hand, implemented mainly command and control policies and showed little interest in the use of tradeable permits. In 1992 an emissions trading scheme for particulate matter (PM10) was installed in Santiago de Chile (Montero *et al.*, 2002). In 1997 the parties of the United Nations Framework Convention on Climate Change (UNFCCC) agreed on the Kyoto

Protocol that mandates a reduction of greenhouse gas (GHG) emissions of about five percent on average until 2012 compared to 1990. GHGs are global pollutants that are expected to contribute to global climate change. With the Kyoto Protocol, two trading schemes were established. First, an international emission trading scheme that allows countries with GHG emission targets (Annex B countries) to trade "Assigned Amount Units" (AAUs) for compliance with the Kyoto Protocol. Second, two project-based offset schemes, Joint Implementation (JI) and the Clean Development Mechanism (CDM). JI allows for crediting of achieved emissions reductions in Annex B countries and the CDM allows for crediting of achieved emissions reductions in Non-Annex B countries (developing and least developed countries). International emissions trading was expected to significantly reduce the costs of compliance under the Kyoto Protocol (Böhringer and Löschel, 2002). In 2011 New Zealand started a domestic GHG treading scheme. In Japan there are local GHG trading schemes in Tokyo and the province of Saitima. In the USA and Canada there are several local GHG trading schemes, like the Regional Greenhouse Gas Initiative (RGGI) which is active since 2008. In 2012, Australia introduced a carbon pricing scheme which will likely be converted to an emissions trading scheme in 2015. In early 2013, an emissions trading scheme for GHG emissions was introduced in California.

21.4. The EU Emissions Trading Scheme

The European Emissions Trading Scheme (EU ETS) currently is the largest GHG regulation scheme worldwide (Kossoy and Ambrosi, 2010). About 11,000 industrial installations are covered by the scheme, located in 30 countries (the EU27, Liechtenstein, Norway and Iceland). The first trading phase, scheduled for 2005 to 2007, was intended as a test phase and was characterized by generous free allocation of permits. In the second phase, scheduled for 2008 to 2012, free allocation was reduced, but free allocation of at least 90% of permits is obligatory for EU Member States. With the start of the third trading phase from 2013 to 2020, allocation rules will be changed. Free allocation will be completely phased out for electricity producing facilities. The remaining facilities will receive free allocation based on product benchmarks. Free allocation shall be phased out entirely until 2027. The overall aim of the EU ETS is to achieve a reduction of GHG of 21% versus 2005 until 2020. To achieve the targeted emissions reduction, the overall cap on emissions will be decreased by 1.74% each year, starting in 2013.

21.4.1. *Origin and design of the EU ETS*

Ellerman *et al.* (2010) discuss the origin of the EU ETS. We focus on the most important milestones here to illustrate the political process of establishing the EU ETS. Initially the EU was in opposition to a market based approach in the Kyoto Protocol and planned to install a community wide carbon and energy tax. First efforts to install a community wide carbon and energy tax were made in 1992, but the proposal was finally withdrawn in 1997. One reason for this was the resistance of EU countries to shift tax competencies to the European level. Rather the EU defined internal country specific emissions reduction targets in the Burden Sharing Agreement in 1998 to meet the Kyoto obligations and welcomed a quantity based approach. In 2000 the Green Paper on Emissions Trading was released, summarizing options to implement an emissions trading scheme in Europe. The ETS Directive was drafted in 2001 and finally adopted in 2003. Since the Emissions Trading Directive focused on the Europe-wide trading scheme, the European Commission drafted a separate Directive to link the EU ETS to the project based offset mechanisms of the Kyoto Protocol, defining the options for using permits from the CDM and JI in the EU ETS in addition to domestic permits, called European Emissions Allowances (EUAs). The Linking Directive was agreed in 2004. Because there were different positions on how to include CDM and JI in the EU ETS, the ultimate decision to define quantity restrictions for the use of offset permits from the CDM or JI was left to the member states.

The ETS Directive (Directive 2003/87) defines the regulated activities and the type of regulated emissions in Annex I. From 2005 onwards the Directive stipulated only the regulation of carbon dioxide (CO_2) from combustion installations, mineral oil refineries, coke ovens, metal ore roasting and sintering, pig iron and steel production, cement and clinker production, glass manufacturing, ceramic and bricks manufacturing as well as pulp and paper production. By doing so, many branches of the European economy remain outside the regulation of the trading scheme, like the transportation sector or the agricultural sector. Those sectors are often referred as "non-traded sectors". In general the EU ETS covers roughly 50% of CO_2 emissions in Europe and approximately 40% of GHG emissions, respectively. A downstream approach was chosen since pure upstream regulation in the European Emissions Trading Scheme would have effectively lead to double taxation of many products and would have partly removed the fiscal autonomy of the member states. EU member states have autonomy with regard to fiscal policy and many countries already had taxes, e.g. on fuel, in place. The regulatory framework set out in the ETS Directive is hence motivated by practical concerns and the requirements of the political architecture of the European Union, rather than concerns of economic efficiency and environmental effectiveness of the scheme. However, the ETS directive explicitly considers an extension of the EU ETS to the transport sector for the future, recognizing its importance in contributing to greenhouse gas reductions.

The ETS Directive states that each member state shall develop a "National Allocation Plan" (NAP) for each trading period based on "objective and transparent criteria", listed in Annex III of the Directive. The NAPs define the amount of freely allocated EUAs to each installation in the respective member state. The EU Commission has the right to reject a NAP. The allocation of permits in the EU ETS is characterized by free allocation (grandfathering) in its first two trading periods. In the first period from 2005–2007, at least 95% of permits had to be allocated for free, according to the Directive. For the period 2008–2012, at least 90% of permits must be allocated for free. Permits have to be handed out to installations by the national authorities by 28 February each year. Firms have to surrender permits for compliance by 30 April each year. While the Directive does not allow for borrowing of permits in general, the allocation rule actually enables firms to borrow permits from their next year's free allocation for current compliance. This one-year ahead implicit borrowing increases flexibility of firms and help to reduce potential distortions on the markets for permits due to excess demand around the compliance date. To ensure compliance with the EU ETS, the Directive defines a fine, posed to regulated firms that do not surrender a sufficient amount of permits. For each emitted ton of carbon dioxide for which a firm has not surrendered a permit it is liable to a financial penalty of EUR 40 for the period 2005–2007 and EUR 100 from 2008 onwards. Paying the fine does not release firms from the obligation to surrender the respective permits.

Information about the freely allocated EUAs, the verified emissions and surrendered permits of regulated installations are available to the public. The Community Independent Transaction Log (CITL) database provides the information online. In addition, each national authority in the member states runs a registry, where regulated installations and their emissions data can be viewed online by the public. Each person or firm is allowed to hold permits under the premise that it has an account at a national authority's registry. Within the registries, exchange of permits is documented, so that the person or firm holding a permit can be identified. Member states have to report to the EU Commission each year, first to collect emissions data and second to control the general implementation of the trading scheme in the member states. Annex IV and V of the ETS Directive defines principles for monitoring, reporting and verification (MRV) of emissions. Detailed MRV rules were developed for each technology since 2003. In Germany the competencies are shared between the federal states (Länder) and the German Emissions Trading Authority (DEHSt). The Länder are responsible for the permission of installations and the validation and approval of the monitoring plans which describe the methods used for calculating CO_2 emissions. They also randomly check installation-specific issues in the emission reports. A verifier validates and verifies the emission reports by installations and reports the amount of emissions to the DEHSt. The DEHSt checks the emission reports more closely to ensure compliance and is responsible for enforcement and sanctioning.

21.4.2. *First and second trading phase (2005–2012)*

Ellerman *et al.* (2010) provide a detailed analysis of the first trading phase of the EU ETS (2005–2007). In general the first trading phase was characterized by generous free allocation of permits. It is important to note that European Emissions Allowances (EUAs) from the first trading period could not be transferred to the subsequent periods. This underlines the test-character of the first trading period. While firms had to fully comply with the schemes rules like specified in ETS Directive, EUAs from the first period were only valid for compliance in the first trading period. The EU ETS was scheduled as a decentralized trading scheme, combining the schemes of the EU member states. Ellerman *et al.* (2010) points out that the EU ETS shows a much higher degree of decentralization as any previous trading scheme, like the U.S. SOx trading scheme or the Nox budget trading program. Each member state had to develop its own National Allocation Plan (NAP). The overall cap, consisting of the sum of capped emissions from the member states, was ex ante unknown. In addition, when the NAPs were developed, countries could possibly treat installations differently, meaning that installations could receive a different amount of free allocation in different countries, which might raise fairness issues. Since the first trading period did not seek to reduce emissions in the first place, but rather phase-in the trading scheme, the cap was planned to be set close to business-as-usual (BAU) emissions (Ellerman *et al.*, 2010). However, exact data on BAU emissions were not available, e.g. in Germany (Matthes and Schafhausen, 2007), or were not reliable, as in the case of Sweden (Zetterberg, 2007). Hence authorities in charge for developing the NAPs partly relied on voluntary submitted emissions data from industry, which set incentives for the industry to over-report.

As a consequence, there was a net surplus of allowances in the first trading period which was ex-ante unknown to regulated firms. When allowance trading started in 2005 prices were between EUR 20 and EUR 30 per ton of CO_2, showing relatively high volatility. When the first year of trading ended and the first verified emissions data in the EU ETS were published, prices dropped sharply from EUR 29.43 on 24 April 2006 to EUR 10.90 on 2 May 2006. Regulated companies became aware of the allowance surplus in the scheme, when verified emissions where published for the first time. This event is known as “compliance break”. It outlines the importance of the test phase of the EU ETS when the scheme was introduced. Towards the end of the trading phase, prices for emissions allowances eligible for the first phase converged to zero. The reason for that is simply that permits from the first trading period were not eligible for compliance in the second trading period and the scheme was oversupplied with emissions permits.

With the start of the second trading phase in 2008, also the commitment period to the Kyoto Protocol started. The overall cap was decreased in the EU from 2,299 $MtCO_2e$ per annum in the years 2005 to 2007 to an annual

Table 21.1. CERs und ERUs usage for compliance in the EU ETS (million permits).

	2008	2009	2010	Total (2008–2010)
EU-wide				
CERs (from CDM)	82.5	77.9	116.9	277.3
ERUs (from JI)	0.05	3.2	20.1	23.35
Germany				
CERs (from (CDM)	23.7	26.0	33.4	83.1
ERUs (from JI)	0	0.67	4.2	4.87

Source: CITL (2011), CDC (2011).

maximum of 2,083 $MtCO_2$ in the years 2008 to 2012. Free allocation was slightly decreased compared to the first trading period. The EU member states were asked to allocate a minimum of 90% of emissions permits for free to regulated installations. From 2008 onwards, permits from the Clean Development Mechanism (CDM) and Joint Implementation (JI) were also eligible for compliance for installations regulated under the EU ETS. While permits from CDM (Certified Emission Reductions — CERs) and JI (Emission Reduction Units — ERUs) were only moderately used in the start of the second trading phase, usage speeded up gradually in the years 2009 and 2010. However, the usage of permits from CDM and JI is limited in the EU ETS. Each EU member state individually decides how much permits can be used for compliance. In Germany, for example, installations can surrender permits from CDM and JI up to 22% of received free allocation. Table 21.1 summarizes overall usage of permits from CDM and JI in Europe and Germany.

21.4.3. *Perspectives on emissions trading in Europe from 2013 onwards*

With the start of the third trading phase of the EU ETS in 2013, several changes will apply to the EU ETS. New sectors will be regulated. The biggest one is the aviation sector, which will already be added to the scheme in 2012. In 2013 new stationary installations from sectors producing bulk organic chemicals, hydrogen, ammonia and aluminum will be opted in to the EU ETS. While only CO_2 emissions were covered by the scheme until 2012, N_2O emissions from the production of nitric, adipic and glycolic acid

Table 21.2. Sectors and allocation principles in the EU ETS form 2013 onwards.

Activity/Sector	Allocation rule
Electricity Generation	No more free allocation
Heat	Free allocation
Industrial Sectors and Subsectors which are deemed to be exposed to a significant risk of carbon leakage (EU, 2010b)	100% free allocation based on benchmarks from 2013 to 2020
Industrial Sectors (remaining)	80% free allocation based on benchmarks in 2013, falling to 30% in 2020

Source: EU (2011)

production as well as per fluorocarbons from the aluminum sector will be covered by the scheme (EU, 2010b). The Netherlands and Austria included some N_2O emissions from specific installations already in the second trading period of the EU ETS. The most relevant change for existing installations from 2013 onwards will be the revised free allocation rules. Auctioning will be the principle of allocation from 2013 onwards. In practice auctioning will progressively replace free allocation from 2013 onwards and grandfathering shall be completely phased out until 2027 (see Table 21.2). Over the third trading period from 2013 to 2020, about half of the issued European Emissions Allowances shall be auctioned (EU, 2010a). Also the way how free allocation is organized will change. From 2013 onwards, free allocation will be based on product specific performance measures. To do so, the EU Commission developed product benchmarks. In practice, free allocation will be orientated on the performance of the ten percent most energy efficient installations in a certain product group (see Table 21.3). This implies that less efficient installations will receive far less free allocation from 2013 onwards.

Apart of the reduced amount of freely allocated permits, the EU-wide amount of issued permits ("cap") will be decreased annually by a linear factor of 1.74%. In practice this will result in an annual reduction of the overall EU-cap of 37.4 $MtCO_2$ for stationary sources. The cap for the year 2013 is determined at 2,039 $MtCO_2$. An amount of 107 $MtCO_2$ for newly opted in sources in 2013 is included in the 2013 cap (EU, 2010b). The EU, however, will determine the exact cap finally in 2013. Until 30 September 2011, the national authorities of the EU member states shall

Table 21.3. Benchmarks and their estimated coverage of emissions.

Activity/Sector	Benchmark
Electricity Generation	None
Heat Generation	Heat-benchmarks (approx. 20% of emissions)
Industry	Product benchmarks (approx. 75% of emissions)
	Fuel benchmarks (approx. 5% if emissions)
	Process emissions (approx. less than 1% of emissions)

Source: EU (2011).

work out the plans for revised free allocation from 2013 onwards and submit their "national implementation measures" (NIM) to the EU commission. The commission is expected to review the NIMs in the course of 2012. Hence, it is likely that the concrete figures for free allocation will be available at best a few months before the start of the third trading period in March 2013. This practice leaves firms with a relatively high degree of uncertainty.

21.5. The German Experience — Evidence from the KfW/ZEW CO_2 Barometer

The Centre for European Economic Research is annually surveying all German companies regulated by the EU ETS on their behavior under the trading scheme in a common project with KfW Bankengruppe. In the survey, topics like abatement efforts, trading activities and issues related to carbon management and organizational responds of firms to the EU ETS are addressed. The firm-level survey data are merged with emissions data from the Community Independent Transaction Log (CITL), allowing an identification of larger and smaller emitters based on verified official data from the EU. The data are aggregated from the installation to the firm-level. In the CITL there are 1,668 regulated installations identified for Germany. After aggregation, 816 firms can be identified. From the survey additional information on the firms are available, e.g. the number of employees and the main activity based on the NACE classification. The firm-level data gathered at ZEW in the years 2009 to 2011 describes the firm-behavior of German firms in the EU ETS and gives empirical evidence on the transformation towards the third trading period of the EU ETS.

21.5.1. *Abatement*

For the years 2009 and 2010, 63% of surveyed German firms said that they already had implemented CO_2 abatement activities since the start of the EU ETS in 2005. Small and Medium Sized Companies (<250 employees) and small emitters (<25,000 tCO_2 emissions p.a.) are less active in abatement then larger companies or larger emitters. In March 2010, 57% of emitters planned to implement abatement activities until the end of 2012. In March 2011, 65% of companies stated that they plan abatement from 2013 onwards, implying slightly increasing overall abatement activities in the third trading period. Abatement activities carried out between 2005 and 2010 were in most cases not scheduled to achieve CO_2 reductions in the first place. Abatement mostly occurred as a side-effect of other measures, including general restructuring and optimization of production processes. 25% of companies state, that from 2013 onwards activities with the primary reason of CO_2 abatement will be scheduled, implying an increasing impact of the EU ETS on economic activities in Germany, caused by the stricter rules in the EU ETS and higher carbon price from 2013 onwards.

One important determinant of CO_2 abatement is the technical lifecycle of existing installations under regulation. While existing installations might be optimized to reduce emissions, the achievable reductions are limited. One reason is the non-existence of end-of-pipe technologies for carbon dioxide abatement. Under existing regulation firms might take carbon emissions into account when it comes to new investment in machinery, replacing existing installations. As a consequence, considerable emissions reductions might be achieved by replacing existing installations by more energy efficient ones. Based on that, German firms were asked about the remaining average life-cycle of existing installations in their machinery portfolio. 82% of emissions stem from installations with an average remaining technical life-cycle of 15 to 20 years. Accordingly, carbon dioxide abatement through investment in energy efficient new machinery might only occur to a relatively large extent in the long run (See Table 21.4).

Table 21.4. Average remaining technical life-cycle of existing installations in Germany.

Average remaining life-cycle (evaluated in 2011)	% of companies	% of current emissions
Up to 5 years	8%	0.6%
5 to 10 years	20%	4.8%
10 to 15 years	19%	3.3%
15 to 20 years	28%	81.9%
20 to 30 years	14%	7.3%
30 to 40 years	1%	<0.02%
More than 40 years	10%	2.1%

Source: Löschel *et al.* (2011).

Current abatement activities are mostly based on process optimization. In March 2011, 63% of surveyed firms said that emissions reductions were achieved in this way. 59% also invested in energy efficient technology. Fuels-switch was used by 26% of respondents and renewable energy sources by 25%. One option to reduce emissions is, off course, the reduction of output by firms. For the year 2009, during the financial and economic crisis, 20% of firms stated that they had achieved emissions reductions by reducing output. The statement corresponded to the overall figures for CO_2 emissions in Europe, which dropped in 2009 compared to previous years. After the German economy recovered in 2010, emissions rose again. As a consequence, only 9% of firms reported the reduction of output as a reason for declined emissions for the trading year 2010.

21.6. Allowance Trading

Trading of allowances was carried out in a similar frequency by regulated companies in Germany in the trading-years 2009 and 2010. 51% of surveyed firms traded allowances in 2009 and 54% in 2010 respectively. About two third of companies only trade once per year, the remaining firms trade more than once per year. Very large emitters also choose to trade daily or weekly. In contrast small emitters (<25.000 tCO_2) usually trade once per year or, in rare cases quarterly or biannual. Surprisingly, there is no significant difference in trading activities of firms with an overall surplus or net short position of freely allocated allowances. This gives evidence for intra-firm or intra-trust exchange of permits and borrowing. In fact, 77% of firms with more than one regulated installation prefer to exchange permits internal, rather than using markets or exchanges in the first place. Only 23% of firms usually clear internal short/long positions generally via the market. Most companies prefer to trade allowances via intermediaries, like carbon funds or banks. The second most frequent way of trading are over the counter (OTC) transactions. Only a relatively small number of firms trade directly at exchanges. The preferred way of trading is dependent on the annual emissions levels. Smaller emitters tend to trade via intermediaries or OTC. Only relatively large emitter directly trade at exchanges.

21.6.1. *Companies response to changes in the EU ETS from 2013 onwards*

Some of the regulated companies are particularly struggling with the uncertainty of planning that stems from the lengthy drafting process of the actual reforms. Indeed, as many as 72% of the interviewed companies in Germany have tried to estimate the volume of the new allocation from 2013 by March 2011. However, the information base for such estimations remains rather limited. It seems that approximately 63% of all interviewed companies expect an insufficient allocation of certificates from 2013. To be compared, just 27% of all respondents suffered from insufficient allocation in 2010. As a consequence, many companies will have to expect substantial additional expenditures from 2013 due to emissions trading. While the additional expenditures for smaller emitters will be limited, bigger emitters will face additional expenditures of several hundreds of thousands of euros. Very large emitters even have to expect additional costs amounting to millions of euros, especially in the energy sector where the most reductions to the free allocation will be made.

Indeed, for a transformation towards a low greenhouse gas European economy, it is necessary to set clear incentives for CO_2 reductions by means of a price signal. This fact is widely accepted within the German economy. However, given the extensive reforms, the companies must be sufficiently informed and need enough time for an adequate preparation for the changes to come. This is clearly a weak point of the European climate policy. The interviewed companies stated that they would need an average of

20 months between the announcement of the actual free allocation from 2013 and the enactment of the regulations in order to be perfectly prepared for the changes. This seems reasonable, since many companies will have to plan for substantial additional costs. In reality, the preparation phase for the companies is likely to be way shorter. It depends on the assessment procedure that the EU commission will apply to the propositions for free allocation made by the German Emissions Trading Authority, but most likely the allocation will be known only a few months ahead.

21.7. Conclusion

Emissions trading offer the chance for Europe to achieve its emissions reduction commitments in an economically efficient way, meaning at least costs. In addition to that, regulated companies are provided with a great amount of flexibility. Firms with high costs of carbon dioxide abatement are allowed to buy additional emissions allowances, while companies with relatively low costs of carbon dioxide abatement have incentives to reduce emissions. In the curse of that, the overall emissions reduction target can be achieved with certainty, because of the Europe-wide fixed amount of allowed emissions. Based on the German experience, firms in general respond very well to the regulatory scheme. Most firms already have started to implement CO_2 abatement measures and abatement activities are expected to increase in the future. The EU emissions trading scheme was introduced gradually, starting with a "test phase" from 2005 to 2007, a period of moderate regulation from 2008 to 2012 and increasing regulatory pressure from 2013 onwards. This way to introduce the trading scheme actually turned out to be of great use for regulated firms as well as for policy makers. Firms were able to adapt to the new regulation and policy makers were able to adjust the scheme based on gathered experience and data. Despite the fact that emissions trading plays a key role for Europe to efficiently reduce its CO_2 emissions, some criticism remains. Most importantly, firms are confronted with ongoing reforms in the EU ETS and climate policy in Europe. Uncertainty about future regulation actually can hamper investment in new technology and has a negative impact on the overall efficiency of the emissions trading scheme.

References

Baumol, W. J. and Oates, W. E. (1971). The use of Standards and Prices for Protection of the Environment. *The Swedish Journal of Economics*, 73(1), 42–54.

Böhringer, C. and Löschel, A. (2002). Assessing the Costs of Compliance: The Kyoto Protocol. *European Environment*, 12(1), 1–16.

CDC (2011). "The Monthly Bulletin on the European Carbon Market." Tendances Carbone No. 58, Caisse des Dépôts, May 2011.

CITL (2011). "EU ETS Emissions Data from the Community Independent Transaction Log." http://ec.europa.eu/environment/ets/

Crocker, T. D. (1966). The Structuring of Atmospheric Pollution Control Systems. In H. Wolozin (ed.), *The Economics of Air Pollution*. New York: Norton, pp. 61–86.

Dales, J. H. (1968). *Pollution, Property and Prices*. Toronto: University of Toronto Press.

Downing, P. B. and White, L. J. (1986). Innovation in Pollution Control. *Journal of Environmental Economics and Management*, 13, 18–29.

Ellerman, D., Convery, F. J. and de Perthuis, C. (2010). *Pricing Carbon — The European Union Emissions Trading Scheme*. Cambridge: Cambridge University Press.

EU (2010a). "Emissions Trading: Questions and Answers on the EU ETS Auctioning Regulation." MEMO/10/338, Brussels, 16 July 2010.

EU (2010b). "Emissions Trading: Questions and Answers concerning the Second Commissions Decision on the EU ETS Cap for 2013." MEMO/10/513, Brussels, 22 October 2010.

EU (2011). "Questions and Answers on Draft Commission Decision on Free Allocation Rules for the ETS." MEMO/11/258, Brussels, 27 April 2011.

Goulder, L .H., Parry, I. W. H. and Burtraw, D. (1997). Revenue-raising Versus other Approaches to Environmental Protection: The Critical Significance of Pre-existing Tax Distortions. *RAND Journal of Economics*, 28(4), 708–731.

Hahn, R. and Hester, G. (1989). Where Did All the Markets Go — An Analysis of EPAs Emissions Trading Program. *Yale Journal of Regulation*, 6, 109–153.

Hanemann, M. (2009). The Role of Emission Trading in Domestic Climate Policy. *The Energy Journal*, 30 (Special Issue 2), 79–114.

Heindl, P. (2011). "The Impact of Informational Costs in Quantity Regulation of Pollutants: The Case of the European Emissions Trading Scheme." ZEW Discussion Paper No. 11–040, Mannheim.

Howe, C. W. (1994). Taxes Versus Tradable Discharge Permits: A Review in the Light of the U.S. and European Experience. *Environmental and Resource Economics*, 4, 151–169.

Kossoy, A. and Ambrosi, P. (2010). States and Trends of the Carbon Market 2010, World Bank, Washington.

Löschel, A., Brockmann, K. L., Heindl, P., Lutz, B. and Schumacher, J. (2011). "KfW/ZEW CO_2 Barometer 2011 — Hoher Anpassungsbedarf im EU Emissionshandel ab 2013 — deutliche Defizite bei der Vorbereitung in den Unternehmen." KfW Bankengruppe, Frankfurt am Main.

Matthes, F. C. and Schafhausen, F. (2007). Germany, In A. D. Ellerman, B. K. Buchner and C. Carraro, (ed.). *Allocation in the European Emissions Trading Scheme: Rights, Rents and Fairness*. Cambridge: Cambridge University Press.

Montero, J. P., Sánchez, J. M. and Katz, R. (2002). A Market-Based Environmental Policy Experiment in Chile. *Journal of Law and Economics*, 45(1), 267–287.

Montgomery, D. W. (1972). Markets in Licenses and Efficient Pollution Control Programs. *Journal of Economic Theory*, 5(3), 395–418.

Newell, R. G. and Pizer, W. A. (2003). Regulating Stock Externalities under Uncertainty. *Journal of Environmental Economics and Management*, 45(2), 416–432.

Nordhaus, W. D. (2006). After Kyoto: Alternative Mechanisms to Control Global Warming. *American Economic Review*, 96(2), 31–34.

Nussbaum, B. D. (1992). Phasing Down Lead in Gasoline in the U.S.: Mandates, Incentives, Trading and Banking. In T. Jones and J. Corfee-Morlot (eds.), *Climate Change: Designing a Tradable Permit System*. Paris: OECD, pp. 21–34.

Pigou, A. C. (1920). *The Economics of Welfare*. London: Macmillian and Co.

Roberts, M. J. and Spence, M. (1976). Effluent Charges and Licenses under Uncertainty. *Journal of Public Economics*, 5(3–4), 193–208.

Pizer, W. A. (2002). Combining Price and Quantity Controls to Mitigate Global Climate Change. *Journal of Public Economics*, 85(3), 409–434.

Tietenberg, T. (2006). *Emissions Trading — Principles and Practice*, 2nd ed. Washington: Resources for the Future.

Weitzman, M. L. (1974). Prices vs. Quantities. *The Review of Economic Studies*, 41(4), 477–491.

Zetterberg, L. (2007). Sweden. In A. D. Ellerman, B. K. Buchner and C. Carraro (ed.), *Allocation in the European Emissions Trading Scheme: Rights, Rents and Fairness*. Cambridge: Cambridge University Press.

Chapter 22

LOW-CARBON ECONOMIC MECHANISIM

*Thomas Gries**

22.1. Introduction

Due to the threat of climate change and the increasing awareness of many political leaders of the threat, the discussion of a low-carbon economy has gained increasing importance in recent years. In order to get a general idea of the stylized facts of CO_2 emissions, we start with an international comparison of the most important actors. Naturally, as carbon emissions are closely linked to economic activities, high income economies are also major polluting economies and hence have to take responsibility for addressing this issue. However, with Asia developing so successfully, the problem has spread beyond high income economies to include others. Today large emerging economies such as China or India also substantially contribute to CO_2 emissions. While the fact of successful growth in these economies must be appreciated, we must be aware of the externalities, especially with respect to carbon emissions and climate change. Therefore the first step in a discussion of a low-carbon economy is to look at some stylized facts on emissions and energy use in terms of an international comparison. As we cannot give a full discussion either of the current global situation or of the perspectives for the next decades, we would like to refer to a fairly comprehensive study of this issue by the OECD (2008a). For our purposes we do not need to go that far; instead, we choose to compare four economies: the three largest high-income economies — the United States, Japan and Germany — and the

*Thomas Gries, Professor at the Center for International Economics, University of Paderborn, Germany.

largest emerging economy, China. As the international comparison intends to examine current activities, we also only look at current emission flows. We disregard the historical accumulation of emissions, which is important for other discussions.

22.2. CO_2 Emissions and Efficiency Indicator

CO_2 *emissions and economic activity:* Looking at CO_2 emissions of the economies it is evident that there is a link between market size and emissions. However, the strength of that link can vary, exemplified by the fact that Germany actually reduced its carbon emissions even though its economy underwent substantial growth during the period indicated.

CO_2 *efficiency*: Another way to examine international CO_2 emissions is to look at the aggregate CO_2 efficiency of economies. Figure 22.1 illustrates CO_2 per GDP measured in units of purchasing power parity. This very broad and aggregate measure gives a general indication of an economy's pollution output per unit of GDP. However, a number of problems are associated with this kind of efficiency measure. First, economies go through different stages of development. Today's advanced economies like the U.S., Japan and Germany generate 70% of their income in the service sector, which is neither energy nor emissions intensive. Therefore the state of development of an economy can influence energy consumption and pollution levels. Second, different economies are located in different climate zones. For

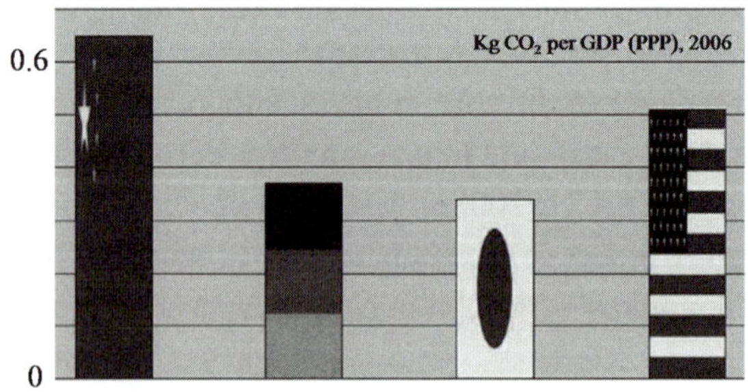

Fig. 22.1. CO_2 efficiency: CO_2 per GDP (PPP) (2006).
Source: IEA.

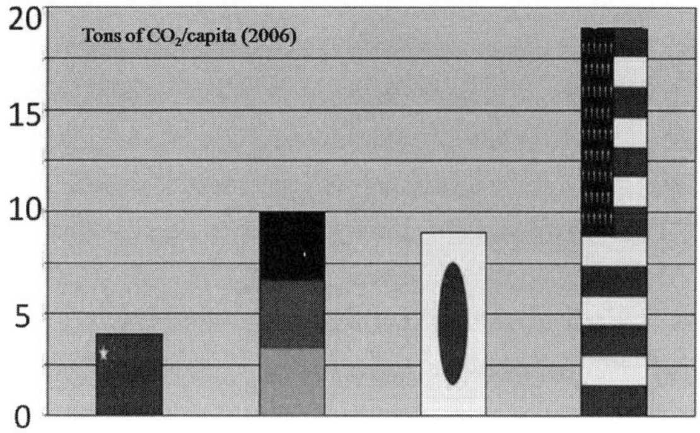

Fig. 22.2. CO_2 efficiency: CO_2 per capita (2006).
Source: IEA.

example, Germany is far further north than the other economies in the comparison, causing it to consume more energy for heating. In turn, it has higher pollution levels than the other economies under review. Hence a direct comparison must always be conducted with care.

CO_2 *efficiency*: Looking at CO_2 emissions per capita we can clearly recognize a difference (a) between emerging and advanced economies and (b) across advanced economies (see Fig. 22.2). While China's emissions per capita is still lower than that of Germany and Japan due to its lower per capita income, the United States, which has a similar income and wealth structure as Germany and Japan, has per-capita pollution levels that are twice as high as those of Germany and Japan.

22.2.1. *Energy mix and renewable resources*

Total primary energy mix:

Looking at the energy mix (see Fig. 22.3), it is evident that CO_2 emissions are impacted not only by the level of economic activity, but also by how resources are used. While the energy mix in the advanced economies is similar, China's energy mix has a clear focus on coal, the source of over 60% of its primary energy. Surprisingly, China has a relatively high share of renewable energies which is due to its strong hydropower sector.

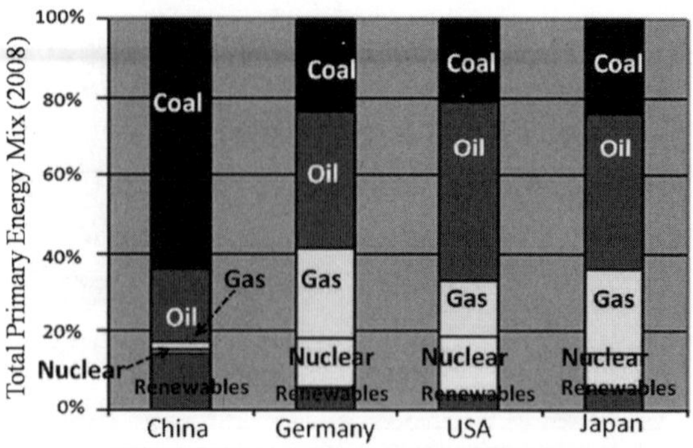

Fig. 22.3. Energy mix: Total primary energy mix (2008).
Source: IEA.

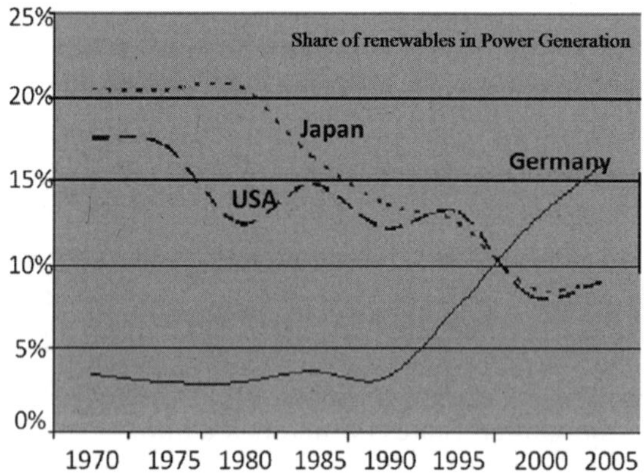

Fig. 22.4. Share of renewables (1972–2006).
Source: IEA.

Share of renewables in power generation: Figure 22.4 shows that renewables play a very small role in Germany's power generation industry. The reason for this is that in most countries, renewable energy traditionally means hydroenergy. Unlike the U.S. and Japan, Germany lacks the geography needed to develop a significant hydropower sector. Until the 1990s, therefore, renewable forms of energy accounted for less than 5% of the power generation industry. This clearly changed in the 1990s with the arrival of wind, solar and bio energy.

22.3. Emissions and Renewable Energies in Germany

22.3.1. *Policies and instruments — background*

After the oil crisis in the 1970s Germany became increasingly aware of the limitations of exhaustible resources and the benefits of a pleasing environment. Today there is widespread awareness of environmental issues both among politicians and the general public. The political system, too, reflects these developments. The Green Party, which was founded in the late 1970s, is a firmly established political organization. All political parties have included defined environmental policies in their programs. Therefore there is less resistance to environmental policy instruments provided they do jeopardize the competitiveness of German industry.

Looking at the policy mix we can identify three categories of different instruments. First, the traditional instrument of *direct regulation* is essentially a "command and control" approach. Second, *market instruments* work through price incentives. The third type comprises *informative and voluntary instruments* such as PR campaigns and environmental incentives for firms or environmental projects. All three categories are strongly represented in the German policy mix. More recently, German policy has increasingly attempted to switch to instruments conform to market conditions. These widespread and much discussed instruments are taxes, subsidies and certificate trading. The idea of using taxes to correct inefficient market prices due to negative externalities goes back to Pigou (1920). The Pigou tax corrects market failure and can even lead an economy back to an efficient first-best solution. However, owing to major information problems first-best solutions are not likely to be achievable, so a discussion about second-best solutions ensued. As a major benchmark in further development Pigou's idea was later extended by Baumol and Oates (1971) and Baumol and Oates (1988). Their standard price model is still one of the most important theoretical instruments in the discussion of implementing environmental policies,[1] and is included in almost all modern textbooks on environmental issues.[2] The approach has been extended beyond pure price correction and the lump-sum redistribution of tax revenue by the government to include a discussion

[1]For a recent discussion see OECD (2006a).

[2]See Oates (1996), Pearce (2000), Pearce and Turner (1990) or Baumol and Oates (1988).

of the double dividend of tax policy.[3] The double-dividend hypothesis is based on the idea that price correction not only impacts on market efficiency, but also promotes the investment of tax revenue in environmental projects and environmental technology research. However, this hypothesis is hard to verify since there is no explicit link between certain tax revenue and expenditures in German tax policy. A second important instrument, conform with market conditions, is the introduction of emissions trading.[4] The system defines a maximum level of pollution for an aggregate level of an economy, a region or even an industry, hence allowing a certain level of pollution that is defined according to requirements and policy goals. Once the total level of pollution is defined potential polluters can compete for government-issued permits to emit certain amounts of pollutants. As long as there are no other externalities between the firms this competition will also lead to an efficient market allocation of pollution, since tradable permits direct pollution allowances towards the most efficient form of use. The third category is a category of informative and voluntary instruments, which raise awareness of benefits among the public and among businesses. Recent OECD studies have described a number of programs of this kind that were implemented in advanced economies.[5]

A number of comprehensive recent studies, in particular by the OECD (cf. e.g. OECD, 2008b, OECD, 2009c), provide an excellent overview of international policies and best practices for achieving global environmental goals. They take into account all kinds of instruments, the effectiveness of the policies, and in some cases discuss the political economy of implementing environmental policies.

22.3.2. *Development of emissions, fossil fuel savings and renewable energies*

This section examines Germany in more detail. First we look at aggregate developments before analyzing each specific sector in more detail. With respect to the aggregate development of emissions Germany reduced emissions over the last fifteen years by approximately 17%, or around

[3]See e.g. Bovenberg and De Mooij (1994) and Goulder (1995) for a discussion.
[4]These ideas go back to Dales (1968).
[5]Examples of this instrument are given in several OECD studies (OECD 2003a, b, c).

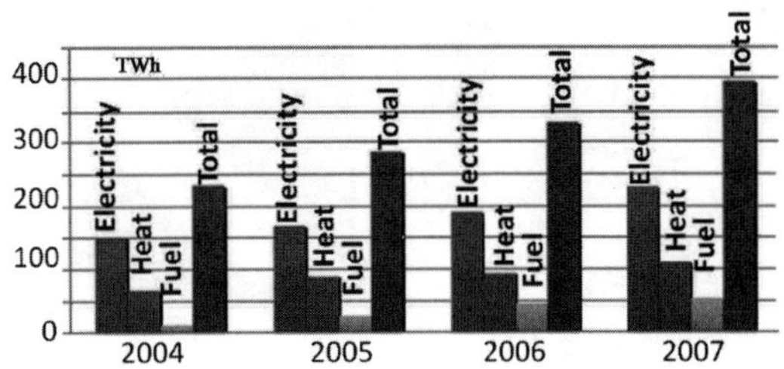

Fig. 22.5. Saved fossil fuel due to renewable energy Germany (2008).
Source: Federal Ministry for Environment, Nature Conservation and Nuclear Safety.

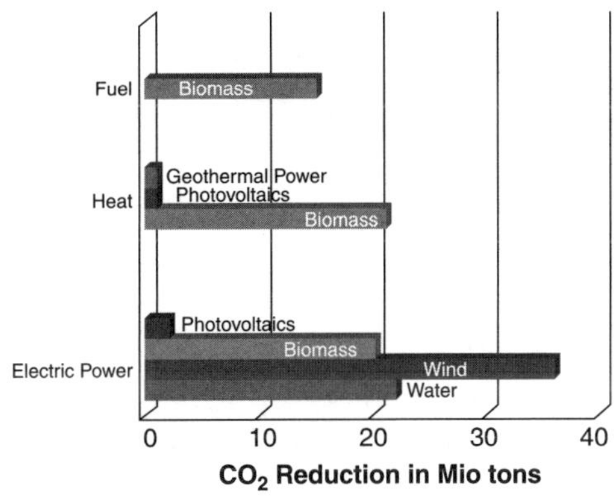

Fig. 22.6. CO_2 reduction due to renewables Germany (2008).
Source: Federal Ministry for Environment, Nature Conservation and Nuclear Safety.

1% per year. While 1% per year does not seem very high, if an economy is able to reduce its emissions slowly but continuously, in the end the total reduction is strong and significant.

Saved fossil fuel due to renewable energy: A significant contribution to this reduction in emissions stems from the rise in renewable energies. Figure 22.5 indicates fossil fuel savings due to renewable energies, broken down by resource use. This picture illustrates that the most important contribution of renewable energy to reducing fossil fuel consumption was in power generation, followed by heating and fuel.

CO_2 *reductions due to renewable resources*: With respect to the different fields of energy use, Fig. 22.6 shows that with the exception of traditional

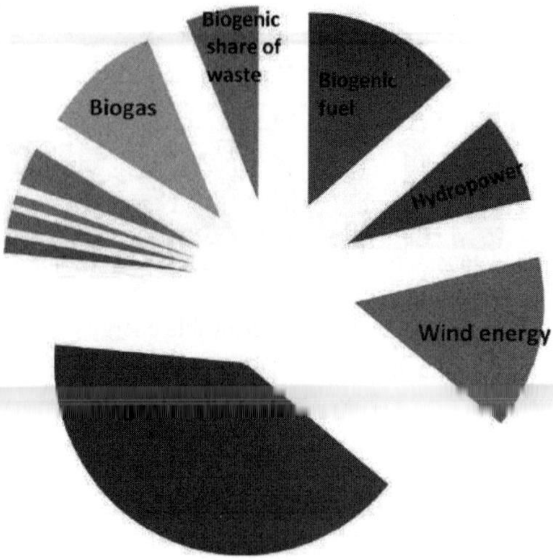

Fig. 22.7. Primary energy mix (renewable energies) in 2008.
Source: Federal Ministry for Environment, Nature Conservation and Nuclear Safety.

hydroenergy, wind and biogenic resources contribute most to power generation. With respect to heating, biomass is by far the largest contributor of renewable energies. The same is true for fuel supply.

Primary energy mix (renewable energies): Figure 22.7 illustrates the mix of renewable energies. The biggest contribution currently comes from biogenic sources, followed by wind energy, hydropower and solar energy.

22.4. Savings and Renewable Energies by Sector

22.4.1. *Emissions by sector*

While Section 22.3 gave a brief overview on aggregate developments, in this section we would like to look at the different sectors and to what extent policy programs to encourage savings and the use of renewable energies have contributed to each sector's change in emissions and energy use. To start with, we look at emissions by sector. Figure 22.8 shows that the power generation industry is by far the biggest polluter in the German economy, followed by mobility, then by all other sectors of industry. At 13% the housing sector is also a significant factor. This holds even more if we

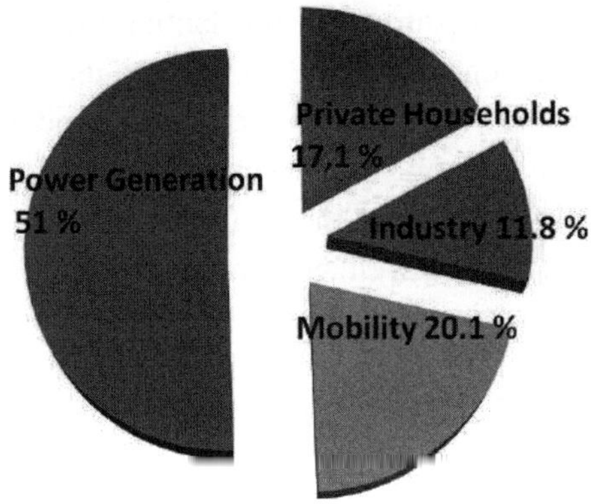

Fig. 22.8. Emissions by sector in 2007 Germany (2008).
Source: Federal Ministry for Environment, Nature Conservation and Nuclear Safety.

consider that households also account for a large share in electric power consumption.

However, the sectors in question have rather different characteristics. There is therefore no general policy or set of instruments that could be applied to all sectors in an equally efficient and effective manner. Policies and instruments need to be adjusted to the characteristics of each sector and to the efficiency and effectiveness with which they apply the instruments.

In the following we discuss for each sector the programs, ideas, and targets for saving energy and emissions, and look at some of the instruments used to put these targets into practice. In each case we first examine energy savings programs before moving on to programs and ideas to promote the use of renewable resources.

22.4.2. *Private households*

The residential sector is highly decentralized. Since most properties are owned by private individuals, there are a large number of homeowners for whom effective instruments must be developed.

Energy savings: The main idea behind most energy savings programs for residential properties is to reduce fossil energy consumption and in turn reduce CO_2 emissions.

The low-carbon refurbishment of residential properties is expected to increase the buildings' energy efficiency in various respects. The idea is to substitute the use of resources by the use of capital, such as thermal insulation of the walls or roofs, or the installation of triple glazing. This enables heating energy to be used more efficiently, eventually reducing the resources used for heating as well as reducing pollution.

As residential properties are typically privately owned, the government has to offer incentives to homeowners to encourage them to take up energy saving capital and technologies. The most common market incentives by the government are favorable loans and direct subsidies. Homeowner implementing energy saving technologies can apply for special low-interest loans or direct subsidies. In Germany the policy-mix is dominated by market incentives. Among the incentives, subsidies are the most common and preferred instrument. As subsidies are nothing but a negative Pigou tax, they are able to push the market towards a more efficient use of resources. However, the distribution of burdens is different. While in the case of taxes, the burden is on the polluter, subsidies shift the burden to taxpayers who contribute part of their income to promote a reduction in pollution. This reversal of the distribution scheme may have something to do with the political economy of policy implementation and asymmetric information and market power. However, the incentive system is a step in the right direction and therefore a potential second-best solution.[6]

Figure 22.9 provides an overview of the potential energy savings in insulated properties. Modern houses consume about 40% less energy per square meter than houses built 60 years ago — an enormous effect. The most effective measure is to install a new and more efficient heating system, followed by thermal insulation and ceiling and roof insulation.

While the graphs give a general impression of the development of energy savings and its potential effects, some studies examine the effectiveness of certain policy instruments in more detail. One such program is the CO_2 Building Restructuring Program that was launched by Germany's Federal Government in cooperation with the Kreditanstalt für Wiederaufbau. The program has undergone several amendments and has run under the name *Energieeffizient Sanieren* since April 2009.

---

[6]For an academic discussion: See Adolf (2008), Sorensen (2007) or Gao *et al.* (2005).

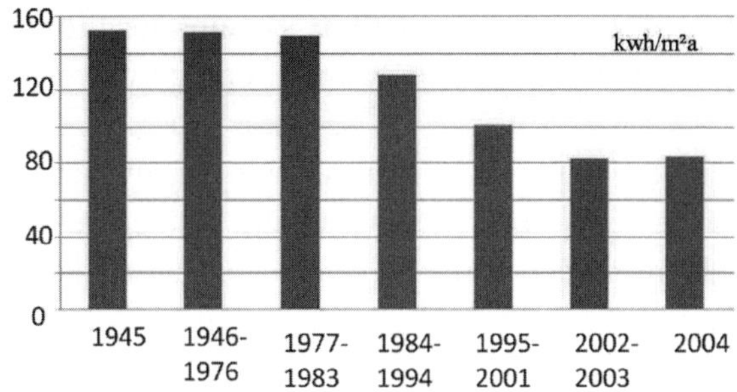

Fig. 22.9. Consumption of energy (kwh/m^2) of houses by year of construction (2007). *Source*: Federal Ministry of Transport, Building and Urban Affairs.

Kuckshinrichs, Hansen and Kronenberg (2009) point out that the CO_2 Building Restructuring Program is an incentive instrument with overall positive effects that will achieve an overall reduction in CO_2 emissions and hence help to prevent environmental damage in the long term. Savings are estimated to reach between € 22 and € 67 per ton of CO_2, a very significant figure (Kuckshinrichs *et al.*, 2009).

For investors, the short-term investment costs are set off by additional benefits in the long-term. For example, tenants in refurbished houses are more likely to accept rent increases. For the government, the benefits are ambivalent. On the one hand, the government has to finance the program. On the other hand, the program increases investments and therefore enhances labor demand, which may lead to an increase in tax revenues and a decrease in social security spending. Eventually, most effects depend on the various elasticities.

Unfortunately these elasticities are very hard to measure and identify. For instance, Kohlhaas (2005) compares marginal elasticity of demand for private households with elasticity of demand in the industry sectors. Private households can only partially substitute energy demand (Boonekamp, 2006). Liu (2004) assumes that elasticity of demand is between −0.03 (short-term) and −0.157 (long-term). However, all these figures are very uncertain and should be interpreted with care.

Renewable energy: The second area of policy action promotes the use of renewable energy. Programs in this field encourage the use of modern heating systems, solar energy, air handling systems, heat from solar energy, photovoltaics, and the substitution of fossil energy by renewable energy sources.

Again, as in the discussion surrounding energy savings, any program to introduce renewable energies in the residential sector must be decentralized. Homeowners must be offered incentives for implementing renewable energy technologies. Therefore, the instruments used are rather similar to the instruments used to promote energy saving. Again, favorable loans and direct subsidies are part of the portfolio. However, there is one additional instrument that leads to a very decentralized power generation by households. Power generated from solar panels can be sold in return for a fixed and highly subsidized feed-in tariff to large power utilities. As the feed-in tariffs are fixed for the long-term, decentralized private households can invest in their own power generation systems and sell any surplus power to the market for a reasonable profit. This is a clear incentive for private energy generating investments. Similar ideas are being discussed and tested for a decentralized electricity supply using gas.

To estimate the effects of these policies we look at Fig. 22.10, which shows the different heating systems in old and new buildings. The first fact to note is that there are virtually no coal-fired heating systems in households any more. As of 1998, there were practically no heating systems that used renewable forms of energy. In 2006, the share of renewable energy heating

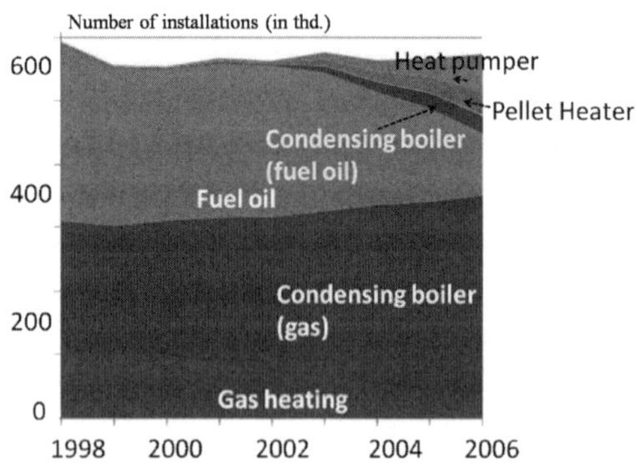

Fig. 22.10. Heating systems in old and new buildings (2007).
Source: Federal Ministry of Transport, Building and Urban Affairs.

systems was still small, but clearly above 10%. Even though the use of fossil resources continues to dominate, the technologies in use today are much more efficient and in far wider use. Solar panels are another example of the implementation of renewable resources in households. Driven by subsidies and market incentives, the number of installed solar panels has increased dramatically since 1990. This is even more surprising since Germany is a northern country geographically.

22.4.3. *Transportation and mobility*

Energy savings: Transportation and mobility is another important sector in terms of CO_2 emissions. It accounts for 20% of all emissions and is hence a major polluter. Here, too, there are two types of policy: energy savings and the promotion of renewable energies. With respect to savings, the programs, ideas and targets aim to encourage individuals to buy fuel-efficient, low-pollution cars. Further, a combination of price incentives for less car use is linked to a well organized transportation system. Public transportation is expected to substitute for individual car use. There are also incentives to encourage car sharing. Among the instruments in use in this sector are heavy fuel taxes and motor vehicle taxes depending on the vehicles' emissions.

All instruments serve to reduce transportation-induced pollution. However, the greatest effects are expected to be generated by direct regulatory measures and tax incentives. High taxes on fuel and strict caps on vehicle emissions are the most important and effective instruments. Fuel taxes in OECD countries are shown in Fig. 22.11. Germany has one of the highest fuel taxes per liter of fuel, which acts as a strong incentive to either use public transport or substitute individual car use with car-sharing schemes or car-pooling. In the medium term consumers may even consider buying smaller, more fuel-efficient cars. The most effective regulatory measure is the emissions check that all cars must undergo every other year. Cars failing to keep within the required limits are taken off the road. While high fuel taxes have been in force in Germany for quite a while, some OECD studies are presently examining the potential influence of taxes on CO_2 emissions of motor vehicles.[7]

[7] OECD (2009a), OECD (2009b).

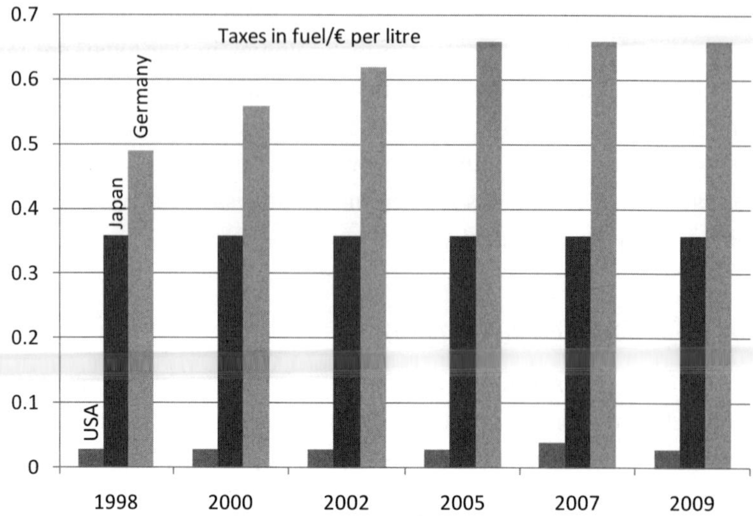

Fig. 22.11. Fuel taxes in OECD countries (# € per litre).
Source: OECD/EEA (2009).

In order to gain an impression of the effect of price incentives on fuel consumption in transportation we take a look at Fig. 22.12, which shows year-by-year growth in fuel consumption in the transportation sector. The 1970s were clearly impacted by the two oil price shocks in 1973 and 1979 — the only years of high oil prices to have negative growth in fuel consumption in the transportation sector. The tables only turned after the introduction of the so-called eco-tax on fuel. The drastic increase in taxes drove up fuel prices, which acted as an incentive to switch to more fuel-efficient cars in the medium-term. Compared to the oil price shocks, which did not curb consumption long-term, the eco-tax was effective because it actually brought about a permanent increase in prices and was not just a short-term shock.

Since the introduction of the eco-tax, demand for mineral oil has been decreasing, some empirical studies claim effectiveness of the eco-tax. Kohlhaas (2005) measured a change rate of between −0.50 (1999) and −1.91 (2005). For fuel in Germany, Graham and Glaiser (2002) assume an elasticity of demand between 0.25 (short-term) and 0.77 (long-term).

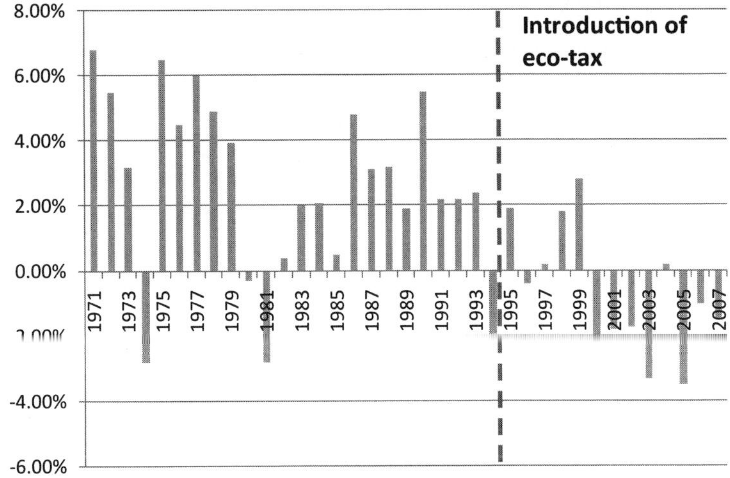

Fig. 22.12. Fuel consumption growth (by year).
Source: UPI Institute (2008).

An international comparison of CO_2 emissions in the transportation and mobility sector shows that Germany is the only country of the three countries in Fig. 22.13 that has been able to keep total consumption in the transportation sector constant. Fuel consumption in the United States increased by 35% in this period, while in Japan it rose almost 20%. Again, this shows that introducing incentives for decentralized actions seems to be a rather effective measure.

Renewable energy: Turning to renewable resources in the transportation sector, there are government policies that promote biofuels and alternative vehicle technologies. Biofuels are an option for car drivers, while alternative vehicle technologies are mostly found in the public transportation sector. Biofuels taxes are lower than other fuel taxes, and alternative vehicle technologies are subsidized via various R&D projects. Very recently, the government announced a plan to spend € 500 million by 2012 on the implementation of alternative vehicles technologies. The aim is to have one million electric cars on the roads by 2020.

Turning to an international comparison of the usage of renewable fuels in transportation we see that Germany is clearly ahead in terms of its experience with alternative fuels (see Fig. 22.14).

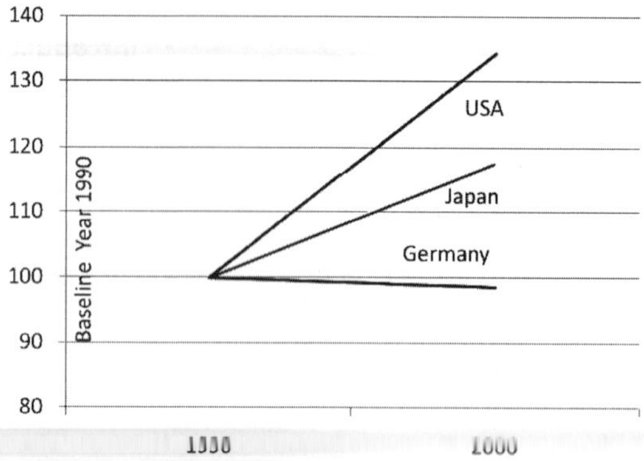

Fig. 22.13. Total growth in CO_2 emissions since 1990.
Source: Institut der deutschen Wirtschaft (2009).

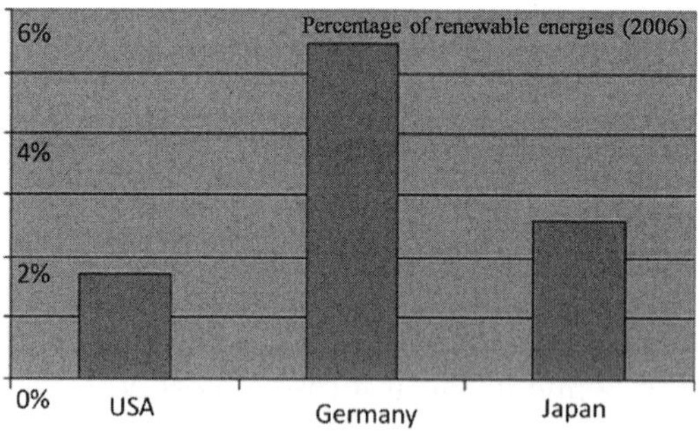

Fig. 22.14. International comparison of renewable fuel use in 2006.
Source: IEA.

22.4.4. *Industry*

Energy savings: In this last section we turn to industry, for which a variety of ambitious programs and targets exist. The power generation sector is part of this sector. Elasticity of energy demand in this sector is assumed to be higher than the elasticity of demand in private households (Kohlhaas, 2005). The aim in this sector is to reduce CO_2 emissions by about 20% by 2020 and cut energy consumption by 20–40%. In agriculture, construction and transport in particular, energy consumption is decreasing (Kohlhaas, 2005). These targets are to be reached via a combination of incentives and regulatory measures. In addition to favorable loans and direct subsidies for energy

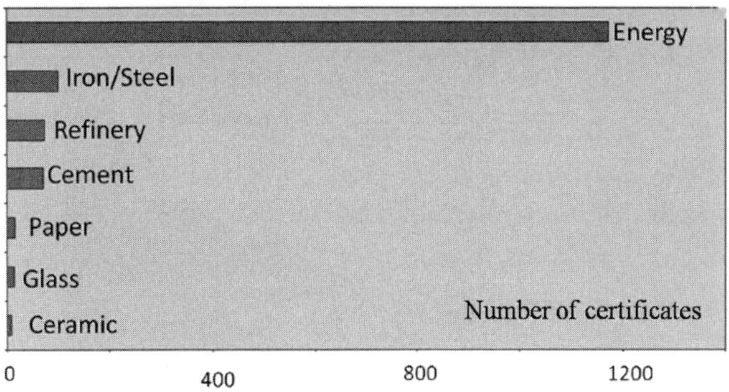

Fig. 22.15. Allocation of CO_2 certificates by industry/million tons (2005).
Source: Federal Ministry for Environment, Nature Conservation and Nuclear Safety; German Emission Trading Agency.

saving investments, emission caps have been introduced. Emissions trading will ensure an optimal distribution of emissions within these limits.

Taking a closer look at all industry sectors, the energy sector is the greatest polluter of all industries. Therefore, instruments should focus on the power generation sector.

The instruments for the industry sector lean more towards direct regulation and tradable permits. As power generation is the greatest polluter in the sector there are far more polluters in this sector than in others. Therefore, both Germany and the EU have introduced tradable permits for the manufacturing and power generation industries.[8] The allocation of CO_2 certificates, shown in Fig. 22.15, indicates that the manufacturing industry accounts for a fairly small share of total pollution. The iron and steel sector, refineries and the cement industry are significant direct polluters. This structure also reflects the general production structure of advanced economies. In these economies more than 2/3 of value added is generated in the service sector, which is naturally less energy-intensive than the manufacturing sector. Advanced economies like Germany hence enjoy an advantage in terms of their eco-balance.

Renewable energy: As already mentioned, the greatest polluter within the industry is power generation. Germany's power generation sector still uses

[8] See e.g. Christiansen *et al.* (2005) or Convery (2009).

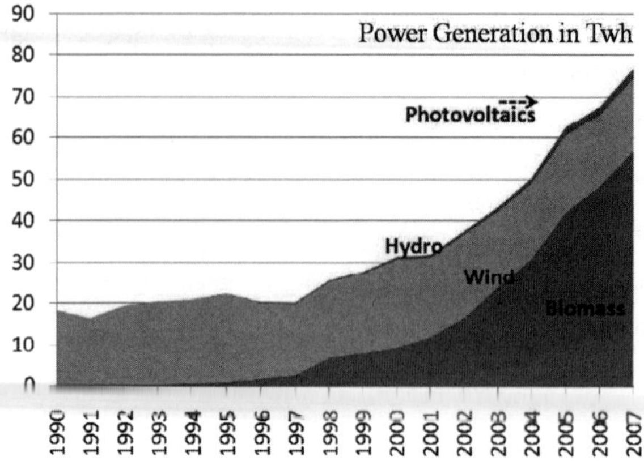

Fig. 22.16. Electric power generation: Renewable energy (2008).
Source: Federal Ministry for Environment, Nature Conservation and Nuclear Safety; German Emission Trading Agency.

50% coal, owing to the fact that its coal mining industry is still significant in certain regions. Gas and Oil accounts for about 13%. However, one reason for the massive incentives to use renewable energies to generate power is the political decision to move away from nuclear power. Today (in 2010), nuclear power accounts for 25% of Germany's power generation sector. By law, the last nuclear power station in Germany is scheduled to shut down in the next decade. Due to the nuclear incident in Fukushima, Japan in 2011, the pace at which Germany moves away from nuclear energy increased. Before the accident in Japan, the last reactor was scheduled to shut down in the 2040s. In the medium- and long-term nuclear power will hence have to be substituted by other energy resources. With CO_2 emissions also slated for a drastic reduction, renewable resources will have to fill this gap, too. Within the 15 years (1990 to 2005) the share of renewable resources in power generation has increased from approximately 3% to more than 16% today. However, the target of 30% by 2030 remains an ambitious goal.

Looking at the mix of renewable energies, there seems to be no growth potential for hydropower. The resources with most growth potential are, according to Fig. 22.16, biomass and wind energy. Photovoltaics also account for a fast growing share; however, its absolute contribution still appears limited.

22.5. Long-Term Scenario and Perspectives

Finally, we would like to provide a long-term outlook concerning the expected use of renewable resources in Germany until 2050. Figure 22.17 illustrates this scenario. The total contribution of renewable resources to all primary energy consumption is expected to double from around 9% today to 18% in 2020, rising to 60% in 2050. According to this scenario its main contribution is expected to be to power generation, where its already respectable share of 15% is anticipated to increase to 50% in 2020, hopefully rising to 80% in 2050. For heating, renewable resources are expected to increase by 2020 to 11% and to almost 50% by 2050. The sector that seems to have most problems moving from fossil fuels to renewable resources is transportation, where the share of renewable resources is expected to increase moderately from 8% today to around 12% by 2020. Between 2020 and 2050 the scenario anticipates one third of fuels to be based on renewable resources.

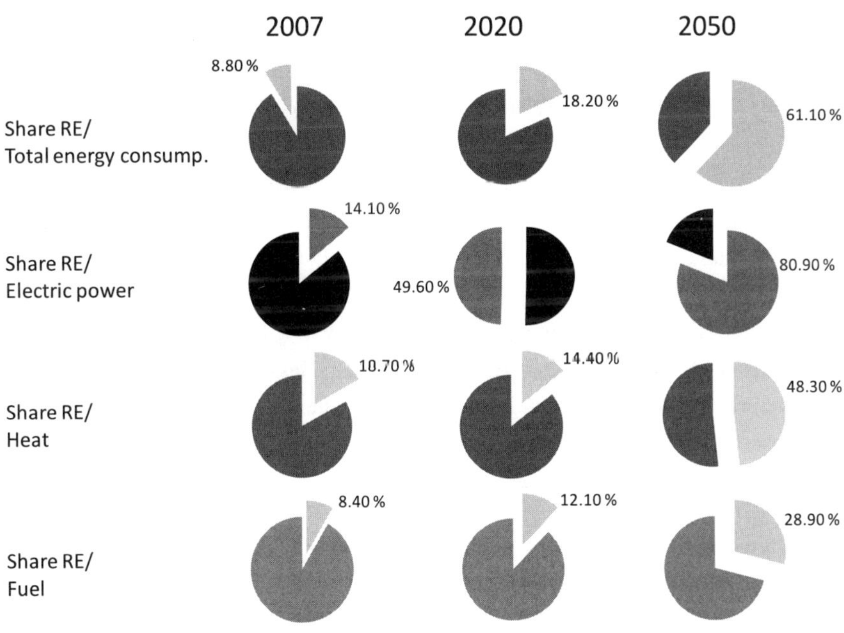

Fig. 22.17. Share of renewable energy until 2050 (2008).
Source: Federal Ministry for Environment, Nature Conservation and Nuclear Safety.

However, even though it is uncertain whether this is a likely scenario, two mechanisms will eventually be responsible for reducing the use of fossil resources and boosting the use of renewables. First, depending on relative prices we will see a shift towards renewable resources that can help to reduce pollution and CO_2 emissions. Second, relative prices are able to direct technical progress towards energy-saving and pollution-reducing technologies. If relative prices can incentivize direct substitution and technical progress, the global economy may be able to reach the targets required to slow down climate change. However, reaching these targets is not just a problem at the global aggregate level — it is also a problem of income and wealth distribution and a fair distribution of burdens across the world's economies.

References

Adolf, J. (2008). Marktwirtschaftliche Instrumente — Königsweg der Klimapolitik? *Wirtschaftsdienst*, 326–333.

Baumol, W. J. and Oates, W. E. (1971). The Use of Standards and Prices for Protection of the Environment. *Swedish Journal of Economics*, 73, 42–54.

Baumol, W. J. and Oates, W. E. (1988). *The Theory of Environmental Policy*, 2nd edition. Cambridge: Cambridge University Press.

Boonekamp, P. G. M. (2006). Price Elasticity, Policy Measures and Actual Development in Household Energy Consumption. A Bottom-up Analysis for the Netherlands. *Energy Economics*, 29, 133–157.

Bovenberg, A. J. and de Mooij, R. A. (1994). Environmental Levies and Distortionary Taxation. *American Economic Review*, 84, 1085–1089.

Christiansen, A. C., Arvanitakis, A., Tangen, K. and Hasselknippe, H. (2005). Price Determinants in the EU Emissions Trading Scheme. *Climate Policy*, 5, 15–30.

Convery, F. J. (2009). Reflections — The Emerging Literature on Emissions Trading in Europe. *Review of Environmental Economics and Policy*, 3, 121–137.

Dales, J. H. (1968). *Pollution, Property and Prices*. Toronto: University of Toronto Press.

Federal Ministry of Transport, Building and Urban Affairs (2007), Available online: http://www.bmvbs.de

Federal Ministry for Environment, Nature Conservation and Nuclear Safety (2008). *Pilot Study 2008 — Advancements in the Strategy of Renewable Energies*, Available online: www.erneuerbare-energien.de.

Gao, W., Madlener, R. and Zweifel, P. (2005). "Promoting Renewable Electricity Generation in Imperfect Markets Price versus Quantity." CEPE Working Paper 45.

German Emission Trading Agency (DEHST) (2005), Available online: http://www.dehst.de.

Goulder, L. H. (1995). Environmental Taxation and the Double Dividend Hypothesis: A Reader's Guide. *International Tax and Public Finance*, 2, 155–182.

Graham, D. and Glaiser, S. (2002). The Demand for Automobile Fuel: A Survey of Elasticities. *Journal of Transport Economies and Policy*, 36, S. 1–26.

Institut der deutschen Wirtschaft (2009). Available Online: http://www.iwkoeln.de

International Energy Agency (2006, 2009). Available Online: http://wds.iea.org/wds/ReportFolders/ReportFolders.aspx?CS_referer=&CS_ChosenLang=en.

Kohlhaas, M. (2005). "Gesamtwirtschaftliche Effekte der ökologischen Steuerreform. Forschungsprojekt im Auftrag des Umweltbundesamts." FuE-Vorhaben Förderkennzeichen 204 41 194. August 2005.

Kuckshinrichs, W., Hansen, P. and Kronenberg, T. (2009). "Gesamtwirtschaftliche CO_2-Vermeidungskosten der energetischen Gebäudesanierung und Kosten der Förderung für den Bundeshaushalt im Rahmen des CO_2-Gebäudesanierungsprogramms." STE Research Report. 04/2009.

Liu, G. (2004). "Estimating Energy Demand Elasticities for OECD Countries — A Dynamic Panel Data Approach." Discussion Papers No. 373 Statistics Norway.

Oates, W. E. (1996). *The Economics of Environmental Regulation*. Cheltenham: Edward Elgar Publishing.

OECD (2003a). "Voluntary Approaches: Two Japanese Cases, Pollution Control Agreements in Yokohama City and Kitakyushu City, 2003."

OECD (2003b). "Voluntary Approaches: Two United States Cases: The Experience of Intel Corporation and Merck Pharmaceuticals in Project XL, 2003."

OECD (2003c). "Voluntary Approaches: Two Danish Cases: The Danish Agreement on Industrial Energy Efficiency, with Examples from the Paper Sector and the Milk Condensing Sector, 2003."

OECD (2006a). "The Political Economy of Environmentally Related Taxes, 2006."

OECD (2008a). "OECD Environmental Outlook to 2030, 2008."

OECD (2008b). "Climate Change Mitigation What do we do? 2008."

OECD/EEA (2009). "OECD/EEA Database on Instruments Used for Environmental Policy and Natural Resources Management."

OECD (2009a). "Incentives for CO_2 Emission Reduction in Current Motor Vehicle Taxes."

OECD (2009b). "The Scope for CO_2-Based Differentiation in Motor Vehicle Taxes."

OECD (2009c). "Ensuring Environmental Compliance: Trends and Good Practices, 2009."

OECD/EEA (2009). "OECD/EEA Database on Instruments used for Environmental Policy and Natural Resources Management."

Pearce, W. A. (2000). *Economics and Environment: Essays on Ecological Economics and Sustainable Growth*. Cheltenham: Edward Elgar Publishing.

Pearce, D. and Turner, K. (1990). *Economics of Natural Resources and the Environment*. Baltimore: John Hopkins University Press.

Pigou, A. C. (1920). *The Economics of Welfare*. London: McMillan.

Sorensen, P. B. (2007). The Theory of Optimal Taxation in the Policy Relevance. *International Tax Public Finance*, 14, 383–406.

Umwelt- und Prognose — Institut e.V. (2008). "Trendwende beim Kraftstoffver brauch stabilisiert sich." Available online: http://www.upi-institut.de/Kraftst offverbrauch.htm

Chapter 23

CARBON KUZNETS CURVE AND ITS TURNING POINTS

Jinjun Xue and Ziqi Lin†*

23.1. Introduction

The 2009 United Nations Climate Change Conference, generally known as the Copenhagen Conference (COP15), was held at the Bella Center in Copenhagen, Denmark. One of its purposes was to establish a decisive global agreement on greenhouse gas emission targets after the year of 2012, which was not discussed in the Kyoto Protocol. During the conference, some countries declared what target they would aim at. Japan, the United Kingdom and Canada stated that they would cut greenhouse gas emissions by 25%, 34% and 25%, respectively, below 1990 levels by 2020. The United States proposed to decrease its greenhouse gas emissions by 17% below 2005 levels by 2020. Similarly, China, the largest greenhouse gas emitter in the world, stated that it would cut per unit of GDP by 40% to 45% below 2005 levels. The targets, however, do not contain any legally binding commitment for the reduction of greenhouse gas emissions. At COP16, held in Mexico from November to December of 2010, BASIC countries (Brazil, South Africa, India, and China) challenged developed countries to take responsibility for achieving quantified goals, to provide developing countries with aid and technology transfer and to protect the rights of developing countries.

There exists a gap in greenhouse gas emissions between developed countries and developing countries and is mainly the result of the gap in income levels between developed countries and developing countries. Hence, this

*Jinjun Xue, Professor of Economics at Graduate School of Economics, Nagoya University, Japan.
†Ziqi Lin, Graduate from Graduate School of Economics, Nagoya University.

chapter systematically regards a developed country as a high-income country and a developing country as a low to middle-income country. This classification method will be discussed later on.

In general, low to middle-income countries, such as China, India and Vietnam, tend to place a priority on economic growth over environmental protection, an approach that is often referred to as "Pollute First, Control Later". On the other hand, most developed countries have already experienced a turning point in the Environmental Kuznets Curve (EKC) and are devoted to reducing their emissions levels. The majority of developed countries, however, still have higher emissions levels than developing countries.

This chapter focuses on emissions levels in high and low-income countries by comparing trends in emissions levels. It also considers the relationship between economic growth and emissions levels with the use of the basic EKC hypothesis. Factors such as population increase, urbanization, industrialization, investment rate, export dependency and energy intensity will be included in the fundamental EKC hypothesis, and will also be integrated into a regression analysis with time-period dummy variables. Thus, this chapter is designed to find how each factor affects emissions levels in high and low-income countries.

23.2. Literature Review

The Environmental Kuznets Curve (EKC) hypothesis is that there is an inverted-U relationship between economic growth and environmental pollution. The EKC hypothesis explains the inverted-U relationship by stating that factors such as changes in the composition of output, the introduction of advanced technology, environmentally friendly regulation and increases in public demand for green products, and how they can cause a decline in pollution associated with increasing per capita income. However, a number of recent studies have argued instead, that an increase in per capita income does not always lead to decreasing, and eventually a negative, marginal change in pollution levels.

Several recent studies have claimed to find empirical evidence of an inverted-U shaped relationship between pollution levels and income levels. These studies are mainly focused on two questions, "Can the EKC hypothesis be validated empirically?" and, if so, "What factors cause the inverted-U relationship between pollution and per capita income?" It is found that certain types of pollution, such as suspended particulate matter (SPM), CO, SO_2 and NO_x, follow an inverted-U shaped relationship. On the other hand, it has rarely been seen that pollutants such as CO_2 follow the inverted-U shaped relationship of the Environmental Kuznets Curve.

Representative research studies on polluting substances, excluding CO_2, are shown below.

In a recent paper, Grossman and Krueger (1995) use cross-country panel data on urban air pollution levels to estimate the point at which atmospheric concentration of SPM and SO_2 experience an EKC. Estimated turning points for the two polluting substances are both under 5000 U.S. dollars (in 1985 U.S. dollars) of per capita GDP.

Similarly, Selden and Song (1994) investigated the EKC using cross-national panel data, collected by the World Resources Institute, on emissions of four important air pollutants: suspended particulate matter, sulfur dioxide, oxides of nitrogen and carbon monoxide. They estimated an analytical regression based on the relationship among per capita emissions, real per capita GDP and population density. Estimated turning points for those pollutants turn out to be higher than Grossman and Krueger's (1995) estimates. In particular, calculated turning point estimates uniformly exceed 8000 U.S. dollars. The difference comes from the fact that Selden and Song do not use data on atmospheric concentration but instead use data on gross emissions levels. It has also been estimated that emissions levels of those substances will have an increasing trend for the next 50 years.

Cole *et al.* (1997) adopts an advanced EKC hypothesis in a number of ways. He employs such factors as carbon dioxide, CFCs and halons, methane, nitrogen dioxide, sulfur dioxide, suspended particulate matter, carbon monoxide, nitrides, municipal waste, energy consumption and traffic volumes. Also, emissions of nitrogen dioxide, sulfur dioxide and suspended particulate matter and energy use by transport sector are integrated into his analysis separately. His research successfully observes a turning point for all pollutants except methane and energy use by transport. As a result, he concludes that turning points for integrated factors are higher than those of Grossman and Krueger's estimates and lower than those of Selden and Song's.

With regard to other research papers associated with energy consumption, Horvath (1997) is notably known as well. He tries to prove the EKC hypothesis for energy consumption per capita using cross-country panel data from 114 countries. He finds that a turning point for energy consumption cannot be identified, although energy consumption per capita increases constantly with increases in income per capita.

Of the other studies, few have researched the EKC hypothesis for carbon dioxide so that a turning point for carbon dioxide and its relationship with income per capita is rarely verified.

There are, however, some studies on the EKC for carbon dioxide. Shafik (1994a) conducts research on the relationship between economic growth and environmental quality, which mainly focuses on local air pollutants, water quality and sanitation. Although carbon dioxide is a monotonically increasing function with economic growth, turning points are observed for SO_2, CO_2, industrial waste matter and dissolved oxygen.

Holtz-Eakin and Selden (1995) examine the relationship between carbon dioxide and economic development by using global panel data provided by the Oak Ridge National Laboratory (ORNL), which suggests that there is a diminishing marginal propensity to emit CO_2 as per capita income increases. In addition, it is suggested that the EKC occurs with the turning point at 35,428 U.S. dollars per capita with a linear regression equation. In contrast, with a log linear equation, it occurs at 8 million U.S. dollars, so that CO_2 is expected to simply increase as income rises. Furthermore, their study indicates that global emissions of CO_2 grow at the annual rate of 1.8%, so that emissions rates will increase in developing countries, where production levels and population levels rise rapidly.

Schmalensee *et al.* (1998) estimate future emissions of carbon dioxide using reduced-form models with national-level panel data for the period of 1950–1990. An inverted-U curve can be observed within the observational period. Developed countries have negative income elasticity of CO_2 emissions per capita, especially if those countries have already achieved a turning point for carbon emissions, meaning that the amount of CO_2 has started to decrease.

The above studies have shown an attempt to verify the EKC hypothesis. This chapter introduces empirical studies on possible factors that explain the existence of the EKC. These possible factors, which are expected to explain the EKC hypothesis, are: change in economic, industrial and social structures, international trade, political freedom, scale of economies, price level, environmental change, institutional change and so on.

A main finding is that quality policies and institution can reduce environmental degradation at low income levels and speed up improvement at higher income levels (Panayotou, 1997). In other words, better environmental regulation and policies can help the EKC flatten or achieve a turning point earlier.

Torras and Boyce (1998) insist that there is a certain relationship between national institutions and turning points in the EKC. Specifically, literacy, political rights and civil liberties are found to have strong effects on environmental quality, especially in low-income countries.

Kaufmann *et al.* (1998) explore the effects of income and the spatial intensity of economic activity on the atmosphere concentration of sulfur dioxide. They conclude that there is a U-shaped relationship between income and atmosphere concentration of SO_2 and an inverted-U relationship between the spatial intensity of economic activity and SO_2 concentrations. In their paper, spatial intensity of economic activity is expected to enhance technologies and policies, and eventually lead to reduced SO_2 emissions.

Suri and Chapman (1998) use pooled cross-country and time-series data to quantify the effects of international trade on the EKC. When industrialized countries export manufactured goods, it makes the slope of the EKC shift upward. In contrast, the EKC curve becomes downward sloping by importing manufactured goods.

Rock (1996) attempts to clarify that, those countries with more export-oriented trade policies have higher pollution intensity per unit GDP than those with more inward-oriented trade policies.

Cole and Elliott (2003) attempts to explain impacts of trade liberalization on the environment, and confirms that it depends, in particular, on whether countries have a comparative advantage in pollution-intensive production, which is mainly determined by a country's endowments and relative environmental regulations.

Similarly, de Bruyn *et al.* (1998b) analyze the EKC hypothesis between energy prices and scale of economies. Neumayer (2002) includes natural factors, such as climatic conditions, the availability of renewable resources and transportation requirements of countries, in his research on the EKC hypothesis.

23.3. Income Gap and Carbon Emissions Gap

23.3.1. *Data*

There are a large number of studies on the EKC that mainly use data from the 1990s. However, this chapter is designed to use long-term data observed between 1960 and 2004; it will discuss what major change the Kyoto Protocol has caused.

This chapter adds 8 more countries to Selden and Song's data set, which originally included 22 high-income countries, 6 middle-income countries, and 2 low-income countries, and systematically divided those countries into two groups: high-income countries and low to middle-income countries. According to a research paper published by the World Bank in 2009, a country that has a per capita income of more than 12,196 U.S. dollars is a high-income country and one that has less than that is regarded as a low to middle-income country. The 23 high-income countries are Australia, Austria, Belgium, Canada, Denmark, Finland, France, Germany, Greece, Ireland, Iceland, Israel, Italy, Japan, Holland, New Zealand, Norway, Singapore, Spain, Sweden, Switzerland, the U.K. and the U.S. The 15 low to middle-income countries are China, Hungary, Egypt, India, Indonesia, the Philippines, Thailand, Portugal, Chile, Colombia, Malaysia, Mexico, South Africa and Turkey.

Data on CO_2 per capita, collected by the World Resources Institute (WRI), will also be incorporated. Data on GDP per capita comes from the online World Development Indicators (WDI) database provided by the World Bank. Its data is based on constant dollar terms in the year 2000.

Seven explanatory variables, including urbanization, population growth, industrialization, export dependency, energy intensity and investment rate, will be integrated into the regression analysis. These variables with the exception of investment rate, are from the WDI database as well. Investment rate comes from the online data set of PWT6.3, collected by the Center for International Comparisons at the University of Pennsylvania.

Also, time-period dummy variables for every 5 years between 1960 and 2004 will be used, and those variables are expressed respectively as 1960dummy, 1965dummy, 1970dummy, 1975dummy, 1980dummy, 1985dummy, 1990dummy, 1995dummy and 2000dummy. Each dummy variable is based on a 5-year period, with, for example 1960dummy representing the time between 1960 and 1964. GDP per capita and CO_2 per capita for every year are shown.

23.3.2. *Panel data*

The data sets used in this analysis are panel data associated with time series from 1960 to 2004. Some data, however, are not completely collected. GDP per capita in Germany in the period between 1960 and 1969 is unknown. Also, GDP per capita of Australia starts from the year of 1965. That is, the data sets are unbalanced panel data because those unknown data fail to be integrated into this analysis. Figure 23.1 shows the scatter plot of GDP per capita (x axis) and CO_2 per capita (y axis) for all observed values.

Fitting a curve to the scatter plot shown above shows a gradual inverted-U shaped curve. The part where the fitted curve increases shows that CO_2 per capita increases with GDP per capita. On the contrary, once reaching a peak, the curve falls to the right, meaning that, GDP per capita increases as CO_2 emissions decrease. Figure 23.1 consists of values that deviate greatly from the fitted curve; these points belong to Australia, the U.S. and Canada. The largest amount of CO_2 per capita in these countries is around 20 metric tons. Similarly, partial values diverge from the trend.

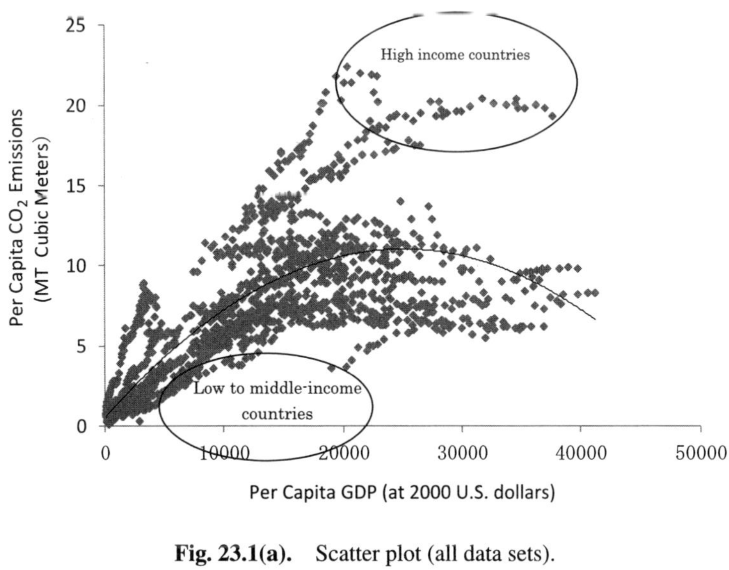

Fig. 23.1(a). Scatter plot (all data sets).

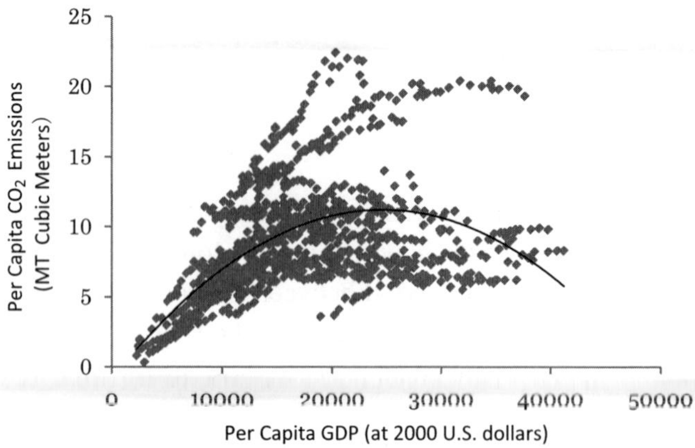

Fig. 23.1(b). Scatter plot (high-income countries).

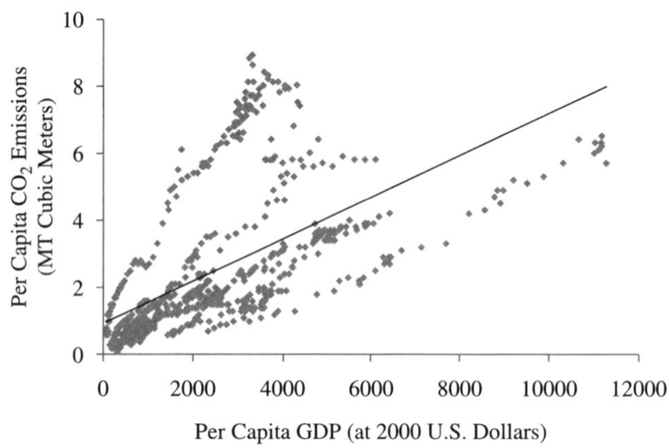

Fig. 23.1(c). Scatter plot (low to middle-income countries).

By considering the relationship between CO_2 emissions and GDP per capita in high-income countries and low to middle-income countries separately, differences appear clearly. The high-income group turns out to have an inverted-U relation, although the fitted curve in the high-income group is a little steeper than that in low to middle-income group. Observed values of the low to middle-income group, however, are plotted in divergent manner, as expected, and increase to the right. Thus, the EKC is not observed in the low to middle-income group. This chapter will further integrate more explanatory variables into the regression analysis.

23.3.3. *Observed trends in CO_2 emissions per capita*

Referring to the panel data shown below, differences in CO_2 emissions between high-income countries and low to middle-income countries appear to be clear. The highest CO_2 emissions per capita in high-income countries is more than 20 metric tons, however, in low to middle-income countries, it is less than 10 metric tons. Figure 23.2 shows trends in CO_2 emissions between 1960 and 2004.

In 2004, the average CO_2 emissions in high-income countries was 10.41 metric tons. Of all values, the U.S. has the largest CO_2 emissions of 20 metric tons, and Sweden has the smallest figure of 6.1 metric tons. In contrast, the average CO_2 in 2004 in low to middle-income countries is 3.43 metric tons, which is smaller by 5.84 metric tons than that of high-income countries. The largest amount of CO_2 is 7.4 metric tons, and the smallest is 0.9 metric tons.

In the period between 1960 and 2004, the average CO_2 in high-income countries goes up from 5.84 metric tons to 10.41 metric tons. By looking at 23 countries' trends in CO_2 emissions, they can be divided into three groups. One is the countries where CO_2 emissions have continued to increase from 1960. Australia, Singapore and Greece have this trend. Second, Germany,

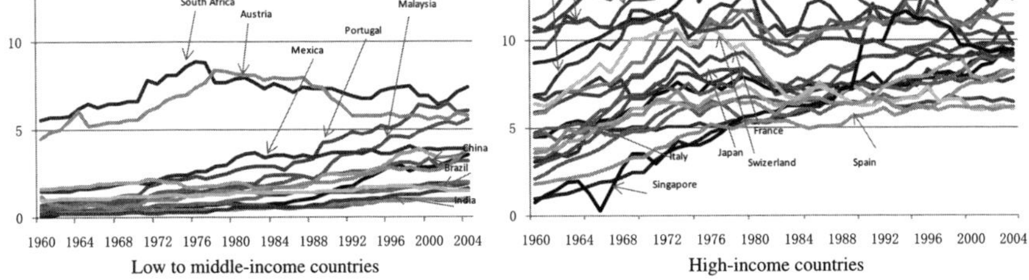

Fig. 23.2. Trends in CO_2 emission per capita (from 1960 to 2004).

the U.K., France and Sweden have a trend where CO_2 emissions decrease through two oil crises. In the third group, CO_2 emissions decrease after each oil crisis and then rise again; this includes such countries as Canada, the U.S., Belgium, Denmark, Holland, Spain, Finland and Japan.

The average CO_2 emissions in low to middle-income countries rose from 1.28 metric tons to 3.48 metric tons. In particular, South Africa, one of the BASIC countries, has a similar trend with high-income countries: after CO_2 emissions peaks at the end of 1970s, it decreases and then increases again. That is, among low to middle-income countries South Africa has a similar trend with high-income countries and relatively higher income. However, other countries in the group of low to middle-income countries do not show much of an increase in CO_2 emissions.

23.4. Carbon Kuznets Curve

23.4.1. *Basic model*

This chapter adopts a static panel data model that is often used to estimate the EKC:

$$E = \alpha 0 + \alpha 1Y + \alpha 2Y2 + \mu 1 \tag{23.1}$$

$$E = \beta 0 + \beta 1Y + \beta 2Y2 + \beta 3\text{Dum} + \mu 2 \tag{23.2}$$

where E is CO_2 emission per capita (metric tons), Y is GDP per capita, and $Y2$ is the square root of GDP per capita. Regression Eq. (23.1) includes only basic explanatory variables and supposes that factors other than GDP do not influence CO_2 emissions. Taking into consideration the price of oil, changes in technology, economic fluctuations, environmental policy and environmental standards, equation (23.2) uses dummy variables for time periods. Time-period dummy variables for every 5 years are, respectively, 1960dummy, 1965dummy, 1970dummy, 1975dummy, 1980dummy, 1985dummy, 1990dummy, 1995dummy, and 2000dummy. Also, the necessary condition associated with the EKC is;

$$\alpha 1 > 0, \quad \alpha 2 < 0, \quad (|\alpha 1| > |\alpha 2|)$$

$$\beta 1 > 0, \quad \beta 2 < 0, \quad (|\beta 1| > |\beta 2|)$$

This chapter tests the EKC hypothesis using the regression analysis shown above by dividing chosen countries into two groups; high-income and low to middle-income.

23.4.2. *The regression results*

Results of the regression analysis are summarized in Table 23.1. Regression Eq. (23.1) is model (1), and Eq. (23.2) is model (2). All models have results for all data sets (all countries), high-income countries and low to middle-income countries.

Model (1) is the result of regressive analysis of random effect and fixed effect which passed F test and Hausman test. H shows that the result supports generally the EKC hypothesis. In particular, it shows that low to middle-income countries have a stronger relationship between GDP and CO_2 emissions than high-income countries, which means the higher the GDP per capita the greater the CO_2 emissions. Therefore, it has become clear that further economic growth in the low to middle-income group brings more CO_2 emissions than in the high-income group. That is simply because low to middle-income countries have relatively inadequate environmental technology, so that energy consumption in the early stage of industrialization is enormous. Referring to the turning point for each group, high income countries have the largest income of 27,805.81 dollars while low to middle-income countries have an income of 17,051.13 dollars. Also, the turning point for all countries is estimated to be 26,940.73 dollars. The greater the income countries have, the higher the turning point is.

The result for high-income countries uses fixed effects and the results for low to middle-income countries use random effects, although the results of Eq. (23.2) are based on the F test and Hausman test. Time-period dummy variables show a 0.01 level of statistical significance. Supposing that the value in 1960dummy is standard, all dummies are estimated as positive values. That is, CO_2 emissions have been increasing over the years. Referring to Fig. 23.3, trends in CO_2 emissions are identical to the trends explained in Section 23.3 of this chapter. In other words, CO_2 emissions decreased after respective oil shocks although CO_2 emissions had an increasing trend over the observed years.

Table 23.1. Results of regression analysis.

Explaining	Equation (1)									Equation (2)								
	All data sets			High income			Low-middle income			All data sets			High income			Low-middle income		
variables	OLS	Fe	Re	OLS	Fe	Re	OLS	Fe	Re	OLS	Fe	Re	OLS	Fe	Re	OLS	Fe	Re
gdp	0.0009***	0.0007***	0.0007***	0.0009***	0.0006***	0.0006***	0.0012***	0.0011***	0.0011***	0.0009***	0.0006***	0.0006***	0.0010[illegible]	0.0006***	0.0006***	0.0012***	0.0010***	0.0010***
	(0.0000)	(0.0000)	(0.0000)	(0.0000)	(0.0000)	(0.0000)	(0.0001)	(0.0000)	(0.0000)	(0.0000)	(0.0000)	(0.0000)	(0.00[illegible])	(0.0000)	(0.0000)	(0.0001)	(0.0001)	(0.0001)
gdp2	−0.0000***	−0.0000***	−0.0000***	−0.0000***	−0.0000***	−0.0000***	−0.0000***	−0.0000***	−0.0000***	−0.0000***	−0.0000***	−0.0000***	−0.000[illegible]	−0.0000***	−0.0000***	−0.0000***	−0.0000***	−0.0000***
	(0.0000)	(0.0000)	(0.0000)	(0.0000)	(0.0000)	(0.0000)	(0.0000)	(0.0000)	(0.0000)	(0.0000)	(0.0000)	(0.0000)	(0.00[illegible])	(0.0000)	(0.0000)	(0.0000)	(0.0000)	(0.0000)
dummy 1965	—	—	—	—	—	—	—	—	—	0.3060	0.3254***	0.3165***	0.33[illegible]	0.4363**	0.4308**	−0.0233	−0.0315	−0.0308
										(0.2675)	(0.1051)	(0.1052)	(0.41[illegible])	(0.1667)	(0.1666)	(0.2427)	(0.0750)	(0.0750)
dummy 1970	—	—	—	—	—	—	—	—	—	0.7612***	0.8812***	0.8619***	0.859[illegible]	1.1899***	1.1778***	0.0643	0.0256	0.0273
										(0.2870)	(0.1103)	(0.1100)	(0.44[illegible])	(0.1871)	(0.1865)	(0.2616)	(0.0783)	(0.0782)
dummy 1975	—	—	—	—	—	—	—	—	—	0.7733***	0.9921***	0.9656***	0.69[illegible]	1.2158***	1.1992***	0.1917	0.1291	0.1316
										(0.2911)	(0.1161)	(0.1155)	(0.45[illegible])	(0.2073)	(0.2063)	(0.2798)	(0.0827)	(0.0825)
dummy 1980	—	—	—	—	—	—	—	—	—	0.2596	0.5358***	0.5040***	−0.1[illegible]5	0.4536**	0.4337*	0.2005	0.1044	0.1076
										(0.2811)	(0.1216)	(0.1208)	(0.45[illegible])	(0.2263)	(0.2248)	(0.2674)	(0.0872)	(0.0869)
dummy 1985	—	—	—	—	—	—	—	—	—	0.1820	0.4823***	0.4456***	−0.3[illegible]9	0.2938	0.2705	0.1955	0.0726	0.0763
										(0.2766)	(0.1278)	(0.1266)	(0.45[illegible])	(0.2497)	(0.2478)	(0.2599)	(0.0904)	(0.0901)
dummy 1990	—	—	—	—	—	—	—	—	—	0.3211	0.6442***	0.6031***	−0.15[illegible]	0.4569*	0.4310	0.3008	0.1143	0.1187
										(0.2724)	(0.1338)	(0.1323)	(0.46[illegible])	(0.2698)	(0.2675)	(0.2456)	(0.0959)	(0.0955)
dummy 1995	—	—	—	—	—	—	—	—	—	0.5630**	0.8827***	0.8370***	0.10[illegible]	0.6297**	0.6011**	0.5051**	0.2371**	0.2424**
										(0.2790)	(0.1415)	(0.1398)	(0.48[illegible])	(0.2962)	(0.2934)	(0.2488)	(0.1032)	(0.1027)
dummy 2000	—	—	—	—	—	—	—	—	—	0.6757**	0.9367***	0.8877***	0.38[illegible]	0.6860**	0.6553**	0.5502**	0.1735	0.1796*
										(0.2818)	(0.1499)	(0.1479)	(0.51[illegible])	(0.3267)	(0.3235)	(0.2461)	(0.1089)	(0.1082)
_cons	0.4360***	1.2970***	1.3054***	−0.5567**	2.0818***	2.1172***	0.2515***	0.1064	0.1095	0.0360	1.3765***	1.3475***	−1.084[illegible]	1.7865***	1.7968**	0.1132	0.2232**	0.2272
	(0.0835)	(0.1262)	(0.4627)	(0.2428)	(0.2209)	(0.7177)	(0.0672)	(0.0695)	(0.4306)	(0.1841)	(0.1443)	(0.4571)	(0.31[illegible])	(0.2994)	(0.7397)	(0.1713)	(0.0882)	(0.4338)
N	1695	1695	1695	1020	1020	1020	675	675	675	1695	1695	1695	102[illegible]	1020	1020	675	675	675
R^2	0.6317	0.5347	0.628	0.3728	0.5136	0.3557	0.4109	0.7511	0.3864	0.6348	0.5654	0.6297	0.38[illegible]7	0.5578	0.3595	0.4182	0.7548	0.3866
F-test	—	310.7***	—	—	307.87***	—	—	568.05***	—	—	331.08***	—	—	335.45***	—	—	562.85***	—
hausman-test	—	—	1.56	—	—	2.27	—	—	0.15	—	—	8.94*	—	—	5.23	—	—	2.49
turning point	24549.79	26973.17	26940.73	24171.78	27841.82	27805.81	8267.99	17078.42	17051.13	24387.64	26507.76	26522.06	2429[illegible]	28326.29	28328.21	8071.78	19184.65	19190.50

Notes:

gdp: GDP per capita (real GDP based on US dollars in year 2000)
gdp2: the square root of GDP
dummy1965: time-period dummy between 1965 and 1969
dummy1970: time-period dummy between 1970 and 1974
dummy1975: time-period dummy between 1975 and 1979
dummy1980: time-period dummy between 1980 and 1984
dummy1985: time-period dummy between 1985 and 1989
dummy1990: time-period dummy between 1990 and 1994
dummy1995: time-period dummy between 1995 and 1999
dummy2000: time-period dummy between 2000 and 2004
R^2: Coefficient of determination
F-test: F-test to check statistical significance
_cons: Constant term
(SE): Standard errors
N: the Number of observations
Turning point: Turning point in the EKC
***: 0.01 level of statistical significance
**: 0.05 level of statistical significance
*: 0.1 level of statistical significance

However, observations in high-income countries do not show statistical significance in dummy1985 and dummy1990, although dummy1980 is statistically significant at the 0.1 level. In the low to middle-income group, dummy1995 is statistically significant at the 0.05 level and dummy2000 is statistically significant at the 0.1 level. From the above, we see that when time-period dummy variables are integrated into the equation, classifying countries according to income level does not lead to superior results.

23.4.3. *Estimation of turning point for each year*

Cross country analysis of the EKC is used for analyzing the relationship between economic growth and environmental quality. Taking the differences in growth process in each country into consideration, periodical peculiarities in the analysis are not really important. In fact, technological advance and environmental policy can contribute to reductions in CO_2 emissions per capita. In this section, this paper analyzes the EKC in each year and changes in estimations of GDP per capita and CO_2 emission using data on CO_2 emission between 1960 and 2004. Regression Eq. (23.1) uses the least squares method, and the result is shown in Table 23.1.

According to the result of the regression analysis, statistical significances between 1960–1969 and 1976–1978 are both 10%, which is regarded as statistically significant. However, the results show better outcomes than that of Osaka's and meet the precondition of the EKC. Also, the estimated turning point (in terms of GDP) first increases, and once it reaches a certain level then stabilizes for a while. Then it starts to increase again. That is, the more economic growth takes place, the higher the income level that can be observed. Estimated CO_2 emissions at the turning point gradually increase between 1960 and 1970, and then reach the largest amount in 1978. After that, it declines a little and stabilizes. As was stated before, this corresponds with the fact that CO_2 emissions decrease after each oil shock.

Estimated income levels at the turning point and CO_2 emissions levels are shown in Fig. 23.3, and demonstrate an apparent trend in the turning point. The income level increases steadily, and CO_2 emissions also increase at first, but then stabilize.

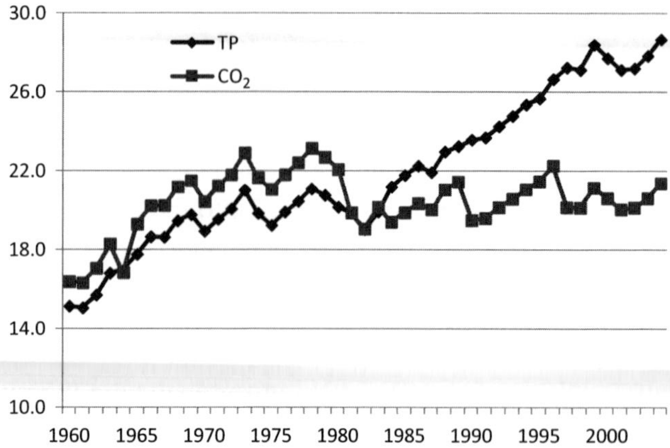

Fig. 23.3. Estimated turning points for each year (with estimated GDP).
Notes: TP: GDP when it reaches turning point in the EKC (thousand dollars); CO_2: CO_2 emission level (metric tons).

Referring to the regression analysis of the whole data, GDP per capita at the turning point is 24,549.79 dollars. Assuming that this GDP per capita is the standard GDP per capita at the turning point, countries that have already achieved the turning point in the EKC are; Canada, Austria, Holland, Finland, the U.K., Singapore, Ireland, Sweden, Denmark, Switzerland, Iceland, the U.S., Japan and Norway. On the other hand, the countries that have not achieved the turning point are; India, Indonesia, the Philippines, Egypt, China, Thailand, Colombia, South Africa, Brazil, Malaysia, Turkey, Chile, Hungary, Mexico, Portugal, Greece, New Zealand, Spain, Italy, Israel, France, Australia, Germany and Belgium.

Observed values in each country are shown in Fig. 23.3. They show that high-income countries have higher income levels and larger CO_2 emission levels. In contrast, low to middle-income countries have lower incomes and smaller CO_2 emission levels. A decrease in CO_2 emissions in low to middle-income countries has not yet been observed. Hence, it is expected that it will take a long time for low to middle-income countries to achieve a turning point in the EKC.

Some countries in the high income group have already achieved the turning point, the rest will achieve the turning point sooner or later.

23.5. The Relationship between Economic Growth and Carbon Emissions

In this section, a stability test is used to examine whether results in the previous section are reliable and six more explanatory variables, including urbanization, population growth, investment rate, export dependency and energy intensity will be integrated into the regression analysis.

23.5.1. *Stability test for the estimated parameter*

In order to conduct a stability test, we use the regression equation,

$$E = \gamma 0 + \gamma YY + \gamma Y2Y2 + \gamma UU + \gamma PP + \gamma IndInd + \gamma \mathrm{II} + \gamma ExEx + \gamma EE + \mu 3 \quad (23.3)$$

$$E = \delta 0 + \delta YY + \delta Y2Y2 + \delta UU + \delta PP + \delta IndInd + \delta \mathrm{II} + \delta ExEx + \delta EE + \delta dumDum + \mu 4 \quad (23.4)$$

where explained variance E is CO_2 emissions per capita (metric tons), Y, which is often used in regression analyses associated with the EKC hypothesis, means GDP per capita, $Y2$ is the square root of GDP, U is the rate of urbanization, P is the population growth rate, Ind is industrialization (the ratio of the industrial sector in GDP over the agricultural sector in GDP, shown by logarithm value), I is the investment rate and Ex (Export/GDP) and EE (Energy consumption/GDP) are export dependency and energy intensity respectively.

Regression Eq. (23.3) includes these additional explanatory variables and Eq. (23.4) includes an additional time-period dummy variable. Also, comparison by income group will be used in order to examine the results separately.

23.5.2. *Results including effects of additional explanatory variables*

The effect of urbanization is measured by the ratio of population in urban areas to the total population. Urbanization is often regarded as one of the factors that is strongly associated with environmental pollution. That is because city life consumes more energy than country life and produces more CO_2 emissions along with that energy consumption. In addition, intensive use of energy in local industry sectors may lead to additional CO_2 emissions. It is, however, sometimes argued that industrialization may contribute to environmental improvement. For instance, establishment of waste processing facilities can improve the environment through economies of scale.

Population increase brings environmental issues such as soil contamination, water pollution, CO_2 emissions and loss of biodiversity and it is also thought that a lower population growth rate has a positive effect upon economic growth. However, in low to middle-income countries, population growth tends to increase because they engage in labor-intensive agriculture and industry.

The industrial sector consumes the largest proportion of energy, and consequently it produces large amounts of greenhouse gases. In the process of economic growth the EKC shows transformation from non pollution-intensive (non energy-intensive) agriculture to pollution-intensive (energy-intensive) manufacturing, and then to less polluting service industries. But this chapter considers agriculture and manufacturing industries, not service industries.

Export dependency, which is the ratio of exports of products and services over total GDP, is integrated into the regression analysis in order to examine the impacts of exports on the environment. Although exports have both positive and negative effects upon the environment, exports increase greenhouse gas emissions, energy consumption and emissions of pollutants along with increases in production levels because of economies of scale. Furthermore, developed countries transfer production facilities to developing countries in order to avoid restrictions on emissions, increasing exports of products from developing countries which results in an acceleration of environmental contamination. On the other hand, it is also true that the transfer of advanced technology and capital, caused by developed countries, transferring their production facilities to developing countries, leads to environmental conservation.

The investment rate is also integrated into the regression analysis to investigate how investment, associated with exports and industrialization, affects the environment.

Finally, regression Eq. (23.4) uses energy consumption per unit of GDP to analyze how energy intensity affects the environment. Improving energy intensity is one of the ways to solve the lack of energy. Here, this chapter will examine whether the EKC observed in the previous section is statistically consistent; how additional variables affect the EKC is also investigated, especially whether the impacts of those additional variables on the EKC are positive or negative.

23.5.3. *The relationship between economic growth and carbon emissions*

Before considering the effects of additional explanatory variables, this chapter examines multicollinearity of each explanatory variable (Table 23.2). If the VIF is over 10.0, there is multicollinearity among those variables. From Table 23.2, the VIF of every additional explanatory variable is lower than 10.0 in all data sets, high-income countries and low to middle-income countries, so there is no multicollinearity among additional explanatory variables.

The result of the regression analysis is shown in Table 23.3. Eq. (23.3) is composed of 6 additional explanatory variables — industrialization, urbanization, population growth, the investment rate, export dependency, and energy intensity — plus GDP, which is the basic explanatory variable, and the square root of GDP. In addition to Eq. (23.3), Eq. (23.4) includes time-period dummy variables.

First, we focus on Eq. (23.3). An inverted U-shaped curve is shown because all data sets and high-income countries have fixed effects, and low to middle-income countries have random effects. The turning point for all data sets is 24,517.95 US dollars, and that of high-income countries is

Table 23.2. Tests for multicollinearity.

	All data sets		High-income		Low to middle-income	
Variable	VIF	1/VIF	VIF	1/VIF	VIF	1/VIF
gdp	22.74	0.043976	34.56	0.028937	15.8	0.063288
gdp2	15.21	0.06575	32.01	0.031242	11.24	0.089004
industrialization	3.11	0.321055	1.98	0.504455	3.37	0.296737
urbanization	2.8	0.357608	1.69	0.592574	2.88	0.347322
population growth	1.94	0.51614	1.4	0.713782	1.96	0.510458
energy intensity	1.59	0.629759	1.3	0.766396	1.69	0.592654
export dependency	1.23	0.812926	1.26	0.791451	1.42	0.706509
investment rate	1.05	0.951449	1.23	0.810303	1.36	0.737852
Mean VIF	6.21		9.43		4.96	

31,245.49 US dollars. However, the turning point for low to middle-income countries is not statistically significant although the correlation of the square root of GDP is negative. The relationship between CO_2 emissions and GDP cannot be an inverted U-shaped curve if they receive any effect from other factors. If this is the case, then it can be said that the EKC hypothesis is not evident for low to middle-income countries. Those counties have difficulty reaching the turning point due to many issues such as technology.

Next consider the relationship between each additional explanatory variable and CO_2 emissions. The data of industrialization uses logarithms, and the result is shown in,

$$\Delta y = (\beta 1/100) \cdot 100\% \Delta x$$

where Δx stands for the change in industrialization, which is the logarithmic ratio of GDP produced from manufacturing over GDP produced from agriculture, Δy stands for the change in CO_2 emissions and $\beta 1$ is the correlation. For example, CO_2 emissions increase by 0.8% with an additional unit of industrialization when $\beta 1 = 0.8$.

The other factors — population growth, urbanization, investment, export dependence, and energy intensity — do not involve logarithms, and the result comes from,

$$\Delta y = \beta 1 \Delta x.$$

Table 23.3. Results of regression analysis II.

	Equation (23.3)								
	All data sets			High-income			Low to middle-income		
	OLS	Fe	Re	OLS	Fe	Re	OLS	Fe	Re
gdp	0.0009***	0.0003***	0.0003***	0.0004***	0.0004***	0.0005***	0.0013***	0.0007***	0.0007***
	(0.0000)	(0.0000)	(0.0000)	(0.0001)	(0.0001)	(0.0000)	(0.0001)	(0.0001)	(0.0001)
gdp2	−0.0000***	−0.0000***	−0.0000***	−0.0000***	−0.0000***	−0.0000***	−0.0000***	−0.0000*	−0.0000
	(0.0000)	(0.0000)	(0.0000)	(0.0000)	(0.0000)	(0.0000)	(0.0000)	(0.0000)	(0.0000)
urbanization	0.0078	0.0416***	0.0442***	−0.0367***	0.0834***	0.0544**	−0.0309***	0.0090*	0.0080*
	(0.0057)	(0.0082)	(0.0080)	(0.0092)	(0.0239)	(0.0210)	(0.0071)	(0.0049)	(0.0049)
population growth	0.6482***	-0.0050	-0.0170	1.1517***	−0.0471	−0.0750	−0.3945***	0.0280	0.0229
	(0.1465)	(0.0667)	(0.0668)	(0.2094)	(0.0991)	(0.0993)	(0.1323)	(0.0462)	(0.0462)
industrialization	3.3783***	0.7345***	0.7311***	6.4452***	1.2887**	1.2121**	2.4139***	0.3337**	0.3701**
	(0.2919)	(0.2544)	(0.2536)	(0.4606)	(0.5915)	(0.5658)	(0.3675)	(0.1569)	(0.1571)
investment rate	−0.0160**	0.0390***	0.0376***	−0.0723***	0.0677***	0.0664***	−0.0151**	0.0096**	0.0097**
	(0.0076)	(0.0064)	(0.0064)	(0.0152)	(0.0104)	(0.0104)	(0.0073)	(0.0043)	(0.0043)
export dependency	−0.0213***	0.0062	0.0019	−0.0446***	0.0118	0.0009	0.0130***	0.0095***	0.0098***
	(0.0047)	(0.0044)	(0.0043)	(0.0053)	(0.0082)	(0.0075)	(0.0033)	(0.0029)	(0.0028)
energy intensity	1.9113***	0.2504*	0.2480*	30.2520***	23.1552***	23.3453**	0.7905***	−0.1569**	−0.1444**
	(0.3176)	(0.1411)	(0.1416)	(1.4674)	(1.4377)	(1.4279)	(0.1656)	(0.0707)	(0.0709)
dummy1965	-	-	-	-	-	-	-	-	-
dummy1970	-	-	-	-	-	-	-	-	-
dummy1975	-	-	-	-	-	-	-	-	-
dummy1980	-	-	-	-	-	-	-	-	-
dummy1985	-	-	-	-	-	-	-	-	-
dummy1990	-	-	-	-	-	-	-	-	-
dummy1995	-	-	-	-	-	-	-	-	-
dummy2000	-	-	-	-	-	-	-	-	-
_cons	−9.1579***	−0.9211	−1.3358*	−15.7470**	−12.4602**	−10.8650**	−3.7055***	−0.8625**	−0.9048*
	(0.8304)	(0.5825)	(0.7320)	(1.1862)	(2.3407)	(2.1712)	(0.8520)	(0.3461)	(0.5085)
N	1241	1241	1241	713	713	713	528	528	528
R^2	0.6913	0.2609	0.6082	0.6221	0.3750	0.4251	0.6256	0.7366	0.3815
F-test	-	298.28***	-	-	205.39***	-	-	325.17***	-
Hausman-test	-	-	27.16***	-	-	22.44**	-	-	11.78
turning point	23194.25	24517.95	24739.62	28146.44	31245.49	32000.64	6960.02	40499.36	41089.67

CO_2 emissions, for instance, increase by 0.8 mt with an additional unit of urbanization when Δx is $\beta 1 = 0.8$.

Next, we analyze the results of the regression over each explanatory variable. All data sets, high-income countries and low to middle-income countries have positive correlation between carbon emissions and urbanization, which is statistically significant. Therefore, carbon emissions increase with urbanization. The index of urbanization for high-income countries is almost 10 times as high as that for low to middle-income countries, so urban areas in high-income countries consume more energy and emit more greenhouse gases than low to middle-income countries.

Regarding population growth, all data and high-income countries show negative correlation with carbon emission and low to middle-income countries show positive correlation with CO_2 emission; however, none are statistically significant. Hence, there may be some problems with the stability of the parameter estimators. Moreover, it seems to be the same pattern as the patterns that environmental problems can be improved through low population growth and increases of income (Table 23.4).

Table 23.4. Correlation coefficients for each explanatory variable.

	Equation (23.4)								
	All data sets			High-income			Low to middle-income		
	OLS	Fe	Re	OLS	Fe	Re	OLS	Fe	Re
gdp	0.0008***	0.0004***	0.0004***	0.0004***	0.0006***	0.0007***	0.0011***	0.0010***	0.0009***
	(0.0000)	(0.0000)	(0.0000)	(0.0001)	(0.0001)	(0.0001)	(0.0001)	(0.0001)	(0.0001)
gdp2	−0.0000***	−0.0000***	−0.0000***	−0.0000***	−0.0000***	−0.0000***	−0.0000***	−0.0000***	−0.0000**
	(0.0000)	(0.0000)	(0.0000)	(0.0000)	(0.0000)	(0.0000)	(0.0000)	(0.0000)	(0.0000)
urbanization	0.0091	0.0673***	0.0683***	−0.0367***	0.1327***	0.0985***	−0.0281***	0.0374***	0.0298***
	(0.0055)	(0.0098)	(0.0088)	(0.0093)	(0.0243)	(0.0216)	(0.0070)	(0.0057)	(0.0055)
population growth	0.6121***	−0.0912	−0.1110*	1.1688***	−0.2417**	−0.2571**	−0.4581***	−0.0888**	−0.0901*
	(0.1479)	(0.0672)	(0.0667)	(0.2161)	(0.1010)	(0.1013)	(0.1492)	(0.0450)	(0.0462)
industrialization	3.7242***	0.8713***	0.9470***	6.4558***	1.5667***	1.4436***	3.1224***	0.5473***	0.6024***
	(0.3078)	(0.2533)	(0.2531)	(0.4669)	(0.5708)	(0.5487)	(0.4065)	(0.1509)	(0.1553)
investment rate	−0.0190**	0.0266***	0.0250***	−0.0780***	0.0541***	0.0531***	−0.0134*	0.0006	0.0033
	(0.0075)	(0.0066)	(0.0066)	(0.0149)	(0.0102)	(0.0102)	(0.0071)	(0.0043)	(0.0044)
export dependency	−0.0188***	−0.0010	−0.0042	−0.0472***	−0.0054	−0.0131*	0.0155***	0.0059**	0.0084***
	(0.0051)	(0.0045)	(0.0043)	(0.0057)	(0.0086)	(0.0077)	(0.0035)	(0.0029)	(0.0028)
energy intensity	1.9277***	0.1690	0.1751	30.0683***	22.3522***	22.4215***	0.7288***	−0.2254***	−0.1933***
	(0.3411)	(0.1390)	(0.1393)	(1.4957)	(1.4222)	(1.4108)	(0.1753)	(0.0665)	(0.0684)
dummy1965	0.7977*	−0.8346*	−0.8632*	0.0000	0.0000		0.2362	−0.2691	−0.2243
	(0.4670)	(0.4849)	(0.4888)	(0.0000)	(0.0000)		(0.3045)	(0.2298)	(0.2372)
dummy1970	−1.0190***	−0.5531	−0.5836	−0.5654	1.1907***	1.0618***	−1.1135***	−0.4910**	−0.3759*
	(0.2921)	(0.4343)	(0.4357)	(0.4055)	(0.3022)	(0.2962)	(0.2847)	(0.2143)	(0.2191)
dummy1975	−1.1918***	−0.5593	−0.6138	−0.0673	1.0306***	0.9452***	−1.1911***	−0.5752**	−0.4272*
	(0.2921)	(0.4414)	(0.4399)	(0.3864)	(0.2490)	(0.2456)	(0.2708)	(0.2211)	(0.2247)
dummy1980	−1.8531***	−1.1356**	−1.2073***	−0.5189	0.2745	0.2276	−1.4813***	−0.8175***	−0.6375**
	(0.3093)	(0.4504)	(0.4461)	(0.3961)	(0.2178)	(0.2155)	(0.3098)	(0.2311)	(0.2335)
dummy1985	−1.9679***	−1.2816***	−1.3722***	−0.6258	−0.0177	−0.0837	−1.5618***	−1.0250***	−0.8151***
	(0.3120)	(0.4604)	(0.4524)	(0.4038)	(0.1819)	(0.1809)	(0.3113)	(0.2394)	(0.2405)
dummy1990	−2.1423***	−1.3122***	−1.4156***	−0.7913**	−0.0839	−0.1414	−1.6915***	−1.1937***	−0.9595***
	(0.3308)	(0.4712)	(0.4602)	(0.4005)	(0.1581)	(0.1577)	(0.3157)	(0.2506)	(0.2508)
dummy1995	−1.8436***	−1.2067**	−1.3127***	−0.1131	−0.0944	−0.1167	−1.7237***	−1.2963***	−1.0428***
	(0.3442)	(0.4821)	(0.4682)	(0.4011)	(0.1308)	(0.1317)	(0.3448)	(0.2596)	(0.2593)
dummy2000	−2.0319***	−1.2359**	−1.3517***	0.0000	0.0000		−2.1894***	−1.6234***	−1.3475***
	(0.3726)	(0.4963)	(0.4794)	(0.0000)	(0.0000)		(0.3379)	(0.2728)	(0.2715)
_cons	−8.2689***	−2.3413***	−2.7853***	−15.1523**	−19.1065**	−16.6605**	−3.6579***	−1.8801***	−1.7577***
	(0.7275)	(0.7506)	(0.8032)	(1.2375)	(2.5350)	(2.3647)	(0.7699)	(0.3546)	(0.4615)
N	1241	1241	1241	713	713	713	528	528	528
R^2	0.6992	0.3001	0.6225	0.6274	0.4311	0.4015	0.6530	0.7763	0.3759
F-test	-	305.85***	-	-	223.19***	-	-	352.47***	-
hausman-test	-	-	38.5***	-	-	25.66***	-	-	40.55***
turning point	23573.61	26555.43	26680.76	27127.09	31083.68	31542.97	7224.70	22962.10	26783.86

All data sets, high and low to middle-income countries have a positive correlation between CO_2 emissions and industrialization, and that is statistically significant. So, further industrialization leads to larger emissions. Investment has the same correlation as industrialization and is thought to cause damage to the environment.

Export dependency shows a positive correlation, but that correlation in low to middle-income countries is statistically insignificant.

The correlation of energy intensity is positive in high-income countries, but negative in low to middle-income countries. On the one hand, energy intensity has a positive correlation with CO_2 emissions in high income countries because it is significant, as the indicator shows, for energy saving. On the other hand, low to middle-income countries have a negative correlation between energy intensity and CO_2 emissions. This is because those countries are still in the process of economic development with mass production, mass consumption and building infrastructure, and they have difficulty reducing CO_2 emissions caused by the process of economic development.

To compare and analyze the details of the correlation between carbon emissions and each parameter in high and low to middle-income countries, this chapter will refer to Table 23.5 shown below.

Table 23.5. The relationship between additional explanatory variables and CO_2 emissions.

Explanatory variables	All data sets	High-income countries	Low to middle-income countries
Urbanization	positive	positive	positive
Population growth rate	×	×	×
Industrialization	positive	positive	positive
Investment rate	positive	positive	positive
Export dependency	×	negative	positive
Energy intensity	positive	positive	negative

×: means that there is a problem in the stability of estimated parameters.

As shown in Table 23.5, the correlation between CO_2 emissions and export dependency in low to middle-income countries is positive while in high income countries it is negative. Low to middle-income countries have a tendency to export lower value-added and energy-intensive products, so that the damage to the environment is more serious if their export dependency is higher. CO_2 emissions also increase from production and transportation. On the contrary, high-income countries mainly export higher value-added and less energy-intensive products, so higher export dependency does not necessarily increase CO_2 emissions like it does in low to middle-income countries. Other findings from Table 23.5 match with the results of previous analyses.

Next we focus on Eq. (23.4). Equation (23.4) integrates dummy variables into Eq. (23.3). From the results, it demonstrates an inverted U-shaped curve, although there is not a big change in correlation between carbon emissions and each explanatory variable. Furthermore, almost all the dummy variables have positive correlation and the observed trends here are different from previous practical analyses, regardless of whether those variables show statistical significance. The examination of this point is a key subject for further experimental analysis.

23.6. Conclusion and Further Study

This chapter attempts to analyze the EKC hypothesis by dividing countries into a high-income group and a low to middle-income group in order to examine the relationship between CO_2 emissions and income inequality. First, it describes previous findings and conducts a review, and then focuses on a comparison of CO_2 emissions in each group. Also, this chapter starts with a basic regression analysis using cross-country data sets, and then integrates additional variables step by step. All results observed in this chapter can be summarized below.

First, low to middle-income countries emit less CO_2 per capita given their level of income, and the amount of CO_2 per capita is even less than that of high income countries in the 1960s.

In addition, many high-income countries have already achieved a turning point in the EKC, but low to middle-income countries have not. That is, CO_2 emissions from low to middle-income countries are expected to increase until they reach the turning point.

According to experimental analysis, without additional explanatory variables, the result implies that both groups prove the EKC hypothesis. With additional variables, the EKC hypothesis in low to middle-income countries is not proven. The relationships between CO_2 emissions and additional variables are summarized in Table 23.5.

The basic regression equation implies that CO_2 emissions have been increasing over the observed years (from 1960 to 2004). However, integrating additional explanatory variables into the basic equation makes differences in each correlation. What it is that makes those differences will be the next topic.

Finally, the turning point in the EKC for all data sets and for the high-income group, is 24,517.95 U.S. dollars and 31,235.49 U.S. dollars, respectively, while the turning point for the low to middle-income group has not been observed in this analysis. Hence, the higher the income level is, the higher the turning point is.

It is suggested that CO_2 emissions in low to middle-income countries increases along with economic development. That is due to inadequate technology and environmental policy associated with CO_2 emissions. However, the increasing rate of CO_2 emissions in the low to middle-income group can be improved through transfers of advanced technology and investment from developed countries because developed countries have also experienced increasing carbon emissions in the process of economic development. Low to middle-income countries should avoid increasing their rate of CO_2 emissions by utilizing support provided by developed countries.

Note: BASIC stands for Brazil, South Africa, India, and China, 4 countries which have been experiencing faster economic growth and increasing carbon emissions. The term "BASIC" was formed in COP15.

References

Brock, W. A. and Taylor, M. S. (2004). "The Green Solow Model." NBER Working Paper No. 10557.

de Bruyn, S. M. (1997). Explaining the Environmental Kuznets Curve: Structural Change and International Agreements in Reducing Sulphur Emissions. *Environment and Development Economics*, 2(4), 458–503.

de Bruyn, S. M., van den Bergh, J. C. J. M. and Opschoor, J. B. (1998b). Economic Growth and Emissions: Reconsidering the Empirical Basis of Environmental Kuznets Curves. *Ecological Economics*, 25(2), 161–175.

de Bruyn, S. M. and Heintz, R. J. (1998a). *The Environmental Kuznets Curve Hypothesis*, Handbook of Environmental Economics. Oxford: Blackwell Publishing Co. pp. 656–677.

Cole, M. A. and Elliott, R. J. R. (2003). "Do Environmental Regulations Influence Trade Patterns? Testing New and Old." Blackwell Publishing Ltd, Wiley Online Library, http://onlinelibrary.wiley.com/doi/10.1111/1467-9701.00567/abstract

Ekins, P. (1997). The Kuznets Curve for the Environment and Economic Growth: Examining the Evidence. *Environment and Planning*, 29(5), 805–830.

Fujita, W. (2003). Diffusion of Environmental Technologies Through Trade: Technology Transfer Issues in International Environmental Policy. *Annual Report of Southeast Asian Studies*, 44, 69–97.

Grossman, G. M. and Krueger, A. B. (1995). Economic Growth and the Environment. *Quarterly Journal of Economics*, 110(2), 353–377.

Holtz-Eakin, D. and Selden, T. M. (1995). Stoking the Fires? CO_2 Emissions and Economic Growth. *Journal of Public Economics*, 57(1), 85–101.

Horvath, R. J. (1997). "Energy Consumption and the Environmental Kuznets Curve Debate." Department of Geography, Sydney NSW: University of Sydney.

Hoshinoa, Y. and Sugiyama, T. (2009). "CO_2 Emissions from Energy Sources Continue to Increase in Countries." SERC Discussion Paper SERC09016, Institute of Social and Economic Research.

Jerrett, M., Eyles, J., Cole, D. and Reader, S. (1997). Environmental equity in Canada: an empirical investigation into the income distribution of pollution in Ontario. *Environment and Planning A*, 29(10), 1777–1800.

Kaufmann, R. K., Davidsdottir, B., Garngam, S. and Pauly, P. (1998). The Determinants of Atmospheric SO_2 Concentrations: Reconsidering the Environmental Kuznets Curve. *Ecological Economics*, 25(2), 209–220.

Kuznets, S. (1955). Economic Growth and Income Inequality. *American Economic Review*, 45(1), 1–28.

Magnani, E. (2000). The Environmental Kuznets Curve, Environmental Protection Policy and Income Distribution. *Ecological Economics*, 32(3), 431–433.

Masuda, M. (2003). "Case Western Countries and Asian Countries — Environmental Kuznets Curve and Population." Demographic Study Group Working Paper, No. 0002, Komazawa University.

Masuda, N. (2007). The Simplified Green GDP and Environmental Kuznets Curve in Japan, South Korea and China. *Hitotsubashi Journal of Economics*, 2(1), 1–23.

Naito, N. (2006). Surveys on the Environmental Kuznets Curve Hypothesis. *Journal of Economics*, 15, 34–55.

Neumayer, Eric (2002). Can natural factors explain any cross-country differences in carbon dioxide emissions? *Energy Policy*, 30(1), 7–12.

Ninomiya, K. (2005). Empirical Analysis of SO_2 Emission Trends and China's Economic Development with Environmental Kuznets Curve. *Research on International Cooperation*, 21(1), 1–14.

Osaka, H. (2001b). "Cross-country Analysis by International Data: Verification of Environmental Kuznets Curve." (in Japanese). Kyushu University Discussion Paper, No. 2001-8.

Panayotou, T. (1993). "Empirical Tests and Policy Analysis of Environmental Degradation at Different Stages of Economic Development." Working Paper WP238, Technology and Employment Programme, Geneva: ILO.

Panayotou, T. and Sungsuwan, S. (1994). An econometric analysis of the causes of tropical deforestation: the case of Northeast Thailand. In Brown, K. and Pearce, D. W. (eds.), *The Causes of Tropical Deforestation: The Economic and Statistical Analysis of Factors Giving Rise to the Loss of Tropical Forests*. London, UK: University College London Press, pp. 192–210.

Panayotou, T. (1995). Environmental Degradation at Different Stages of Economic Development. In I. Ahmed and J. A. Doeleman (eds.), *Beyond Rio: The*

Environmental Crisis and Sustainable Livelihoods in the Third World. The Netherlands: Elsevier Ltd., pp. 13–36.

Panayotou, T. (1997). Demystifying the Environmental Kuznets Curve: Turning a Black Box into a Policy Tool. *Environment and Development Economics*, 2(4), 465–484.

Rock, M. T. (1996). Pollution Intensity of GDP and Trade Policy: Can the World Bank be Wrong? *World Development*, 24(3), 471–479.

Rothman, D. S. (1998). Environmental Kuznets Curves — Real Progress or Passing the Buck? A Case for Consumption-based Approaches. *Ecological Economics*, 25(2), 177–194.

Schmalensee, R., Stoker, T. M. and Judson, R. A. (1998). "World energy consumption and carbon dioxide emissions: 1950–2050." Sloan School of Management, Massachusetts Institute of Technology, Cambridge MA.

Selden, T. M. and Song, D. (1994). Environmental Quality and Development: Is There a Kuznets Curve for Air Pollution Emissions? *Journal of Environmental Economics and Management*, 27(2), 147–162.

Shafik, N. (1994a). Economic Development and Environmental Quality: An Econometric Analysis. *Oxford Economic Papers*, 46(0), 757–773.

Shafik, N. (1994b) Macroeconomic Causes of Deforestation: Barking Up the Wrong Tree? In K. Brown, and D. W. Pearce (eds.), *The Causes of Tropical Deforestation*. London: University College London Press.

Shafik, N. and Bandyopadhyay, S. (1992). "Economic Growth and Environmental Quality: Time Series and Cross-country Evidence." Background Paper for the World Development Report 1992, WPS904, Washington, D.C.: The World Bank.

Stern, D., Common, M. and Barbier, E. (1996). Economic Growth and Environmental Degradation: The Environmental Kuznets Curve and Sustainable Development. *World Development*, 24(7), 1151–1160.

Stokey, N. L. (1998). Are There Limits to Growth? *International Economic Review*, 39(1), 1–31.

Suri V. and Chapman, D. (1998). Economrc Growth, Trade and Energy: Implications for the Environmental Kuznets Curve. *Ecological Economics*, 25(2), 195–208.

Torras M. and Boyce, J. K. (1998). Income, Inequality, and Pollution: A Reassessment of the Environmental Kuznets Curve. *Ecological Economics*, 25(2), 147–160.

Uchiyama, K. (2007). Carbon Dioxide Emissions and Environmental Kuznets Curve: A Verification by Dynamic Panel Data Estimation (in Japanese). *Economics and Business Research*, 27(3), 1–78.

Chapter 24

ECONOMIC GROWTH AND CARBON EMISSIONS

*Lin Guo**

24.1. Introduction

The developing and developed countries have a serious disagreement during climate negotiations about the responsibility for carbon emissions. Developed countries are trying to require that all countries make the same emissions reduction commitments, while developing countries, which are focusing on the historical cumulative carbon emissions and historical responsibility, stress the importance of economic development and different levels of per capita income, in order to advocate that emissions targets be in accordance with historical responsibility and per capita emissions. This disagreement leads to some theoretical issues: (1) Does the income gap between developing and developed countries have something to do with the carbon emissions gap? (2) Should developing and developed countries undertake the same emissions reduction obligations? (3) Do developing and developed countries have the same emissions reduction ability? (4) Can they achieve the same emissions reduction targets?

Based on the Kuznets Hypothesis and Environmental Kuznets Curve (EKC), we first derive the Carbon Kuznets Curve model, then analyze the marginal propensity of emission. Second, we establish an econometric model of the Carbon Kuznets Curve (CKC), analyze the relationship between income level and carbon emissions, and test whether there is an inverted-U relationship between carbon emissions and economic development in countries with different per capita GDP. Third, the econometric model of the CKC with income distribution is proposed in order to find the

*Lin Guo, Research Fellow of Institute of Global Low-Carbon Economy; International University of Business and Economics, China and Nagoya University, Japan.

impact of the income gap between high-income and low-income countries on carbon emissions. Finally, the elastic theory is used to analyze the per capita GDP elasticity and Gini elasticity of carbon emissions which is then used to explain the relationship between the carbon emissions gaps and income gaps among countries.

This chapter contains five sections: Section 24.2 provides an overview of the recent literature. Section 24.3 outlines the theoretical analysis and mathematical derivation of the CKC. Section 24.4 describes the econometric specification for testing the CKC in countries with different income levels, and the income gap elasticity and income elasticity of carbon emissions is discussed for revealing the independent and interaction relationships between carbon emissions and the income gap. Our conclusions can be found in Section 24.6.

24.2. Literature Review

The hypotheses that an inverted-U relationship between growth in per capita income and the income gap was proposed by Kuznets (1955), who argues that income inequality first increases following economic growth, and then decreases when average income reaches a high enough level. Many researchers have conducted empirical analyses to test whether there is such an inverted-U curve. Then in 1994, Grossman and Kruger (1993) proved the emergence of the downward sloping segment of the Environmental Kuznets Curve (EKC). However, the empirical evidence from some studies indicates that there is not an inverted-U relationship between environmental quality and average income.

The existence of such nonlinearities in the relationship between environmental indicators and average income holds implications for the relationship between income inequality and environmental outcomes. Studies focusing on this issue, however, result in ambiguous findings.

Boyce (1994) argues that higher inequality reinforces the power of the rich to impose environmental costs on the poor, affects time preference, and encourages activities that cause environmental degradation. Inequality may reduce the ability of a society to achieve cooperative solutions to environmental problems. Torras and Boyce (1998) tested the effects of various pollutants on the EKC (not including carbon emissions) for differences in within-country income inequality. They found mixed evidence: higher inequality entails higher levels of some pollutants but not others. Other studies support his argument: Marsiliani and Renström (2000) point out that the reason why income inequality may harm the environment is that it may lead to loose environmental policy. Magnani (2000) found that contrary to the EKC hypothesis, moments of the income distribution function, other than the mean, may be important for the emergence of a virtuous path of sustainable growth in high-income countries.

Other studies provide contrary evidence. Scruggs (1998) shows that equality may or may not be necessary to minimize degradation. Under some plausible conditions, greater inequality may even be conducive to lower degradation. He provides two cross-national empirical tests that tend to support the theory that distributional issues do not systematically explain variations in environmental equality.

In our research, we focus on the impact of income gaps on carbon emissions by utilizing the econometric model of the CKC with income distribution. We analyze the lagged effects of income elasticity and income gap elasticity of carbon emissions on carbon emissions, based on the methodology of elasticity economics.

24.3. Mathematical Derivation of the CKC (Carbon Kuznets Curve)

24.3.1. *The CKC and marginal propensity of carbon emissions*

The Environmental Kuznets Curve illustrates an inverted U-shaped relationship between growth in per capita income and environmental degradation. This chapter takes CO_2 emissions as a measure of environmental quality and analyzes the linkage between levels of per capita income and carbon emissions based on a Carbon Kuznets Curve (CKC).

According to the mathematical properties of the inverted U-shaped Environmental Kuznets Curve, we can derive the relationship between the income gap and carbon emissions. The assumption and derivation are as follows:

Assumption: $e_{ij}(y_{ij})$ is the carbon emissions by person i in country j; y_{ij} is the income of person i in country j; $E_j(y_{ij})$ is the total carbon emissions of country j; $\frac{1}{N_j}\sum_{i=1}^{N_j} e_{ij}(y_{ij})$ is the per capita carbon emissions of country j; TE stands for the aggregate level of emissions of all countries (M countries). Quadratic or cubic functions of income per capita have been popular specifications in the past work on the EKC; here we choose a quadratic function for the derivation

$$\begin{cases} e_{ij}(y_{ij}) = \beta_{ij0} + \beta_{ij1}y_{ij} + \beta_{ij2}y_{ij}^2 & (24.1) \\ E_j(y_{ij}) = \sum_{i=1}^{N_j} e_{ij}(y_{ij}) = \sum_{i=1}^{N_j} (\beta_{ij0} + \beta_{ij1}y_{ij} + \beta_{ij2}y_{ij}^2) & (24.2) \end{cases}$$

$$TE = \sum_{j=1}^{M} E_j(y_{ij}) = \sum_{j=1}^{M}\sum_{i=1}^{N_j} e_{ij}(y_{ij}) = \sum_{j=1}^{M}\sum_{i=1}^{N_j} (\beta_{ij0} + \beta_{ij1}y_{ij} + \beta_{ij2}y_{ij}^2) \tag{24.3}$$

Figures 24.1(a and b) show the Carbon Kuznets Curve and the marginal propensity of emissions (MPE :$e'_{ij}(y_{ij}) = \beta_{ij1} + 2\beta_{ij2}y_{ij}$). It is obvious that $y_{ij1} < y_{ij3}$, $\beta_{ij1} > 0$, $\beta_{ij2} < 0$, and that $e'_{ij}(y_{ij})$ is a decreasing function. Therefore, the MPE of low-income persons is higher than that of high-income persons ($e'_{ij}(y_{ij1}) > e'_{ij}(y_{ij3})$).

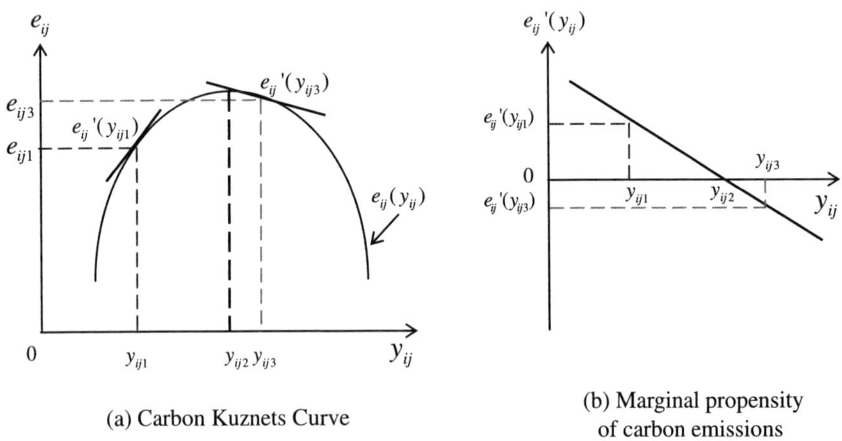

Fig. 24.1. CKC and marginal propensity of carbon emissions (MPE).

The MPE of a country can be specified with the following equation, which is affected by the MPE of all the citizens in this country:

$$E'_j(y_{ij}) = \sum_{i=1}^{N_j}{}' e_{ij}(y_{ij}) = \sum_{i=1}^{N_j} (\beta_{ij1} + 2\beta_{ij2} y_{ij}) = \sum_{i=1}^{N_j} \beta_{ij1} + \sum_{i=1}^{N_j} 2\beta_{ij2} y_{ij}$$

$$= N_j \bar{\beta}_{j1} + 2N_j \bar{\beta}_{j2} \bar{y}_j \bar{y}_j = \frac{1}{N_j} \sum_{i=1}^{N_j} y_{ij}$$

24.3.2. *The relationship between the income gap and carbon emissions*

How can we describe the linkage between the distribution of income and carbon emissions based on the foregoing derivation?

Consider a special case: suppose country j has only two people. The first person's income is y_{1j}, the other's is y_{2j}. If we assume that $y_{1j} < y_{2j}$, then $y_{2j} - y_{1j} > 0$. Also, $\frac{y_{1j}+y_{2j}}{2}$ is the average income of country j. However, if we assume that country j has an equal distribution of income, then the two citizens have the same income ($\frac{y_{1j}+y_{2j}}{2}$). If there is an income gap in country j, the first person's carbon emissions with respect to income is $e_1(y_{1j}) = \beta_{10} + \beta_{11} y_{1j} + \beta_{12} y_{1j}^2$, while the other's is $e_2(y_{2j}) = \beta_{20} + \beta_{21} y_{2j} + \beta_{22} y_{2j}^2$. Consequently, the carbon emissions in the country without the income gap and the country with the income gap are, respectively, $2 \times e_1(\frac{y_{1i}+y_{2j}}{2})$ and $e_1(y_{1j}) + e_2(y_{2j})$.

Suppose the parameters in everybody's emission functions are the same, namely $\beta_{10} = \beta_{20}$, $\beta_{11} = \beta_{21}$, $\beta_{12} = \beta_{22}$. This means that we can ignore individual characteristics in the carbon emissions function. Consequently, $e_1(y_{1j}) + e_2(y_{2j})$ is equal to $e(y_{1j}) + e(y_{2j})$, and $2 \times e_1(\frac{y_{1i}+y_{2j}}{2})$ is equal to $2 \times e(\frac{y_{1i}+y_{2j}}{2})$. The carbon gap can then be calculated as follows:

$$e(y_{1j}) + e(y_{2j}) - 2 \times e\left(\frac{y_{1j}+y_{2j}}{2}\right)$$

$$= e(y_{1j}) - e\left(\frac{y_{1j}+y_{2j}}{2}\right) + e(y_{2j}) - e\left(\frac{y_{1j}+y_{2j}}{2}\right)$$

According to the Lagrange mean value theorem,

$$f(x_1) - f(x_2) = f'(\xi) \times (x_1 - x_2), (x_1 < \xi < x_2),$$

$$\begin{aligned}
&e(y_{1j}) - e\left(\frac{y_{1j} + y_{2j}}{2}\right) + e(y_{2j}) - e\left(\frac{y_{1j} + y_{2j}}{2}\right) \\
&\quad = e'(\xi_1)\left(y_{1j} - \frac{y_{1j} + y_{2j}}{2}\right) + e'(\xi_2)\left(y_{2j} - \frac{y_{1j} + y_{2j}}{2}\right) \\
&\quad = e'(\xi_1)\left(\frac{y_{1j} - y_{2j}}{2}\right) + e'(\xi_2)\left(\frac{y_{2j} - y_{1j}}{2}\right) \\
&\quad = e'(\xi_1)\left(\frac{y_{1j} - y_{2j}}{2}\right) - e'(\xi_2)\left(\frac{y_{1j} - y_{2j}}{2}\right) \\
&\quad = \left(\frac{y_{1j} - y_{2j}}{2}\right)[e'(\xi_1) - e'(\xi_2)]
\end{aligned}$$

Because $y_{1j} < \xi_1 < \frac{y_{1j}+y_{2j}}{2}$, $\frac{y_{1j}+y_{2j}}{2} < \xi_2 < y_{2j}$, $\xi_1 < \xi_2$, and $e'_{ij}(y_{ij})$ is a decreasing function, $[e'(\xi_1)-e'(\xi_2)] > 0$, $(\frac{y_{1j}-y_{2j}}{2})[e'(\xi_1)-e'(\xi_2)] < 0$. This shows that, holding per capita income constant, the emissions of countries with absolute equal income is higher than the emissions of countries with an income gap.

Similarly, at some given initial mean income, decreasing the income gap can also affect carbon emissions. To help see the effects of redistribution on emissions, we postulate a transfer α of income from the second person to the first person. Then the income of the second person is $(y_{2j} - \alpha)$, the income of the first person is $(y_{1j} + \alpha)$, and the total emissions of country j is $e(y_{1j} + \alpha) + e(y_{2j} - \alpha)$. Thus, the carbon emissions gap can be computed by,

$$\begin{aligned}
&e(y_{1j} + \alpha) + e(y_{2j} - \alpha) - [e(y_{2j}) + e(y_{1j})] \\
&\quad = [e(y_{1j} + \alpha) - e(y_{1j})] + [e(y_{2j} - \alpha) - e(y_{2j})] \\
&\quad = e'(\xi_3)(y_{1j} + \alpha - y_{1j}) + e'(\xi_4)(y_{2j} + \alpha - y_{2j}) \\
&\quad = \alpha[e'(\xi_3) - e'(\xi_4)]
\end{aligned}$$

Similarly, because $y_{1j} < \xi_3 < y_{1j} + \alpha \leq y_{2j} - \alpha < \xi_4 < y_{2j}$, then $\xi_3 < \xi_4$, $e'(\xi_3) > e'(\xi_4)$, which reveals that, holding per capita income fixed, narrowing the income gap will increase carbon emissions.

The above mathematical properties of the Environmental Kuznets Curve show that there is a negative relationship between the income gap and carbon emissions, which means trade-offs exist between carbon emissions control and income gap control. So reducing the income gap between countries might well increase carbon emissions, by redistributing income from countries with a low marginal propensity to emit carbon dioxide to those with a high one. Thus fighting poverty by either raising average income or lowering the income gap will exacerbate global warming. Is this story true? We will find the answers through empirical analysis in the following sections.

24.4. Empirical Analysis of the CKC

24.4.1. *The changing trend of carbon emissions*

Using the CO_2 emissions data of World Resources Institute (WRI) from 1950–2005, we can draw the graph of changing trends of carbon emissions in high, middle and low-income countries and the composition of carbon emissions, as shown in Fig. 24.2. From Fig. 24.2(a), we see that carbon emissions in middle-income countries is increasing and tends to exceed high-income countries. Figure 24.2(b) shows that the proportion of carbon emissions in middle-income countries grows fast and begins to account for nearly half of the total carbon emissions in the world.

It is necessary to focus on the relationship between emissions and economic growth in countries with different income levels, and the MPE of countries with different income levels.

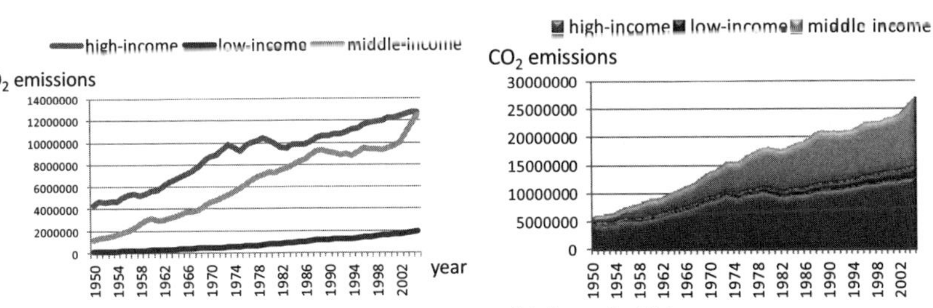

(a) Changing trend of carbon emissions (b) Composition of carbon emissions

Fig. 24.2. Changing trend of carbon emissions by income level.
Source: EarthTrends Database of World Resources Institute (WRI).

24.4.2. *GDP per capita and carbon emissions per capita*

24.4.2.1. *Data*

Data on CO_2 emissions per capita in this section are given by the Earth-Trends database of World Resources Institute (WRI), and data on per capita GDP is offered by the World Development Indicators (WDI) database of the World Bank. Considering availability and comparability, data for 101 countries from 1960 to 2005 is collected. In order to see whether there is an inverted-U relationship in countries with different income levels, the 101 countries have been divided into three categories (high, middle and low-income). According to the four kinds of countries (high, upper-middle, lower-middle, low-income) provided by WDI Database of World bank, we combine upper-middle income countries with lower-middle income countries, and focus on these three kinds of countries.

Figure 24.3 shows scatter plots of per capita CO_2 emissions and per capita GDP of the sample by different categories. Figure 24.3 also shows that the relationship between per capita CO_2 emissions and per capita GDP are obviously different in high, middle and low-income countries. Consistent with the CKC hypothesis, there is an inverted-U relationship between per capita CO_2 emissions and per capita GDP in high-income countries. Per capita CO_2 emissions tend to decline in high-income countries, while they are rapidly increasing in middle-income countries [see Fig. 24.3(c)]. However, we cannot see an inverted-U curve in low-income countries [see Fig. 24.3(d)].

24.4.2.2. *Econometric model*

We estimate an econometric model mainly based on the CKC, using panel data over time across the countries mentioned above. We postulate that all the parameters in the quadratic relationship between log per capita CO_2 emissions and log GDP per capita are linear functions, so the regression models are as follows:

$$\ln perCO_{2it} = \alpha_{10} + \alpha_{11} \ln perGDP_{it} + \alpha_{12}(\ln perGDP_{it})^2 + a_i + e_{it} \quad (24.4)$$

$$\ln perCO_{2it} = \alpha_{20} + \alpha_{21}\_I \cdot category\_2 + \alpha_{22}\_I \cdot category\_3 + \alpha_{23} \ln perGDP_{it} + \alpha_{24}(\ln perGDP_{it})^2 + a_i + e_{it} \quad (24.5)$$

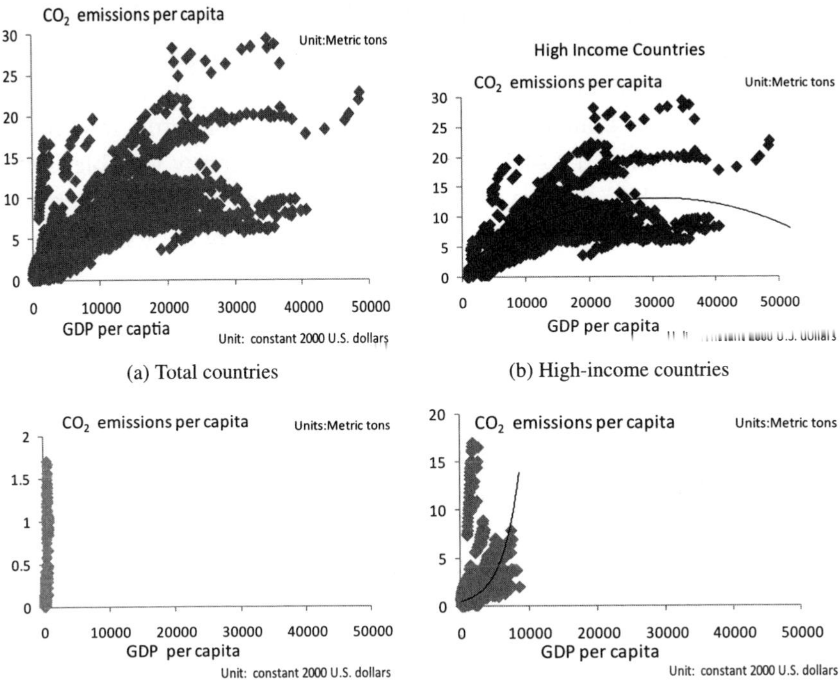

(a) Total countries

(b) High-income countries

(c) Low-income countries

(d) Middle-income countries

Fig. 24.3. Scatter of sample.

Source: World Resources Institute (WRI) and WDI Database of World Bank.

$$\begin{aligned}\ln perCO_{2it} = {} & \alpha_{20} + \alpha_{21}\_I \cdot category\_2 + \alpha_{22} I \cdot category\_3 \\ & + \alpha_{23}\_I \cdot category\_2^{*}(\ln perGDP_{it})^2 \\ & + \alpha_{24}\_I \cdot category\_3^{*}(\ln perGDP_{it})^2 \\ & + \alpha_{25} \ln perGDP_{it} + \alpha_{26}(\ln perGDP_{it})^2 + a_i + e_{it} \quad (24.6)\end{aligned}$$

Equation (24.4) is a popular specification in past work on the EKC, where $\ln per\mathrm{CO}_{2it}$ is log per capita CO_2; $\ln perGDP_{it}$ is log per capita GDP; i denotes the country, and t denotes the time period. The error terms in Eq. (24.4) include a_i and e_{it}. Variable a_i captures all unobservable time-constant factors that affect $\ln per\mathrm{CO}_{2it}$, such as country-specific characteristics; e_{it} is a white noise component.

We introduce two dummy variables, ($I \cdot category\_2$ and $I \cdot category\_3$), into Eq. (24.4) to test whether the categories of countries will affect emissions. This leads to Eq. (24.5), where the $I \cdot category\_2$ stands for middle-income countries, and $I \cdot category\_3$ stands for high-income countries; low-income countries are the base group.

For purposes of comparison, we add two interaction terms into equation (24.4), yielding Eq. (24.6). Adding interacting dummy variables with explanatory variables that are not dummy variables allows for a difference in the second derivative, which results in a difference in the CKC turning point.

24.4.2.3. *Empirical results*

All equations are estimated by OLS, fixed effects and random effects models, as reported in Tables 24.1 and 24.2. Table 24.1 shows results of Eqs. (24.4), (24.5) and (24.6) using all countries (101 countries). Table 24.2 shows results of three models based on Eq. (24.4), using observations of high income countries (33 countries), middle income countries (47 countries) and low income countries (21 countries).

Table 24.1 provides enough evidence for us to prove the CKC hypothesis. According to the results of the fixed effects model based on Eq. (24.4), all the coefficients are statistically significant at the 0.1% level, which means there is an inverted-U relationship between carbon emissions and per capita GDP using a mixed sample including all countries. The fixed effects model based on Eq. (24.5) shows the dummy variable is not statistically significant, which indicates that the initial level of per capita CO_2 emissions does not affect per capita CO_2 emissions. However, the result of the fixed effects model based on Eq. (24.6) gives a statistical significance to both country-category dummy variables. That is to say, the income level will significantly affect the turning point of the inverted-U curve.

Table 24.1 shows that the relationship between per capita CO_2 emissions and per capita GDP are different in different kinds of countries. In middle-income countries, it seems that per capita CO_2 emissions grow at a high speed associated with the increase of per capita GDP, which looks like the back part of a U-shape curve [see Fig. 24.3(c)]. Per capita CO_2 emissions and per capita GDP in low-income countries have no statistically significant correlation with each other [see Fig. 24.3(d)].

Table 24.1. Regression summary.

	Model 4 (Total sample)			Model 5 (Total sample)			Model 6 (Total sample)		
	OLS	Fixed effects	Random effects	OLS	Fixed effects	Random effects	OLS	Fixed effects	Random effects
lnperGDP	2.3404***	2.0503***	2.0525***	1.7218***	2.0503***	2.0352***	[illegible].1771	2.1365***	2.1198***
	(0.0725)	(0.0891)	(0.0874)	(0.0922)	(0.0891)	(0.0883)	(0.1928)	(0.1658)	(0.1656)
$(\text{lnperGDP})^2$	−0.0875***	−0.0629***	−0.0638***	−0.0585***	−0.0629***	−0.0626***	0.0000	−0.1164***	−0.1124***
	(0.0047)	(0.0057)	(0.0056)	(0.0060)	(0.0057)	(0.0057)	(.)	(0.0156)	(0.0156)
_Icategory_2 dummy				0.6518***	0.0000	0.2234	[illegible].0999***	0.0000	−1.8914***
				(0.0423)	(.)	(0.1700)	(0.2111)	(.)	(0.2503)
_Icategory_3 dummy				0.9613***	0.0000	−0.0025	[illegible].3658***	0.0000	−1.0231**
				(0.0621)	(.)	(0.1881)	(0.3810)	(.)	(0.3761)
_Icat2Xlnpe^2 dummy							−0.1005***	0.0639***	0.0583***
							(0.0060)	(0.0056)	(0.0056)
_Icat3Xlnpe^2 dummy							−0.1145***	0.0436***	0.0403***
							(0.0082)	(0.0074)	(0.0074)
Cons.	−12.2943***	−11.5708***	−11.5313***	−9.9706***	−11.5708***	−11.5772***	−7.5541***	−11.8464***	−10.5131***
	(0.2677)	(0.3432)	(0.3412)	(0.3286)	(0.3432)	(0.3468)	(0.5520)	(0.6049)	(0.4888)
No. of Obs.	4646	4646	4646	4646	4646	4646	4646	4646	4646
R^2	0.8197	0.4890	0.8184	0.8299	0.4890	0.8210	0.3400	0.5120	0.7968
F-test	10555.9531***	2174.1313***		5659.3149***	2174.1313***		4060.3192***	1191.0314***	
Wald			4921.05***			4966.6***			5408.31***
Hausman test			4.81			12.63**			67.18***

Note: (SE): Standard error; * $p < 0.05$ (5% level), ** $p < 0.01$ (1% level), *** $p < 0.001$ (0.1% level).

Table 24.2. Regression summary.

	Model 4 (High-income)			Model 4 (Middle-income)			Model 4 (Low-income)		
	OLS	Fixed effects	Random effects	OLS	Fixed effects	Random effects	OLS	Fixed effects	Random effects
lnperGDP	4.7939***	7.0564***	7.0225***	−0.4658	−0.1657	−0.1311	−7.168***	−0.739	−1.117
	(0.3691)	(0.2550)	(0.2546)	(0.2380)	(0.1957)	(0.1960)	(−8.61)	(−0.89)	(−1.35)
$(\text{lnperGDP})^2$	−0.2257***	−0.3454***	−0.3437***	0.0883***	0.1124***	0.1078***	0.803***	0.145	0.183*
	(0.0204)	(0.0142)	(0.0142)	(0.0167)	(0.0141)	(0.0141)	10.73)	(1.92)	(2.43)
Cons.	−23.0203***	−33.6489***	−33.4806***	−1.2991	−4.7087***	−4.7194***	12.67***	−2.617	−1.699
	(1.6605)	(1.1460)	(1.1477)	(0.8381)	(0.6774)	(0.6857)	(5.49)	(−1.15)	(−0.75)
No. of Obs.	1518	1518	1518	2162	2162	2162	966	966	966
R^2	0.4728	0.5984	0.4687	0.4529	0.6311	0.4522	0.5928	0.2230	0.5776
F-test	679.2605***	1105.0463***		893.6685***	1807.4818***		00.8517***	135.3498***	
Wald			2223.57***			3549.39***			302.31***
Hausman test			9.36 **			55.06***			41.98***

Note: (SE): Standard error; * p <0.05 (5% level) ** p <0.01 (1% level), *** p <0.001 (0.1% level).

Taking all model results into consideration, we can safely come to some conclusions. High-income countries have shown a decreasing trend of per capita carbon emissions following the increase of per capita GDP. However, both middle-income and low-income countries have still not shown an inverted-U relationship between economic development and carbon emissions. Middle-income countries are experiencing an ever-increasing trend of carbon emissions, while there is not a clear relationship between per capita carbon emissions and per capita GDP in low-income countries.

24.5. Income Gap and Carbon Emissions Differentials

24.5.1. *Motivation*

Quadratic or cubic functions of income per capita have been popular specifications in related studies, which show the existence of nonlinearities in the cross-country relationship between carbon emissions and average incomes. The inverted-U relationship has implications for the linkage between the income gap and carbon emissions, although those implications seem to have gone unnoticed. There is evidence that inequality between countries has been on an increasing trend since about 1960 (Berry *et al.*, 1991; Pritchett, 1997). This adverse distributional trend may be harmful to carbon emissions control.

This section aims to test the possible effects of the income gap on carbon emissions. We will estimate a model of carbon emissions by introducing an indicator of the income gap between countries into the CKC, so that we can create a link with the original Kuznets Curve. If this can be proved true, the income gap between countries matters to carbon emissions. In that case, the CKC can also be thought of as a reduced-form econometric model, in which the income gap between countries has been solved out or omitted. Furthermore, the economic growth rate may also contribute to the growth of carbon emissions, since both the level of average income and the income gap, may not be neutral in terms of the economic growth rate. An income gap to some extent, improves the efficiency of economic development and accelerates the growth of average income, resulting in the growth of emissions. Thus, the economic growth rate should be taken into consideration when estimating the model.

By adding both an income gap indicator and an economic growth indicator into the model, we will be able to see whether there is a CKC independent of the original Kuznets Curve.

24.5.2. *Data*

In this section, we will discuss the relationship between per capita CO_2 emissions and income gap. Data on per capita GDP in 101 countries from 1960 to 2007 was retrieved from the World Development Indicators (WDI) database of the World Bank. Data on CO_2 emissions per capita in this section was retrieved from the EarthTrends database of the World Resources Institute (WRI). In order to measure the income gap, we use the Gini Index in the following econometric models. The Gini Index is calculated with an equation[1] using per capita GDP in 101 countries from 1960 to 2007. The per capita CO_2 emissions and per capita GDP in all countries in every year is also calculated with the population-weighted data of 101 countries from 1960 to 2007. We compute the economic growth rate of each year by using per capita GDP from the previous year in order to construct a time series data set which covers 48 years.

In order to focus on the income gap, we have omitted population density, technology, resource endowments and other factors which may affect carbon emissions; we only consider the impact of economic growth and income distribution on carbon emissions.

Figure 24.4 shows the scatter plot of per capita CO_2 emissions and per capita GDP, and Fig. 24.5 shows a scatter plot of per capita CO_2 emissions and the Gini index. These two figures suggest that per capita CO_2 emissions have not yet reached the turning point. To some extent, the Gini index affects per capita CO_2 emissions in a positive way, but sometimes there are negative effects.

24.5.3. *Model selection*

As we have discussed before, in order to find the effect of the income gap on carbon emissions, the Gini index should be introduced into the

[1] $\text{Gini} = (1/2\bar{y}_u)[1/n(n-1)\sum_{i=1}^{n}\sum_{j=1}^{i}|y_i - y_j|]$.

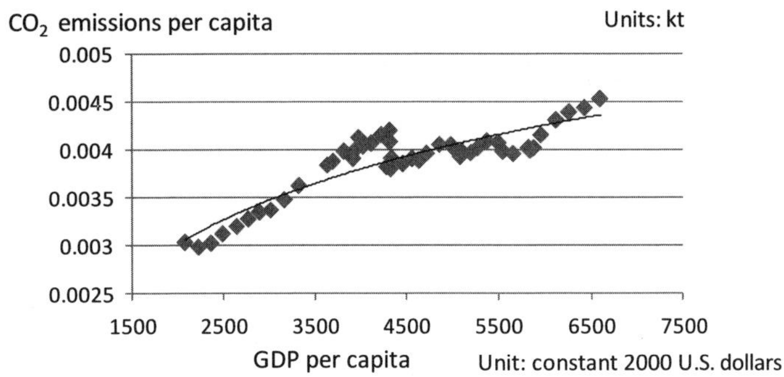

Fig. 24.4. Scatter plot of GDP per capita and CO_2 emissions per capita.

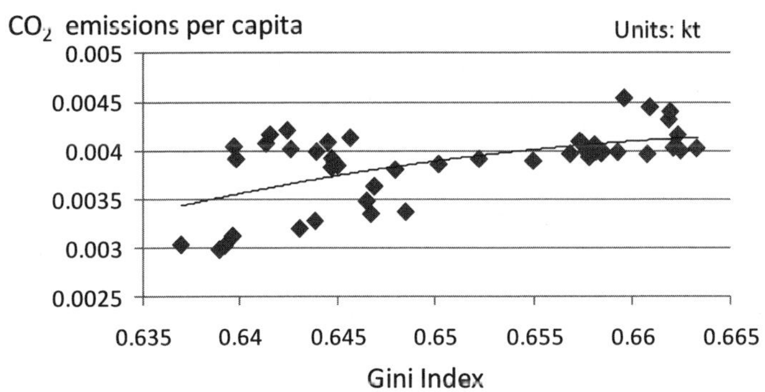

Fig. 24.5. Scatter plot of Gini coefficient and CO_2 emissions per capita.

emissions-income relationship model in as flexible a way as possible so as to allow interaction effects with the other variables, as well as, an independent effect of the income gap on emissions. Therefore, we consider the following four models which were selected from all the models we have tested [see Eqs. (24.7) to (24.10)],

$$\ln perCO_2 = \beta_{10} + \beta_{11} \ln perGDP + \beta_{12}(\ln perGDP)^3 + e \tag{24.7}$$

$$\ln perCO_2 = \beta_{20} + \beta_{21} gini^* \ln perGDP + \beta_{22}(\ln perGDP)^2 + \beta_{23}(\ln perGDP)^3 + e \tag{24.8}$$

$$\ln perCO_2 = \beta_{30} + \beta_{31}(gini)^2 + \beta_{32} gini^* \ln perGDP + \beta_{33} GDPgrowth + e \tag{24.9}$$

$$\ln perCO_2 = \beta_{40} + \beta_{41}(gini)^2 + \beta_{42}gini^* \ln perGDP + \beta_{43}GDPgrowth + \beta_{44}gini^*(\ln perGDP)^3 + e \quad (24.10)$$

Table 24.3 gives the estimation results of the four models. We finally decide to choose model (24.10) on the grounds that it has the highest adjusted R^2 and lowest standard error in all the models.

24.5.4. *The Gini elasticity of carbon emissions*

To further analyze the complex relationship between per capita carbon emissions and the Gini coefficient, in Fig. 24.5, we assume that the effects of economic growth and the income gap on per capita carbon emissions can be simulated by model (24.10). The Gini semi-elasticity of carbon emissions can be calculated according to the estimation results of model (24.10). Because the Gini variable in the model for estimation is in level form, not log form, and per capita CO_2 emissions is in log form, the parameter of the Gini Index can be defined as the Gini semi-elasticity of carbon emissions. Therefore, the first partial derivative of the Gini Index is the percentage change of per capita CO_2 emissions caused by a one unit growth in the Gini coefficient. The Gini semi-elasticity of carbon emissions can be calculated as follows:

$$\frac{\partial \ln perCO_2}{\partial gini} = 2\beta_1 gini + \beta_2^* \ln perGDP + \beta_4(\ln perGDP)^3 = -32.044gini + 2.035 \ln perGDP - 0.006^*(\ln perGDP)^3 \quad (24.11)$$

Figure 24.6 illustrates that the Gini semi-elasticity of carbon emissions is a decreasing function of the Gini Index in any income level. There is a zero semi-elasticity point as we can see from Fig. 24.6. That is to say, above this zero semi-elasticity point, the Gini Index has a positive effect on carbon emissions, while below the zero semi-elasticity point, the Gini Index tends to affect carbon emissions in a negative way. Therefore, the Gini Index may affect carbon emissions in two different ways.

From Eq. (24.11), we can see that the Gini Index semi-elasticity is affected by per capita GDP. The zero semi-elasticity point, or so-called semi-elasticity turning point, interacted by income level (per capita GDP) is

Table 24.3. Model selection.

Variables	Parameters	Model (24.7)	Model (24.8)	Model (24.9)	Model (24.10)
$(\ln perGDP)^3$	β_{12}	−0.009*** (0.002)			
$\ln perGDP$	β_{11}	2.236*** (0.459)			
$(\ln perGDP)^3$	β_{23}		−0.009* (0.004)		
$(\ln perGDP)^2$	β_{22}		0.171*** (0.043)		
$gini^* \ln perGDP$	β_{21}		−0.789*** (0.119)		
$(gini)^2$	β_{31}			−11.825*** (0.907)	
$gini^* \ln perGDP$	β_{32}			0.843*** (0.046)	
GDPgrowth	β_{33}			0.007** (0.002)	
$(gini)^2$	β_{41}				−16.022*** (1.991)
$gini^* \ln perGDP$	β_{42}				2.035*** (0.512)
GDPgrowth	β_{43}				0.007** (0.002)
$gini^* (\ln perGDP)^3$	β_{44}				−0.006*** (0.003)
cons.	β_{10}	−18.725*** (2.521)			
	β_{20}		−7.945*** (1.145)		
	β_{30}			−5.163*** (0.188)	
	β_{40}				−7.593*** (1.054)
N		48	48	48	48
Adjusted R^2		0.868	0.932	0.940	0.946
Std. error of the estimate		0.03657	0.02623	0.02466	0.02349
F-test		156.038***	216.684***	247.161***	205.592***

Note: (SE): Standard error; * $p < 0.05$ (5% level), ** $p < 0.01$ (1% level), *** $p < 0.001$ (0.1% level).

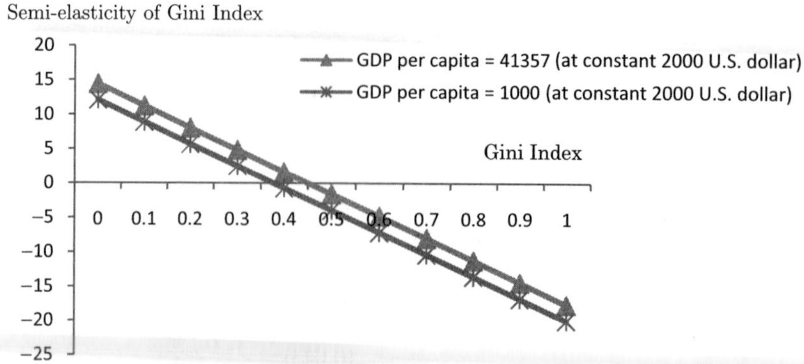

Fig. 24.6. The Gini elasticity of carbon emissions.

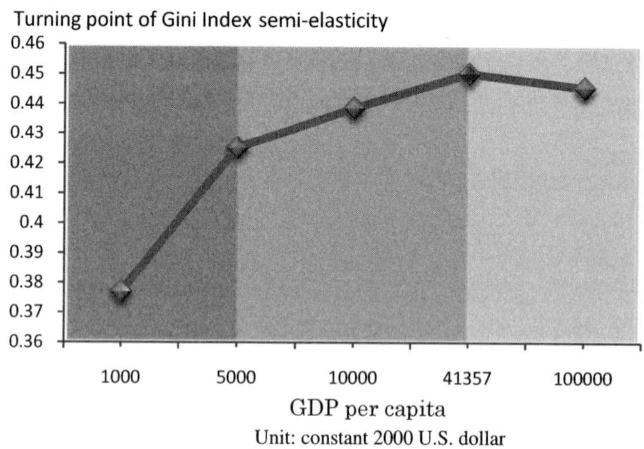

Fig. 24.7. Turing point of the Gini semi-elasticity under different GDP levels.

given in Fig. 24.7. The zero semi-elasticity point grows with per capita GDP. At a higher income level, the Gini Index at the zero semi-elasticity point will be higher, which means the positive effect of the income gap on carbon emissions continues to work for a longer period. Therefore, the higher per capita GDP is, the stronger the impact of the income gap on carbon emissions will be.

Taking all the foregoing analysis into consideration, the impact of the income gap on carbon emissions can be summarized in two ways: first, the effect of the income gap on carbon emissions changes from positive to negative [(see Fig. 24.6)]. Second, the effect of the income gap on carbon emissions is affected by per capita GDP. A higher level of per capita GDP will strengthen the positive effects of the income gap on carbon emissions, thereby enhancing the strength of the influence, extending the influence period and substantially increasing per capita carbon emissions [(see Fig. 24.7)]. That is to say, the income gap at a higher income level will make the situation even worse.

24.5.5. *Income elasticity of carbon emissions*

In model (24.10), another import variable that significantly affects carbon emissions is per capita GDP, the indicator of income. To help see the influence of per capita GDP on emissions and that influence interacted by the income gap indicator, the Gini Index, we calculate income elasticity of carbon emissions by Eq. (24.12), and calculate the income elasticity at three given Gini Index levels by Eqs. (24.13) to (24.15)

$$\frac{\partial \ln perCO_2}{\partial \ln perGDP} = \beta_2 gini + 3\beta_4 gini^*(\ln perGDP)^2$$

$$= 2.035gini - 0.018gini^*(\ln perGDP)^2 \tag{24.12}$$

$$\frac{\partial \ln perCO_2}{\partial \ln perGDP} = 0.2035 - 0.0018(\ln perGDP)^2 \quad (\text{gini} = 0.1) \tag{24.13}$$

$$\frac{\partial \ln perCO_2}{\partial \ln perGDP} = 1.0175 - 0.009(\ln perGDP)^2 \quad (\text{gini} = 0.5) \tag{24.14}$$

$$\frac{\partial \ln perCO_2}{\partial \ln perGDP} = 1.8315 - 0.0162(\ln perGDP)^2 \quad (\text{gini} = 0.9) \tag{24.15}$$

Figure 24.8 shows the changing trend of the income elasticity of carbon emissions associated with income growth at three specific levels of the Gini Index. It is clear that income elasticity decreases faster at a higher Gini Index (gini = 0.9) than at a lower Gini Index (gini = 0.5). Income elasticity changes little when the income gap is small enough, as seen in the trend of income elasticity at the lowest Gini Index level (gini = 0.1). This evidence implies that before reaching the turning point of the CKC, the higher the income gap is, the higher the income elasticity of carbon emissions is at any given per capita GDP level. Thus, holding per capita GDP constant, we can

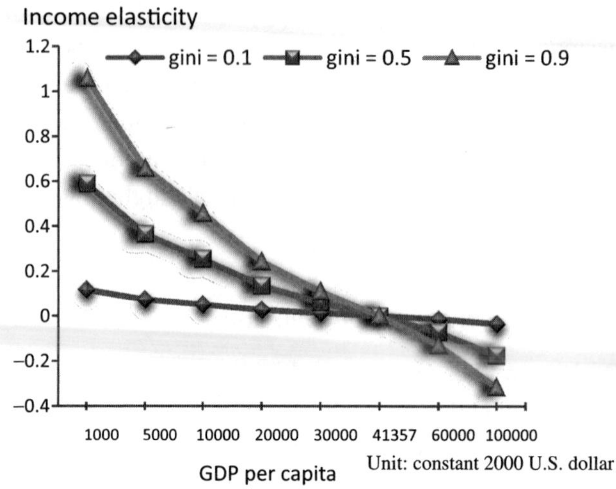

Fig. 24.8. Effects of Gini on income elasticity.

infer that a growing income gap will undoubtedly be accompanied by an increase of income elasticity, which will lead to even more CO_2 emissions.

Similarly, the evidence shown in Fig. 24.9 implies that the income elasticity of emissions, holding per capita GDP constant, is an increasing function of the Gini Index before reaching the turning point and a decreasing function of Gini Index after reaching the turning point. So the specification indicates that a higher income gap will strengthen the income elasticity and eventually result in more emissions. Meanwhile, a higher income gap can also constrain the environmental renovation ability of economic growth.

To sum up, the effects of the income gap on income elasticity of emissions are two-fold: first, the wider the income gap, the greater the income elasticity and additional emissions; second, the effects of the income gap on income elasticity tend to decline. Before reaching the turning point, the elasticity is much higher than it is after reaching the turning point, which means the environmental governance of economic growth is constrained by the income gap.

24.5.6. *Discussions and implications of the model*

From the foregoing analysis based on model (24.10), we can see that the impact of the income gap on per capita carbon emissions has

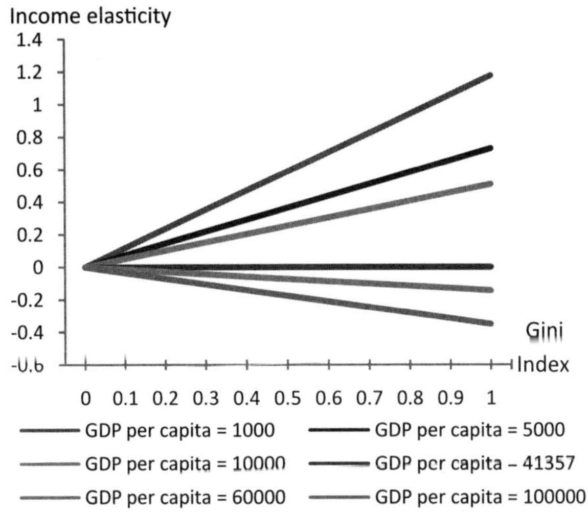

Fig. 24.9. Sensitivity of income elasticity on the change of Gini.
Note: All units are at constant 2000 U.S. dollar.

a conversion trend from positive (increasing emissions) to negative (decreasing emissions). The implications of these complex effects can be summarized in the following three aspects.

24.5.6.1. *Positive (increasing) direct effects of income inequality on carbon emissions*

A positive impact of income inequality on carbon emissions indicates that a widening income gap entails an increase in carbon emissions. The possible reasons for the positive effects are: first, middle and low-income developing countries are eager to expand economic development in order to narrow the income gap with high-income, developed countries, which may reduce the willingness to pay for environmental protection and lead to resource depletion and environment deterioration. As the proportion of middle- and low-income developing countries account for about 70% of the sample, this leads to a decline in the average willingness to pay for environmental protection worldwide, accelerating the increase in carbon emissions. Second, a significant proportion of middle- and low-income countries lack the ability to pay for environment improvements due to their low-income level; even if they desire to improve the environment, they have to develop the economy at the price of environmental degradation due to the shortage of funds, technical backwardness and expensive cost of environmental governance. Third, the widening income gap among countries will undoubtedly lead to a worldwide unbalanced development of environmental quality. Per capita carbon emissions of high-income developed countries are declining [Fig. 24.3(b)], while per capita carbon emissions in middle- and low-income developing countries are rapidly increasing [Figs. 24.3(c and d)]. As a result of the large share of middle and low-income developing countries in the world, the average per capita carbon emissions of the world as a whole is doomed to rise.

24.5.6.2. *Positive (increasing) indirect effects of income inequality on carbon emissions*

The indirect effects of income inequality on carbon emissions take place via economic growth. A widening income gap, to some extent, promotes the economic growth of middle- and low-income developing countries, resulting in more carbon emissions due to the extensive economic growth. On the other hand, a growing income gap entails unbalanced environmental governance, reducing the total demand for high-tech environmental goods, subsequently weakening environmental governance.

24.5.6.3. *Negative (decreasing) effects of income inequality on carbon emissions*

An acceptable explanation for the negative effects of income inequality on carbon emissions is that efforts to reduce income inequality tend to result in the growth of carbon emissions.

The way to develop traditional industries for narrowing the economic gap with developed countries will definitely lead to a sharp rise in environmental pollution. That is because the rapid growth of industry simultaneously increases energy consumption and industrial pollution in the initial stage of traditional industrialization. Environmental awareness in middle- and low-income developing countries is at a low level, so environmental protection often gives way to economic growth. As a matter of fact, reducing the income gap comes with a growth in per capita carbon emissions.

The process of narrowing the income gap between countries is often accompanied by industry transformation, and technology diffusion. For instance, industry transfer between countries on the global value chain has made China into the world's factory. In this industry transfer process, middle- and low-income developing countries bear the high pollution and high emissions associated with traditional manufacturing, which causes the ever-increasing amount of carbon emissions. Consequently, before the elimination of the income gap between countries, industry transfer is often accompanied by the transfer of polluted areas and high-carbon regions. The area of environmental pollution and carbon emissions spreads significantly and results in an increased average level of pollution and carbon emission.

24.6. Conclusions

First, the mathematical properties of the Environmental Kuznets Curve show that there is a negative relationship between the income gap and carbon emissions, which means trade-offs exist between carbon emissions control and income gap control. However, the empirical analysis shows that there is a significant overall positive relationship between the income gap and carbon emissions, with a negative correlation only in certain intervals. Therefore, we need to amend and expand the Environmental Kuznets Curve in order to analyze the relationship between carbon emissions and income inequality.

Second, high-income countries have shown a decreasing trend of per capita carbon emissions with the increase of per capita GDP. However, both middle- and low-income countries have not shown an inverted-U relationship between economic development and carbon emissions. Middle-income countries are experiencing an ever-increasing trend of carbon emissions, and there is no clear relationship between per capita carbon emissions and per capita GDP in low-income countries.

Third, the existence of an income gap between countries will intensify and expand the growth of per capita carbon emissions, delay the arrival of the Kuznets Curve turning point and weaken the beneficial effects of economic growth on the environment.

Fourth, according to the elasticity analysis and given the big economic development gap in the world, globalization and liberalization of the international economy will promote the worldwide transfer of industries, which greatly accelerates the industry upgrading of developed countries and shifts the high pollution and high emissions industries to developing countries. In this way, it leads to the worldwide expansion of environmental pollution and carbon emissions, and to a decline in world average environmental quality. If this were true, it may induce a suggestion that per capita carbon emissions reduction in high-income countries is partly a result of carbon emissions transfer.

Fifth, the significantly delayed impact of the income gap on carbon emissions shows that this impact is not a temporary effect, but a long-term effect (theoretically 7 years). This implies that income inequality between high-income countries and low-income countries may lead to an unbalanced environmental governance capacity, making the beneficial effects of economic growth on the environment ones that cannot be manifest for a long period of time.

Therefore, middle- and low-income countries can no longer accept the heavy industry or energy-intensive industry transfer and high-income countries should provide financial support and technical support to help middle- and low-income countries develop a low carbon economy so as to enhance global environmental quality, reduce the growth rate of carbon emissions and promote international cooperation toward achieving carbon reduction agreements.

References

Berry, A., Bourguignon, F. and Morrission, C. (1991). Global Inequality and its Trend since 1950. In L. Osberg (ed.), *Economic Inequality and Poverty: International Perspectives*. New York: M.E. Sharpe.

Boyce, J. K. (1994). Inequality as a Cause of Environmental Degradation. *Ecological Economics*, 11, 169–178.

Grossman, G. and Krueger, A. (1995). Economic Growth and the Environment. *Quarterly Journal of Economics*, 110, 353–377.

Kuznets, S. (1955). Economic Growth and Income Inequality. *American Economic Review*, 1, 1–28.

Magnan, I. E. (2000). The Environmental Kuznets Curve: Environmental Protection Policy and Income Distribution. *Ecological Economics*, 32, 431–443.

Marsiliani, L. and Renström, T. I. (2000). "Inequality, Environmental Protection and Growth." *Center Working Paper*, Tilburg University, The Netherlands.

Ravallion M., Heil, M. and Jalan, J. (2000). Carbon emissions and income inequality, Oxford Economic Papers, 52, 651–669.

Scruggs, L. A. (1998). Political and Economic Inequality and the Environment. *Ecological Economics*, 26, 259–275.

Torras, M. and Boyce, J. K. (1998). Income, Inequality and Pollution: A Reassessment of the Environmental Kuznets Curve. *Ecological Economics*, 25, 147–160.

Index